# Quantitative Hydrogeology

*Groundwater Hydrology for Engineers*

The "R'statts," well diggers since the tenth century in the R'hir wadi, Algeria. At the beginning of this century these men were still able to dig and maintain the wells, diving in the dark more than 80 m down to find the water, "this marvellous element which can bring life to all things" (the Koran). (Photo by Roger Viollet.)

# Quantitative Hydrogeology

## Groundwater Hydrology for Engineers

### Ghislain de Marsily
PARIS SCHOOL OF MINES
FONTAINEBLEAU, FRANCE

Translated by
*Gunilla de Marsily*

 1986

## ACADEMIC PRESS, INC.
**Harcourt Brace Jovanovich, Publishers**
Orlando   San Diego   New York   Austin
London   Montreal   Sydney   Tokyo   Toronto

COPYRIGHT © 1986 BY ACADEMIC PRESS, INC.
ALL RIGHTS RESERVED.
NO PART OF THIS PUBLICATION MAY BE REPRODUCED OR
TRANSMITTED IN ANY FORM OR BY ANY MEANS, ELECTRONIC
OR MECHANICAL, INCLUDING PHOTOCOPY, RECORDING, OR
ANY INFORMATION STORAGE AND RETRIEVAL SYSTEM, WITHOUT
PERMISSION IN WRITING FROM THE PUBLISHER.

ACADEMIC PRESS, INC.
Orlando, Florida 32887

United Kingdom Edition published by
ACADEMIC PRESS INC. (LONDON) LTD.
24–28 Oval Road, London NW1 7DX

Library of Congress Cataloging in Publication Data

Marsily, Ghislain de.
    Quantitative hydrogeology.

    Translation of: Hydrogéologie quantitative.
    Bibliography: p.
    Includes index.
    1. Hydrogeology—Mathematics.    I. Title.
GB1001.72.M37M3713    1986      551.48      85-9149
ISBN  0–12–208915–4  (hardcover) (alk. paper)
ISBN  0–12–208916–2  (paperback) (alk. paper)

PRINTED IN THE UNITED STATES OF AMERICA

86 87 88 89       9 8 7 6 5 4 3 2 1

ENLARGED AND TRANSLATED FROM THE ORIGINAL FRENCH EDITION
ENTITLED Hydrogéologie quantitative, PUBLISHED BY MASSON, PARIS.
© MASSON, EDITEUR, PARIS, 1981.

# Contents

# Foreword

Understanding the factors that control the flow of water through soils and rocks has led to the development of one of the most important fields in the earth sciences: groundwater hydrology. Over the years, an impressive body of theory and practice has been developed. It was recognized very early that, although the basic concepts of fluid mechanics must still hold, the nature of underground flow paths must be understood in order to develop meaningful solutions to groundwater problems. This is not an easy task because the complexities of the geologic process cause many variations in natural flow systems.

Early treatises on the subject of groundwater hydrology usually started with the simplest flow systems: isotropic and homogeneous porous media. In this way, it was possible to develop analytic solutions through the application of rigorous mathematical methods. These have provided valuable insights into what to expect in the field, but as the groundwater hydrologists gained experience, it became clear that these solutions were often inadequate.

Groundwater systems are not often very homogeneous over distances of any practical significance. In the case of an aquifer in a porous medium (sandstone or limestone), variability of the flow properties in both the horizontal and vertical directions is a result of the vagaries of sedimentation. And there is another problem that the groundwater hydrologist has had to face. Discontinuities of one kind or another are commonly present in rock systems. These may create boundaries that limit the flow regime or produce unusual flow conditions internal to the system. The presence of fractures or joints in the rock mass can have a profound effect on the flow regime. The groundwater hydrologist has thus been forced to develop methods of analyzing fluid flow in rock systems with very complex geometries.

The spatial variability of the flow regime due to the geologic processes at work has led to another problem that must be treated in the field of hydrogeology. The flow field will usually extend for significant distances outside the area where a detailed record of hydraulic properties has been developed through drilling and testing. Thus, the flow regime will be adequately known in one part of the total flow system and less and less well known as distances extend to more remote regions.

In general, the spatial variability of the controlling parameters of a groundwater system (aquifer thickness, hydraulic head, permeability, transmissivity, storage, etc.) is not purely random; it can often be shown that some kind of correlation exists in the spatial distribution of these parameters. The problem then is to develop through the methods of geostatistics an appropriate expression for the spatial correlations that exist. Kriging can be used as a method for optimizing the estimation of a regionalized variable, i.e., a hydraulic property that is distributed in space and measured at a network of points. When the uncertainties associated with these estimates are too high, the problem should be considered as a stochastic rather than a deterministic process. Stochastic partial differential equations can then be used to analyze the problem.

The widespread pollution of groundwater with various chemicals and toxic wastes has necessitated the study of chemical transport. Solutes are diffused and advected during this process, and the dissolved (or suspended) species may or may not react with the rock matrix through which the groundwater solutions flow. Considerable variation in the microscopic and macroscopic velocities within the complex flow paths of the rock mass results in dispersion of the dissolved species.

It is not surprising that numerical solutions to groundwater problems are often necessary to solve the complex problems that can be encountered in the field. A tremendous effort has been expended in developing accurate numerical methods of handling both the flow and transport equations. Sometimes the semianalytical approach is used when it is possible to first obtain a solution in the Laplace transform domain and then apply numerical methods of inversion to obtain the final solution. Numerical methods have been developed using either finite differences or finite elements. More recently, boundary elements of boundary integral methods have been proposed for solving the flow equations.

As the application of fluid mechanics to the problems of water moving through rock systems has progressed, the science of groundwater hydrology has slowly emerged. The complications of the geologic environment indicate the need for care in the application of established theories and an awareness of the validity of the assumptions that one must make. Realistic solutions to groundwater problems must involve a combination of the right application of physical principles with a mature insight developed through experience in the laboratory and field.

This approach is very well demonstrated in this book. Emphasis is placed on the fundamental properties of porous and fractured rocks that control the flow regime. The basic equations and methods of their solution for single- and multiphase flow are presented in detail. An extension of the treatment of flow equations to include transport is included. The author was one of the first to recognize the value of the geostatistical and stochastic approaches in hydrogeology and he has developed a thorough treatment of these topics. Numerical methods of solving the flow and transport equations are also reviewed. The reader will find this a very comprehensive treatment. The appearance of *Quantitative Hydrogeology* represents a significant step forward for this field.

*Berkeley, California*                                    PAUL A. WITHERSPOON

# Preface

This book attempts to combine two separate themes: a description of one of the links in the chain of the water cycle inside the earth's crust, i.e., the subsurface flow, and the quantification of the various types of this flow, obtained by applying the principles of fluid mechanics in porous media. The first part is the more descriptive and geological of the two. It deals with the concept of water resources, which then leads us on to other links in the cycle: rainfall, infiltration, evaporation, runoff, and surface water resources. The second part is necessary in order to quantify groundwater resources. It points the way to other applications, such as solutions to civil engineering problems, including drainage and compaction, and solutions to transport problems in porous media, including aquifer pollution by miscible fluids, multiphase flow of immiscible fluids, and heat transfer in porous media, i.e., geothermal problems. However, the qualitative and the quantitative aspects are not treated separately but are combined and blended together, just as geology and hydrology are woven together in hydrogeology.

This book is intended for engineers with a mathematical background. They will find a fairly detailed description of the physical processes occurring in porous and fractured media followed by the development of the basic flow and transport equations for both steady and transient states. Outlines are given of the methods for solving these basic equations as well as the most common analytic expressions used in handling practical problems. Basic geologic structures containing groundwater and the hydrologic processes occurring within them are described together with practical methods for measuring the relevant physical parameters of the media. We normally give orders of magnitude of these parameters for various rock types in order to provide an initial data base for solving practical problems. The International System of units (SI, i.e., metric) is used throughout. The appendix contains complete definitions and conversion factors for nonmetric units.

Geologists who do not want to burden themselves with the mathematics will find a complete description, given in simple terms, of the assumptions and conditions required for applying the equations and formulas, which is not always found in other treatises.

Apart from the classical basic flow equations, the main chapters concern (i) transport phenomena and pollution problems, (ii) the stochastic definition of medium parameters for addressing the problem of spatial variability, which includes a chapter on kriging as applied to hydrology, and (iii) the principle of numerical techniques, finite differences, integrated finite differences, and finite elements, which nowadays are prerequisites for solving practical groundwater problems.

A great effort was made to keep the book short. As in a quote from Goethe in a letter to a friend, "If I had had more time, the letter would have been shorter." The developments have been kept as brief as possible and the strict selection of the material is guided by the criterion of practical applicability.

This book was originally written in French as lecture notes for the students in engineering at the Paris School of Mines. It was later extended during a sabbatical that the author spent at the Department of Hydrology and Water Resources at the University of Arizona, Tucson. The help and advice of colleagues both at the Paris School of Mines (M. Armstrong, J. P. Delhomme, A. Dieulin, P. Goblet, P. Hubert, P. Iris, E. Ledoux, G. Matheron, and H. Pelissonnier) and at the University of Arizona (S. N. Davis, L. Duckstein, T. Maddock, D. E. Myers, S. P. Neuman, and E. Simpson) were greatly appreciated. Professor James Philip O'Kane of University College, Dublin, kindly reviewed the manuscript and helped to improve it immensely. Professor Paul A. Witherspoon (University of California, Berkeley) has often acted as a guide for the author's own research and has contributed the foreword. Professor Daniel F. Merriam (Wichita State University, Kansas) has provided constant encouragement from the earliest stages of the book. Professors Lynn W. Gelhar (Massachusetts Institute of Technology, Cambridge) and Alan L. Guijahr (New Mexico Institute of Mining and Technology, Socorro, New Mexico) were very helpful guides to the new stochastic theories during their sabbaticals in Fontainebleau and later as well.

This book is dedicated to Dr. Richard E. Jackson (Environment Canada, Ottawa), who strongly urged its translation from the French and with whom the author believes he shares the irresistible fascination for the magic of a lost paradise: the Arizona desert.

# Table of Notation

## 1. Usual Mathematical Symbols

| | |
|---|---|
| $\mathbf{U}$ | A bold face letter: a vector, in one, two or three dimensions with components $U_x$, $U_y$, $U_z$. |
| $^T\mathbf{U}$ | Transpose of vector $\mathbf{U}$ |
| $U$ | A nonboldface letter: a scalar |
| $\mathbf{x}$ | Position vector in one, two, or three dimensions |
| $\mathbf{T,K,D}$ | Tensors |
| $da/dx$ | Derivative of variable $a(x)$; also material or substantial derivative of $a$ (see p. 102) |
| $\partial a/\partial x$ | partial derivative of variable $a(x,y,z)$ |
| div $\mathbf{U}$ | $\partial U_x/\partial x + \partial U_y/\partial y + \partial U_z/\partial z$: the divergence of vector $\mathbf{U}$ |
| grad $h$ | $(\partial h/\partial x, \partial h/\partial y, \partial h/\partial z)$: the gradient of the scalar $h$, i.e., a vector with the given components |
| $\nabla^2 h$ | $\partial^2 h/\partial x^2 + \partial^2 h/\partial y^2 + \partial^2 h/\partial z^2$: the Laplace operator on the scalar $h$; also a scalar number |
| $\mathbf{U} \cdot \mathbf{V}$ | $U_x V_x + U_y V_y + U_z V_z$ is the scalar product of two vectors |
| $e^a$ | Exponential function |
| $\exp(a)$ | Exponential function |
| $E(a)$ | Expected value of the random variable $a$ |
| log | The logarithm base ten |
| ln | The Naperian logarithm |
| $\langle a \rangle$ | The spatial average of $a$ (see p. 41) |
| $\Sigma_{i=1}^{n} a_i$ | The summation $a_1 + a_2 + \ldots + a_n$ |

## 2. Other Symbols Used

Several symbols have different meanings in different chapters. For each one we give the page where the symbol is defined, or the chapter where the symbol is used.

| | | |
|---|---|---|
| $K_1$ | Kinetic constant of a linear chemical adsorption | 259 |
| $\ell$ | Length of an object | |
| $m$ | Constant expected value of a stationary random function $Z$ | Chapter 11, 288 |
| $m(x)$ | Weighting function for spatial averaging | 42 |
| $m(x)$ | Expected value of a nonstationary random function $Z$ | 309 |
| $M$ | Atomic weight of an element or molar mass of a gas | 266, 210 |
| $\mathbf{n}$ | Vector normal to a surface, oriented outward | |
| $p$ | Pressure of a fluid | 39 |
| $p_c$ | Capillary pressure | 29 |
| Pe | Peclet's number in hydrodynamic dispersion | Chapter 11, 237 |
| Pe | Numerical Peclet number in digital models | 389 |
| $pF$ | Suction potential | 30 |
| $q$ | Volumetric flow rate per unit volume of a porous medium (source or sink term) | 50 |
| $Q$ | Constant flow rate of a well | Chapter 7, 8 |
| $Q$ | Source or sink term in the transport equation | 252 |
| $r$ | Radius of air–water interface | 29 |
| $r$ | Distance to the origin in polar coordinates | 149 |
| $R$ | Radius of action of a well | 149 |
| $R$ | Retardation factor in the transport equation | Chapter 11, 256, 260, 267 |
| $R_a^*$ | Rayleigh's number | 283 |
| $R_e$ | Reynolds' number | 65, 74 |
| $R_i$ | Rotation matrix for fracture systems | 70 |
| $R_r$ | Relative roughness of a fracture | 66 |
| RF | Random function | Chapter 11 |
| RV | Random variable | Chapter 11 |
| $s$ | Drawdown in an aquifer | 144 |
| $s, s_i, s_w$ | Saturation in a porous medium | 26, 207, 276 |
| $s_k$ | Skin effect of a well | 156 |
| $S$ | Storage coefficient of a confined aquifer | 111 |
| $S_s$ | Specific storage coefficient of a confined aquifer | 108 |
| $S_{sp}$ | Specific surface area of a porous medium | 22 |
| $\sinh(a)$ | Hyperbolic sine $(e^a - e^{-a})/2$ | |
| $t$ | Time | |
| $T$ | Half-life of a radionuclide | 265 |
| $T$ | Transmissivity of an aquifer | 73 |
| $\mathbf{T}$ | Transmissivity tensor of an aquifer | 73 |
| $\tanh(a)$ | Hyperbolic tangent $[\sinh(a)/\cosh(a)]$ | |
| $u = 4Tt/r^2s$ | Theis' dimensionless time | Chapter 8, 163 |
| $\mathbf{u}$ | Velocity vector of a fluid | 39 |
| $\mathbf{u}^*$ | Mean microscopic velocity in a porous medium | 50, 237 |
| $\mathbf{u}'$ | Mean fictitious velocity in a porous medium | 239 |
| $\mathbf{u}_s$ | Velocity vector of the solid in a deformable porous medium | 100 |
| $\mathbf{u}_s^*$ | Real mean velocity vector of a solid in a deformable porous medium | 101 |
| $\mathbf{u}_\sigma$ | Velocity vector of the fluid-solid interface in a deformable porous medium | 47 |
| $\mathbf{U}$ | Darcy's velocity vector in a porous medium (or filtration velocity vector) | 59 |

# Chapter 1

# The Water Cycle

Precipitation (rainfall and snow) falling on the surface of the earth accounts for almost all of the water entering into the soil. We shall study the case of rainfall and snow separately and follow up with other types of recharge.

When rain falls on the ground, three processes are set in motion: (1) wetting of the soil and infiltration, (2) surface runoff, and (3) evaporation.

## 1.1.  Wetting and Infiltration

In most countries where it rains, the ground contains a significant amount of water in normal conditions. A usual profile of the quantity of water versus the elevation is given in Fig. 1.1.

This moisture content is obviously dependent on the porosity and the permeability of the soil. Below a certain elevation $N$, the water content no longer increases with depth. The soil is said to be saturated: all the empty spaces (pores) in the soil contain water. This water is said to belong to the water table aquifer, or phreatic aquifer. The term aquifer will be further defined in Chapter 6. The water table is the surface, at elevation $N$, constituting the upper limit of the aquifer.

Above the elevation $N$, the soil is said to be unsaturated, as the empty spaces in the soil contain both water and air simultaneously. The relationship

1

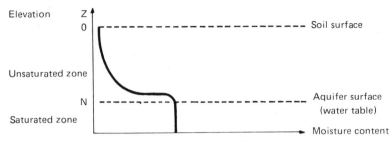

**Fig. 1.1.** Typical moisture profile in a soil.

between the two is discussed in Chapter 2. It is sufficient to note here that the water is, on the whole, subjected to the forces of gravity in the saturated zone and, furthermore, to the capillary forces (which very soon become the most influential) in the unsaturated zone.

Water falling on the soil surface begins by moistening its upper layer (a few centimeters). The resulting moisture profile is shown in Fig. 1.2. This increase in moisture on the surface does not necessarily cause an immediate significant vertical flow. The water is retained as in a sponge.

As the water content continues to increase, the water spreads downward and moistens a deeper zone. If rain continues long enough, the moistening will be progressively greater and eventually cause infiltration, i.e., an inflow into the aquifer. However, this process is very slow: depending on the depth at which the aquifer is situated and the permeability of the soil, it may take a week, a month, or several months for the water to reach the aquifer.

In the case of a continuous and large flux of water at the surface (e.g., extremely long rainfall, or artificial recharge), a decrease in the infiltration rate can be observed: from an initially large value, the infiltration flux decreases to a much smaller value. The mathematics of infiltration will be described in more detail in Chapters 2 and 9.

In temperate zones, a first-order estimation of the height of water naturally infiltrated into the aquifer is about 300 mm/yr, i.e., 10 liters $s^{-1}$ $km^{-2}$.

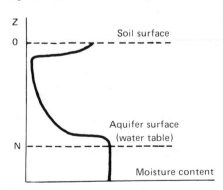

**Fig. 1.2.** Typical moisture profile during a storm.

## 1.2. Surface Runoff

If the rain is heavy, the soil is unable to absorb the water. After the first moments, when the uppermost layer of the soil is moistened, an excess of water appears on the surface (Fig. 1.3).

The upper layer of the soil is saturated in a zone of no great depth, and the moisture does not spread fast enough for the falling rain to be absorbed. Thus a film of water may move around on the ground surface. This is what we call runoff. One even distinguishes, somewhat artificially, between pure surface runoff and "hypodermic flow," which takes place in the first few centimeters of the soil or the vegetation. This runoff moves along the line of the steepest slope of the ground and feeds the natural drainage network of the soil: ditches, brooks, rivers, etc. It gathers up solid particles through erosion, which gives rise to the transport of solids in streams.

In cases where the ground is almost completely impermeable (urban areas or zones with outcrops of rocks of very low permeability) the runoff appears almost instantaneously, as soon as the water has filled the first hollows in the ground (e.g., puddles).

Finally, it is worth noting that the vegetation creates a screen for the above-mentioned mechanisms: the first of the rain is caught up by the trees and grass, and this may prevent a slight rainfall from starting the wetting process.

## 1.3.  Evaporation

Even during the rainfall a large portion of the water immediately evaporates. Indeed, the moisture in the atmosphere is rarely at the saturation point even during a thunderstorm. Once the rain stops, this evaporation continues and gradually dries the water caught up by the vegetation or remaining on the surface. It does, of course, continue on the surface waters (streams, lakes) and on the ground surface.

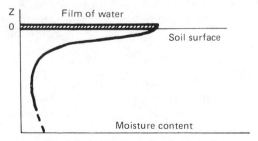

**Fig. 1.3.** Appearance of surface runoff.

Evaporation also continues inside the ground itself. Because of the existence of an air phase in the unsaturated zone, this evaporation might occur simultaneously over the entire profile and even extend nearly as far as the water table itself: however, because the mechanism by which this moisture is extracted into the air phase is so slow (diffusion toward the surface), the evaporation at the soil surface is the dominant phenomenon when the soil is not extremely dry. The water in the soil is "sucked up" and ascends by capillarity to the surface, where it evaporates.

The power of the atmosphere to extract water from the soil decreases with the moisture content of the soil: the smaller this is, the more the water is bound to the ground by capillarity and the more energy is needed to extract it. The effect also depends on the power of the atmosphere to cause evaporation, i.e., on the temperature, the wind, and the exposure to the sun. In the summer, when this evaporation is intensive, the atmosphere generally takes back all the moisture received by the profile during a storm. Eventually there is no infiltration into the aquifer. Figure 1.4 shows a succession of characteristic moisture profiles in the soil in summer and winter, illustrating the seasonal differences. In summer, when there is no rainfall, evaporation at the surface causes water to move upwards from the water table to the surface by capillarity, but the deeper the water table, the smaller the flux.

In practice, it is accepted that the loss through evaporation from the aquifer becomes negligible, even in tropical or arid zones, when the aquifer is situated at a depth of 10–15 m below the ground surface. We shall explain this capillary rise in Chapter 2.

Another phenomenon plays a role similar to that of the evaporation on the ground: plant transpiration. The roots of the plants are able to take up water from the soil in the unsaturated zone, or even in the saturated zone, if it is near (some trees have roots 10-m long or longer).

This transpiration thus gradually reduces the moisture content of the soil. Below a certain border-line value of moisture content, the plants are not able to extract water from the soil: this is the wilting point, which varies from one species to another. This is generally expressed as suction or tension in bars, rather than in moisture content (see the study of the unsaturated zone in Chapter 2).

The two phenomena, evaporation and transpiration, are generally treated together without distinction under the term "evapotranspiration," and it is this quantity which one tries, with difficulty, to estimate or measure.

### 1.3.1. *Empirical Estimation*

Several empirical formulas that give the evapotranspiration have been developed. They are based on climatological measurements (temperature,

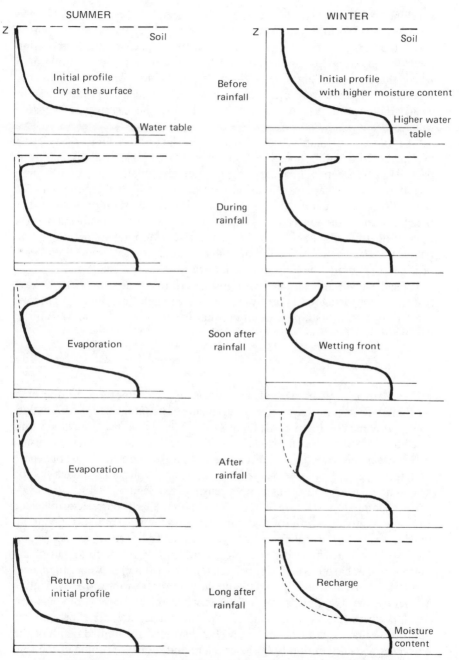

**Fig. 1.4.** Changes in the moisture profile of the ground after a rainfall.

sunshine, wind velocity, etc.). We give, in Appendix 1, as an example, the formulas of Thornthwaite, Turc, and Penman. They estimate a monthly "potential" evapotranspiration (called $ET_P$) that represents the evaporating power of the atmosphere observed on the ground in a plant-covered area where there is at all times sufficient water in the soil for the needs of the vegetation. If there were a shortage of water, the real evapotranspiration (called $ET_R$) would be a function of the $ET_P$ and the quantity of available water.

As a first approximation, one imagines that the upper layer of the soil (the first meter, for example) constitutes a reservoir, the readily available supply in the soil (RAS), the maximum capacity of which is estimated (usually 100 mm). In this reservoir, evapotranspiration may occur freely at the potential $ET_P$ rate. When it is empty, the evapotranspiration can only feed on the precipitations of the given month. When it is full, the excess moisture generates infiltration towards the aquifer. During a given month, one calculates the balance of the rainfall, the $ET_P$, and the reserve in the RAS, which makes it possible to compute the $ET_R$ and the infiltration into the aquifer.

Table 1.1 gives an example, for a given year, from the Camlibel (Turkey) plateau region, using the Thornthwaite formula to calculate the $ET_P$. Thus we can estimate, as a first approximation, that infiltration is 37 mm/yr and the real evapotranspiration 392 mm/yr.

## 1.3.2. *Measurements*

There are also direct methods for measuring the evapotranspiration on a portion of the ground, based on measurements of its energy balance (radiative, convective, and conductive heat flux). The term for latent heat of evaporation given by the balance is converted into water mass [see, in particular, Choisnel (1977)]. But this method is painstaking and, at the moment, it is applicable only to very small surfaces of the order of 1 m². Projects are underway for extending its use by means of remote sensing [see Seguin (1980)].

Another method is the use of a lysimeter. This is a large barrel (e.g. diameter 1 m, depth 2 m) filled with soil and buried in the ground so that its top is at the same elevation as the ground surface. Vegetation can be grown on and around the lysimeter. One can either measure the infiltration flux seeping out of the bottom of the barrel (which can be reached by a tunnel) or weigh the barrel regularly, thus determining rainfall and daily evapotranspiration. Lysimeters are very expensive and not very precise: when the soil dries up, voids can develop where the soil is in contact with the barrel, creating short-cuts for infiltration that do not exist in nature. Furthermore, the limited depth of the barrel causes a discontinuity in the pressure and moisture profile of the soil, which is not present in nature (see Chapter 2).

**Table 1.1**

Estimation of Evapotranspiration and Infiltration

| | Jan. | Feb. | March | Apr. | May | June | July | Aug. | Sept. | Oct. | Nov. | Dec. | Annual |
|---|---|---|---|---|---|---|---|---|---|---|---|---|---|
| Mean temp. (°C) | 6 | 8.2 | 13.1 | 18.3 | 23.1 | 27.6 | 29 | 29.9 | 26.7 | 21 | 14.7 | 8.7 | 18.9 |
| $ET_P$ (mm) | 5.2 | 9.6 | 30.8 | 65.8 | 118.3 | 171.5 | 189.5 | 190.6 | 133.2 | 75.4 | 31.4 | 10.5 | 1031.8 |
| Rainfall (mm) | 49.9 | 38.1 | 48.7 | 47.9 | 58.3 | 38.1 | 8.7 | 5.7 | 17.6 | 28.4 | 36.4 | 51.1 | 428.9 |
| RAS (mm) | 90 | 100 | 100 | 82.1 | 22.1 | — | — | — | — | — | 5 | 45.6 | — |
| Infiltration (mm) | — | 18.5 | 17.9 | — | — | — | — | — | — | — | — | — | 36.4 |
| $ET_R$ (mm) | 5.2 | 9.6 | 30.8 | 65.8 | 118.3 | 60.2 | 8.7 | 5.7 | 17.6 | 28.4 | 31.4 | 10.5 | 392.2 |

Evaporation is a very important phenomenon in hydrology. Indeed, if we try to assess the entire hydrologic cycle on the planet, we find the mean figures in the accompanying tabulation (Budiko *et al.*, 1962).

| | |
|---|---|
| Depth of water fallen on dry land | 720 mm |
| Evapotranspiration | 410 mm (57%) |
| Stream and groundwater flow toward oceans | 310 mm (43%) |
| Direct evaporation from oceans | 1250 mm |
| Depth of water fallen on oceans | 1120 mm |

The figures nearly balance, if one takes into consideration that the oceans take up 70% of the surface of the earth and the continents 30%:

*Excess of precipitation* on land in relation to evapotranspiration: $310 \times 0.3 = 93$ mm;

*Deficit in precipitation* on the oceans as compared to the evaporation: $130 \times 0.7 = 91$ mm.

The infiltrated water circulating in the aquifers (which is our main concern in this book) flows out and is eventually found in the streams, which it feeds even when there is no rainfall; this recharge from the underground medium to the surface flow net is called baseflow, as opposed to stormflow, which includes a component of surface runoff.

This is the reason that engineers working in surface hydrology often call the real evapotranspiration "the flow deficit": it is indeed that part of the precipitation that does not eventually find its way into the streams.

In spite of the great variations in the rainfall, which depends on the geographic location, the altitude, the year, etc., and the large variety of runoff, infiltration, and evapotranspiration mechanisms, this deficit, strangely enough, does not vary a great deal; in temperate climates, it is, on the average, on the order of 470 mm/yr.

## 1.4. Snow

The precipitation that falls in the form of snow has a fate similar to that of rain, but with a time lag. At the outset, there is no wetting, infiltration, or runoff. Evaporation occurs as sublimation of the snow. When the snow melts, infiltration and runoff begin.

The infiltration rate is generally higher because the recharge of water into the soil is slower than in the case of rain. However, if the soil is constantly frozen at a certain depth (permafrost), a large portion of the water becomes runoff and may carry away the top layer of the soil, which is not frozen (mud slide).

## 1.5. Schematization of the Hydrologic Cycle

In Fig. 1.5, the elements of the hydrologic cycle described above are schematically summarized, based on Eagleson (1970). Some figures should be added, as follows.

### 1.5.1. *Statics*

Estimation of the volumes of water available in the world is given in the accompanying tabulation.

| | | |
|---|---|---|
| Oceans | 1320 million km$^3$ | 97.20% |
| Snow and ice | 30 million km$^3$ | 2.15% |
| Groundwater at a depth of less than 800 m | 4 million km$^3$ | 0.31% |
| Groundwater at a depth of more than 800 m | 4 million km$^3$ | 0.31% |
| Unsaturated zone | 0.07 million km$^3$ | 0.005% |
| Fresh water lakes | 0.12 million km$^3$ | 0.009% |
| Salt water lakes | 0.10 million km$^3$ | 0.008% |
| Rivers | 0.001 million km$^3$ | 0.0001% |
| Atmosphere | 0.013 million km$^3$ | 0.001% |

### 1.5.2. *Dynamics*

The annual volume of precipitation in the world may be estimated at 0.5 million km$^3$, i.e., about 0.04% of the volume of water on the earth, or, again, 40 times the volume of water vapor in the atmosphere. This implies a very fast renewal of this atmospheric moisture: the average time the water vapor "spends" in the atmosphere is only 9 days.

## 1.6. Different Branches of Hydrology

The study of the water cycle or hydrology in its wider sense is usually divided into three separate disciplines: meteorology, surface hydrology, and hydrogeology.

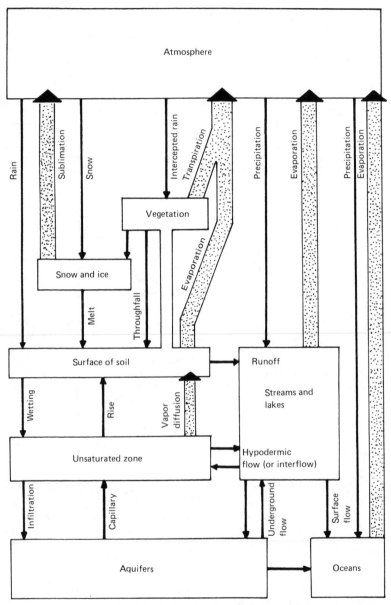

**Fig. 1.5.** Hydrologic cycle. After Eagleson (1970).

*Meteorology or climatology* comes first in the study of the water cycle. It has several aspects: (1) composition and general circulation of the atmosphere; (2) energy balance of the atmosphere; (3) precipitation, rainfall and snow, snowmelt, artificial rain; and (4) evaporation and evapotranspiration.

The random nature of the climate results in a great variability, on different levels of time and space, of the precipitation, which is the first link in the chain of the hydrologic cycle. This precipitation is consequently studied from a statistical viewpoint, which is also used in the following links in the chain.

*Surface hydrology* is concerned with flow in the hydrographic network. It may be studied with several aims in mind:

(1) Evaluation of available resources, either in their natural state or after development (dam), and the calculation of the reservoir volume necessary to ensure a given flow.

(2) Forecasting of flood risks and the works required to control them (drainage network, retarding basin). Very often the works (dams) have to fulfill several simultaneous and often contradictory needs: a reservoir to control floods must be emptied as fast as possible, and this is directly antagonistic to the objective of a reservoir meant to increase flow at low water. Hence the difficult management problems attached to multipurpose installations.

In hydrology, two methods are commonly used:

(1) The stochastic method: because of the variability of rainfall, streamflow is studied as a random variable.

(2) The deterministic method: the process of runoff and infiltration is studied from a physical deterministic viewpoint (flow equations) based on an impulse assumed to be known, rainfall, on which the entire variability is concentrated.

The basin may be represented as a black box in which its components are lumped together, which one studies according to the theory of systems analysis (Fig. 1.6).

On the other hand, one may study the watershed from a physical point of view by considering all the physiographic parameters of the medium. *Groundwater hydrology* or *hydrogeology* is our main concern in this book.

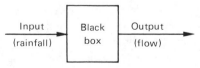

**Fig. 1.6.** Black box system.

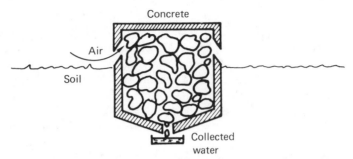

**Fig. 1.7.** Condensation of atmospheric water vapor.

### 1.7. Other Possible Origins of Groundwater

Most of our groundwater is part of the hydrologic cycle described above and is called cyclic water. However, there are other ways in which the ground may be recharged with water:

*Condensation of atmospheric water vapor in the empty spaces of the ground.* This is the equivalent of morning dew on the surface. This phenomenon can be of some importance and is usually called "occult precipitation." In ancient times, it was reported that the city of Theodosia in Crimea was supplied with water from a huge pile of rocks linked to seven fountains. Experiments made in Montpellier (France) have given (based on Geze, 1967) a flow of 2 liters/day for a pile of 5 m$^3$ of rocks (see Fig. 1.7). This is also cyclic water.

*Juvenile water.* This water has its origin deep down. A granitic magma expells a small amount of water when it cools. It has been calculated that a magma of 1000 m thickness, containing 5% water in weight, gives rise to a flow of the order of 25 liters min$^{-1}$ km$^{-2}$ a figure that should be compared to 10 liters s$^{-1}$ km$^{-2}$ (the order of magnitude of infiltration in temperate climates of cyclic waters). It is therefore usually negligible (Geze, 1967).

*Fossil water.* This is cyclic water dating from a more humid period in the Quaternary period. A good example is the Sahara desert, which contains large amounts of fresh water infiltrated a few thousand years ago. However, a minor recharge still occurs during exceptional storms (about once every 30 yr). Another case of fossil water is connate water, generally saline, which dates back to the formation of the sediments.

*Thermal water.* This is mostly cyclic water which follows complicated paths, is heated at depth and then ascends toward the surface by way of thermal springs.

*Chapter 2*

# Rock Porosity and Fluid–Solid Relations in Porous Media

Most rocks and soils naturally contain a certain percentage of empty spaces, which may be occupied by water or fluids. This is what is known as their porosity. These voids must be distinguished from their interconnecting pathways, which allow fluids to circulate through them: this second property, the permeability, is examined in Chapter 4. Suffice it to say that porosity is a necessary, but in itself insufficient, condition for permeability.

In the study of porosity, we distinguish between (1) the existence of voids and their geometry proper, defining the total porosity, and (2) the manner in which the fluid is distributed in these voids and the ensuing fluid–solid relations, which enable us to define the kinematic (or effective) porosity.

Finally, we describe how to measure the porosity and the fluid pressure in the pores.

## 2.1. Total Porosity

(a) *Granulated rocks.* Most rocks are constituted by solid mineral particles, more or less tightly stuck together, forming a skeleton around which empty spaces remain. These are *porous media* in terms of fluid mechanics. For example, sand and sandstone have a total porosity that may reach 30%. However, even rocks that are generally thought to be solid have a certain porosity; examples are limestone, dolomite (particularly secondary), and even crystalline and metamorphic rocks (from 1 to 5%).

Clays belong to a separate category. Their constituent elements resemble thin shavings and are organized into "sheets," which are stacked in parallel layers separated by variable intervals where a fluid might lodge. This gives the clays, in particular, the property of swelling in the presence of water. Furthermore, we shall see that this water is strongly linked to the solid clay particles. All the same, the percentage of voids may be very high, on the order of 40% and even up to 90%, in unconsolidated marine red clays.

(b) *Fractured rocks.* Fracturing is a special case of voids in solid rocks. Because of tectonic movements, e.g., faults, fissures, joints, cracks, openings along bedding planes almost all rocks in the earth's crust are fractured. These fractures are generally oriented in at least two (generally three or four) main directions, which cut up the rock into blocks (Fig. 2.1).

We then have a network of fractures, more or less interconnected, which may create voids in the rock, if the fractures are not sealed by some kind of deposit (clay, calcite, quartz, etc.). In this case, we talk about fracture porosity, as opposed to the interstitial porosity already mentioned. Moreover, these two types of porosity may coexist (sandstone, limestone, etc.).

(c) *Definition.*

$$\text{Total porosity } \omega = \frac{\text{volume of the voids}}{\text{total volume of the rock}}$$

Soil mechanics also uses

$$\text{Void ratio } e = \frac{\text{volume of the voids}}{\text{volume of the solid}}$$

**Fig. 2.1.** Typical fracture network.

Both are of course dimensionless. We shall always use $\omega$, but one may pass from one to the other by

$$e\omega = e - \omega$$

i.e.

$$\omega = \frac{e}{e+1}, \qquad e = \frac{\omega}{\omega - 1}$$

(d) *Representative elementary volume or random functions: Definition of the local properties of a porous medium.* The notion of porosity is easy enough to understand but, on reflection, it poses some problems if we want to define it with precision. We shall discuss them here while keeping in mind that the following applies to other properties of the porous medium as well, such as permeability.

There are two accepted ways of defining the local properties of a porous medium: the notion of the representative elementary volume (REV) and that of the random functions (RF, which is also expressed as "ensemble average"). We shall see that these two notions implicitly influence any description of the spatial variations of the hydrogeological parameters.

The whole problem stems from the fact that the notions of porosity and permeability, which are notions concerning points in an equation with partial derivatives, for instance, cannot be defined or measured at single points, since a porous medium is a conglomeration of solid grains and voids. Below a certain scale of volume, porosity and permeability have no physical significance.

The REV method consists in saying that we give to one mathematical point in space the porosity or permeability of a certain volume of material surrounding this point, the REV, which will be used to define and possibly measure the "mean" property of the volume in question. Consequently, this concept involves an integration in space. It is obviously the first method that comes to mind. Behind it lies the idea of a sample, which is collected and from which the relevant property is estimated by measurement. More exactly, the size of the REV is defined by saying that it is

(1) sufficiently large to contain a great number of pores so as to allow us to define a mean global property, while ensuring that the effect of the fluctuations from one pore to another are negligible. One may take, for example, 1 cm$^3$ or 1 dm$^3$.

(2) sufficiently small so that the parameter variations from one domain to the next may be approximated by continuous functions, in order that we may use the infinitesimal calculus, without in this way introducing any error that may be picked up by the measuring instruments at the macroscopic scale, where meters and hectometers are the usual dimensions.

This is, incidentally, a bit like the problem in fluid mechanics of passing from the "corpuscular" scale to that of the "particle of matter." It should be

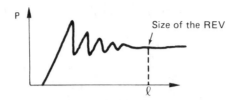

**Fig. 2.2.** Definition of the REV.

noted that in a fractured medium, the size of the REV may be quite astonishingly large so as not to satisfy the second hypothesis of "continuous functions" on the scale of the measuring instruments.

The size of the REV (measured, for example, by one of its characteristic dimensions $l$, such as the radius of the sphere or the side of the cube) is generally linked to the existence of a flattening of the curve that connects the studied integral property $P$ with the dimension $l$ (Fig. 2.2). However, nothing allows us to assert that such a flattening always exists. The size of the REV thus stays quite arbitrary.

Other important objections that can be made to this conception of the porous medium are of two kinds. First, it is very badly suited to the treatment of discontinuities in the medium. When, in a thought experiment, the REV is moved across a discontinuity, the studied property is subjected to a continuous variation (Fig. 2.3). This sometimes poses problems of how to correctly represent boundaries or limits between two media. Second, the most important objection is that it gives no basis for studying the structure of the property in space. The most that can be said is that the spatial variations of the studied property must be smooth in accordance with the same thought process as above concerning the discontinuities.

Marle (1967) has suggested a more rigorous conception of spatial integration. In order to achieve this, he proposes the use of an integrable nonnegative weighting function $m(\mathbf{x})$ such that its integral, when extended over the whole space, is equal to 1; this weighting function would not necessarily have a bounded support. The macroscopic magnitude $\langle a \rangle(\mathbf{x})$ will then be defined

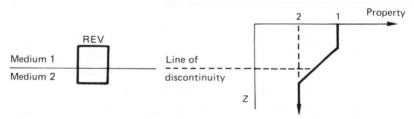

**Fig. 2.3.** Definition of the properties of a discontinuous medium using the REV.

from the local microscopic magnitude $a(\mathbf{x})$ by a convolution extended over the whole space of $a$ by $m$:

$$\langle a \rangle(\mathbf{x}) = \int a(\mathbf{x} + \mathbf{x}')m(\mathbf{x}') \, d\mathbf{x}'$$

where $\mathbf{x}$ stands for the coordinates in three-dimensional space $(x_1, x_2, x_3)$. For the study of porosity, we choose an indicator $a(\mathbf{x})$. If the point $\mathbf{x}$ is in a pore, $a(\mathbf{x}) = 1$; if it is in a grain, $a(\mathbf{x}) = 0$.

Furthermore, Marle suggests that this definition be generalized to the properties "$a$" that are not continuous in the whole space* and that can be described by distributions. The convolution is then taken in the sense of distributions.

This method has the advantage of making the function $\langle a \rangle$ continuous and indefinitely differentiable, even if $a$ is not, by a suitable choice of $m$. If the problem of the size of the REV is eliminated, that of the choice of the weighting function still remains arbitrary. However, with the help of this weighting function it is possible to establish the connection between this method and the second, which we examine in the following.

The random functions (RF) method is a more powerful concept. It consists in saying that the studied porous medium is a realization of a random process. Let us try to visualize the concept. Suppose that we create in the laboratory several sand columns, each filled with the same type of sand. Each column represents the same porous medium but is somehow different from the others. Each column is a "realization" of the same porous medium, defined as the ensemble of all possible realizations (infinite in number) of the same process.

A property like porosity can then be defined, at a given geometrical point in space, as the average over all possible realizations of its point value (defined as 0 in a grain and 1 in a pore). One speaks of "ensemble averages" instead of "space averages." For the sand columns just described, it is obvious that the ensemble average (or expected value) of these point porosities will be identical to the space average defined by taking the column itself as the REV. Furthermore, this ensemble average will be the same for any point of the column. We will define later the conditions necessary for this to be true.

In more general terms, a property $Z$ will be called a random function (RF) $Z(\mathbf{x}, \xi)$ if it varies both with the spatial coordinate system $\mathbf{x}$ and with the "state variable" $\xi$ in the ensemble of realizations. Then $Z(\mathbf{x}, \xi_1)$ is a realization of $Z$; $Z(\mathbf{x}_0, \xi)$ is a random variable, i.e., the ensemble of the realizations of the RF $Z$ at $\mathbf{x}_0$; and $Z(\mathbf{x}_0, \xi_1)$ is the single value of $Z$ at $\mathbf{x}_0$ for realization $\xi_1$. To simplify the notations, the variable $\xi$ is generally omitted.

---

* For example, a surface density of the adsorbed matter on the fluid–solid interface.

If we want to find a less abstract example of a random porous medium, we can recall the countless alluvial plains and fans (debris cones) descending over several thousands of kilometers from the Andes to the coasts of Peru and Chile, which are created by erosion of the same materials under the same conditions and consequently constituted by the same kind of deposits. There, we have a very large number of realizations of the "same" medium.

The immense advantage of the stochastic approach is that one can study other statistical properties of the porous medium in the ensemble of realizations than just the expected value. One very often uses the variance (called dispersion variance, see Chapter 11) of the property, which characterizes the magnitude of the fluctuations with respect to the mean, and the autocovariance (or simply covariance), which characterizes the correlation between the values taken by the property at two neighboring points in space.

However, when studying a given porous medium, there will be only one realization of the conceptual random medium. Some assumptions are necessary to make this concept useful. The most common are stationarity and ergodicity.

*Stationarity* assumes that any statistical property of the medium (mean, variance, covariance, a higher-order moment) is stationary in space, i.e., does not vary with a translation. It will be the same at any point of the medium. *Weak stationarity* refers to a medium where only the first two moments are stationary: if $Z(\mathbf{x})$ is the studied property, $\mathbf{x}$ being the coordinates in one, two, or three dimensions, then the random function (RF) $Z(\mathbf{x})$ satisfies:

(1)  *expected value*

$$E[Z(\mathbf{x})] = m \qquad \text{not a function of } \mathbf{x}$$

(2)  *Covariance*:

$$E[(Z(\mathbf{x}) - m)(Z(\mathbf{x} + \mathbf{h}) - m] \qquad \begin{array}{l}\text{not a function of } \mathbf{x}, \\ \text{but a function only of the lag } \mathbf{h}, \\ \text{a vector in two or three dimensions.}\end{array}$$

By developing, and labeling this covariance $C(\mathbf{h})$,

$$C(\mathbf{h}) = E[Z(\mathbf{x}) \cdot Z(\mathbf{x} + \mathbf{h})] - m^2 \qquad (2.1.1)$$

By definition,

$$C(0) = E[(Z(\mathbf{x}) - m)^2] = \sigma_Z^2$$

is the variance of $Z$.

In more rigorous terms, strong stationarity means that all the probability distribution functions (pdf) of the random function $Z(\mathbf{x})$ are invariant under translation, whether we consider one point $p(Z(\mathbf{x}))$ or $n$ points $p(Z(\mathbf{x}_1), \ldots, Z(\mathbf{x}_n))$.

*Ergodicity* implies that the unique realization available behaves in space with the same pdf as the ensemble of possible realizations. In other words, by observing the variation in space of the property, it is possible to determine the pdf of the random function for all realizations. This is called the "statistical inference" of the pdf of the RF $Z(x)$. We will see in Chapter 11 how it can be done.

In the vocabulary of stochastic processes, a phenomenon that is "stationary" and "ergodic" is called *homogeneous*. We would then use "uniform" to describe a medium in which some property does not vary in space. Geologists traditionally call it "homogeneous."

Other less stringent hypotheses can also be defined, e.g., stationarity of increments of $Z$. These will be defined in Chapter 11.

Marle (1967) compares this method to the one based on spatial integration and shows that the stochastic definition may be regarded as the limiting case of an integral definition when the porous medium is assumed to be infinite, ergodic, and stationary and the weighting function does not have a bounded support. As a matter of fact, spatial integration in an infinite volume reproduces the mathematical expectation over all possible realizations, if the medium is indeed stationary and ergodic. We shall use, in turn, these two methods for defining the properties of porous media.

Other approaches can also be used to define the properties of porous media. One is that of composite materials (Beran, 1968) and, has been applied to porous media by Dagan (1979, 1981, 1982b).

(e)   *Porosity and grain size.*   If one studies *conceptual* porous media, which consist of a cluster of spheres with the same diameter, one can show that there are six possible ways of packing the spheres, resulting in porosities of 26%, 30%, 40%, and 48%. These are, of course, independent of the size of the spheres.

In the case of spheres of different sizes, the porosity is always lower because, as Houpeurt (1974) states, if one reasons on the basis of the large grains, one can say that the small ones take up part of the pores that may exist between them. Conversely, if one reasons on the basis of the small grains, one can say that any large grain gives a greater compactness by its mere presence.

In the case of nonspherical grains, this tendency to lower the porosity is to a certain extent compensated by the irregularities in the shape of the grains, which prevent them from being tightly pressed together.

For unconsolidated media (e.g., sands), one tries to establish the size distribution of the grains in the medium. In general, the wider this distribution is, the smaller the porosity.

A grain-size analysis of the medium, by sieving for example, is represented by the grain-size curve, which gives the percentage (by volume or usually by

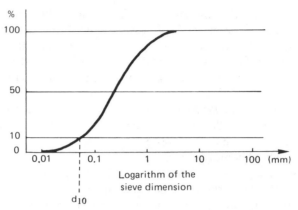

**Fig. 2.4.** Grain-size curve of an unconsolidated medium.

weight) of the elements passing through a sieve, with holes of a given size (Fig. 2.4). Effective grain-size $d_{10}$ is the sieve-dimension at which 10% of the elements of the medium are smaller than $d_{10}$. It is accepted that $d_{10}$ is the most important parameter among those governing the permeability properties of a porous medium (see Chapter 4).

However, it is always necessary to measure the porosity of the medium without destroying the arrangement of its grains (see Section 2.3). We know that the porosity varies with the arrangement of the grains from the study of packed spheres and that this arrangement is a function of the consolidation, or compression, of the medium.

Figure 2.5 (from Bear, 1972) gives a few examples of grain-size curves and a classification of the terms used, according to the International Society of Soil Sciences: gravel, sands, silts, clays.

(f) *Surface porosity.* Using a section of the porous medium, we can define the total surface porosity

$$\omega_s = \frac{\text{surface area of the voids on the section}}{\text{total surface area of the section}}$$

If the size and distribution of the voids are entirely random, the surface porosity is independent of the orientation of the studied section. Furthermore, it has the same value as the volume porosity. To be certain of this, one only needs to integrate the surface porosity on an elementary length at right angles to the section plane. It is, of course, necessary to choose volumes and surfaces of the order of the REV.

The situation is, however, not at all the same if the distribution of the voids is not random, but follows a law of organization resulting from the deposition

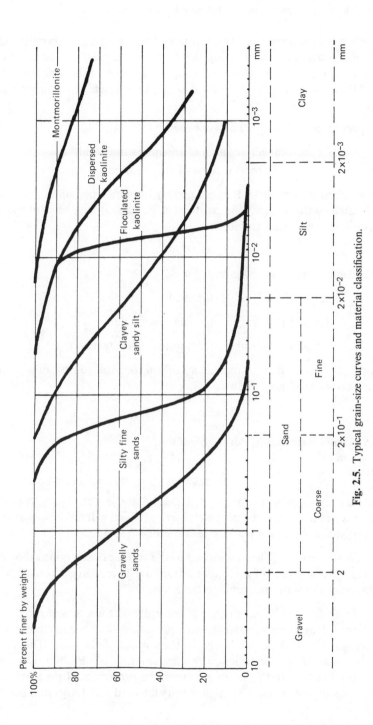

**Fig. 2.5.** Typical grain-size curves and material classification.

process when the medium was formed. It would be possible to determine such directional surface porosities using a "texture analyzer," a device that determines geometrical properties of images. However, this is not usually done, and the surface porosity is assumed isotropic and identical to the volume porosity.

(g) *Specific surface area.* This is defined by

$$S_{sp} = \frac{\text{total surface area of the interstitial voids}}{\text{total volume of the medium}}$$

with dimensions of $(\text{length})^{-1}$. It varies greatly from one medium to another. The more divided the medium, the greater it is. For example, spheres of radius $R$ in a cubic arrangement exhibit

$$S_{sp} = \pi/2R$$

Here are a few orders of magnitude for $S_{sp}$:

$1.5 \times 10^4 \, \text{m}^2/\text{m}^3$    for sand

$1.5 \times 10^5 \, \text{m}^2/\text{m}^3$    for fine sandstone

$1.5 \times 10^9 \, \text{m}^2/\text{m}^3$    for montmorillonite (clay)

This parameter is of great importance for the phenomena of fluid–solid relations that we shall now discuss.

## 2.2. Fluid–Solid Relations in Porous Media

### 2.2.1. *Water-Saturated Media*

First, we shall discuss two-phase media: solid and water. Apart from the water that is a constituent part of rock minerals (which will not be discussed), one must distinguish between adhesive water and free water.

(a) *Adhesive water.* This is attached to the surface of the grains through the influence of the forces of molecular attraction. These forces decrease with the distance of the water molecule to the grain:

(1) A first adsorbed layer, which is of the order of a few tens of molecules thick (about 0.1 $\mu$m), corresponds to an orientation of the water molecules with a bipolar H—OH perpendicular to the surface of the solid. The stress created by these forces of attraction reaches several $10^{12}$ Pa, but decreases rapidly with distance. In this adsorbed layer, the properties of the water are greatly changed: strong viscosity, high density (around 1.5). Large numbers of

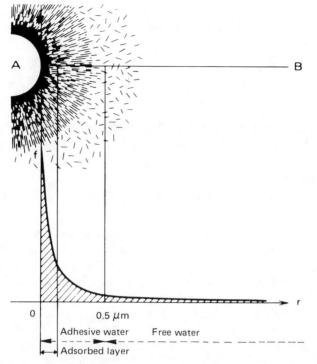

**Fig. 2.6.** Structure of the adhesive water. [From Polubarinova-Kochina (1962).]

ions, mainly cations, may be retained (adsorbed) by joint attraction of the water and solid molecules. We shall return to this in Chapter 10.

(2)  A transition zone, between 0.1 and 0.5 $\mu$m, contains water molecules that are still subjected to a nonnegligible attraction and stay immobile.

(3)  Beyond this, the forces of attraction are negligible, and the water is said to be free.

It is obvious that this limit of 0.5$\mu$m is somewhat arbitrary and varies from one medium to another. Figure 2.6 illustrates the variation of the forces of attraction of the water molecules and their orientation in the vicinity of a solid grain. For the adsorbed water layer in contact with a solid particle, the curve shows the variation of the force of attraction on particles along the radial section AB.

This phenomenon of adsorption of water molecules and ions is linked to the specific surface area of the medium and is especially strong in clay media, where it greatly reduces the possibility of water and ions circulating; this leads us to the definition of the kinematic porosity of a porous medium.

(b)  *Free water.*  We have already defined this. It is the water that is outside the field of attraction of the solid particles and that can be displaced—as opposed to the adhesive water—by gravity or pressure gradients (see Chapter 4).

(c)  *Kinematic porosity of a saturated medium.*  From the point of view of fluid displacement, the adhesive water may be viewed as part of the solid. The empty volume, where the water can circulate, is smaller than the total porosity: it defines the kinematic or effective porosity of a saturated medium. It must be understood, however, that this definition of porosity is already linked to the concept of fluid circulation and not to the percentage of the volume taken up by the fluid phase.

Other phenomena besides adhesion have a limiting effect on the kinematic porosity.

(1)  The existence of unconnected pores. These are "bubbles" of liquid occurring in the solid phase. As their liquid cannot circulate, these voids are of no account in the kinematic porosity. In Section 2.3, we shall see that certain methods for measuring the porosity, based on the impregnation of the porous medium by a fluid, exclude the unconnected pores.

The most common example is that of secondary dolomite, i.e., dolomite formed after the deposition by diagenetic transformation of calcite into dolomite. This transformation is accompanied by a shrinking with angular crystallization of the dolomite. The total porosity is high, 20–30%, but the kinematic porosity is low, because the pores are often not interconnected.

(2)  The existence of dead-end pores, as in Fig. 2.7. The water contained in the cul-de-sac is almost motionless. Only the water in the "pipes" of the medium circulates. Thus, these pores are excluded from the kinematic porosity, but they do play a role when we study the mechanisms of compressibility or of solute transport in porous media.

(3)  On an even larger scale, a fractured rock in which water circulates only in the fractures has a kinematic porosity linked to the volume of these fractures, even if the unfractured rock matrix is porous. Hence, a fractured granite, which has a total matrix porosity of 1–2%, may have a kinematic porosity of less than 1%, because the matrix itself has very low permeability.

Fig. 2.7.  Dead-end pore.

Thus, we define kinematic porosity as

$$\omega_c = \frac{\text{volume of water able to circulate}}{\text{total volume of rock}}$$

(d)  *Comments: Consequences for the tracing of water.*  We shall see later that one can link the kinematic porosity to the velocity of water circulating in the ground, hence the idea of adding a tracer to the water to measure its velocity by *in situ* experiments.

This gives rise to a number of problems, because it is necessary to choose a tracer that will not be adsorbed by the layer of adhesive water or onto the grain surface. However, even if this problem is solved (e.g., by using a tracer of the water molecule, such as tritium, $^3$H), there remains another mechanism of interaction between the tracer and the immobile water. Indeed, our description of circulating water phase/immobile water phase corresponds to a certain microscopic scale of observation of the phenomena, that of fluid layers and flow.

On a molecular scale, things change. There may be a continual exchange of molecules from one phase to the other through molecular Brownian motion: for example, a circulating molecule may become immobilized in the course of its progress, while another one that was originally immobile, may be set in motion. From the point of view of fluid circulation, nothing is visible, but the idea of tracing the route taken by an individual water molecule is bereft of meaning. Are we right to distinguish between two water molecules that we have no physical means of telling apart? Hence, a molecule which was in the position A at the instant $t_0$, and in position B at the instant $t_1$ may very well have progressed along the route ACB (Fig. 2.8) and been "exchanged" in C for a molecule which was initially adhesive, and to which it has communicated its energy. Is there any meaning to the question "Is the molecule in B the one that was originally in A?"

In a porous medium, the progress of a water molecule may be much more complex than the microscopic image leads us to suppose and may make the concept of tracing useful only in ascertaining the circulation velocity of a substance dissolved in the water. This problem will be discussed in Chapter 10.

A

C

B

**Fig. 2.8.** Water molecule trajectory in a porous medium.

### 2.2.2. *Unsaturated Media*

This problem is complicated by the existence of a third phase, air, as well as the water and solid phases.

(a) *Moisture content and volumetric saturation.* The moisture content $\theta$ in a REV is defined by the ratio

$$\theta = \frac{\text{volume of water}}{\text{total volume of the REV}}$$

and the volumetric saturation by the ratio

$$s = \frac{\text{volume of water}}{\text{total pore volume}}$$

where $\theta$ may vary from 0 to the total porosity $\omega$, and $s$ from 0 to 1 or from 0 to 100%.

(b) *Air–water relationships for different moisture contents.* We observe that in a soil containing both air and water the free water "wets" the solid grains, i.e., surrounds them, whereas the air tends to stay in the middle of the voids. Thus, we have the following descriptions for various moisture contents.

(1)   Soil close to maximum water saturation (Fig. 2.9).

(*a*)   The water phase is continuous and may circulate under the influence of gravity. This is called "funicular" or gravitational water.
(*b*)   The air phase is discontinuous and does not circulate. It may reach 10–15% of the porosity, even in a medium said to be saturated, close to the water table of the aquifer. The imprisoned air bubbles can only pass through the contractions in the small channels connecting the pores with each other if there is a sufficiently strong pressure gradient in the water phase.

(2)   Soil at "equilibrium water saturation" or at its "capillary retention capacity" (Fig. 2.10).

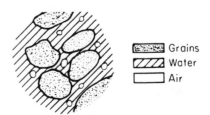

|  | Grains |
|---|---|
|  | Water |
|  | Air |

**Fig. 2.9.** Wet soil.

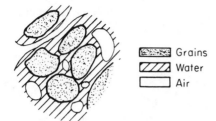

Grains
Water
Air

**Fig. 2.10.** Soil at its retention capacity.

(a) The water phase is still continuous, but it no longer circulates under the sole influence of gravity. In agronomy, one says that the ground has reached its "field capacity" a few days after a rainfall when the water that can percolate by gravity has left the soil profile. The term specific yield or drainage porosity ($\omega_d$) is used to describe the part of the porosity that can be drained by gravity, i.e., the difference between the moisture content of the saturated medium and that attained at the equilibrium saturation. Note that pressure is transmitted through the continuous water phase and that, as a consequence, the equilibrium saturation varies, in principle, with the elevation above the water table of the point being considered (to be discussed further).

(b) The air phase is also continuous, but does not generally circulate.

(3) Weakly saturated soil (Fig. 2.11).

(a) The water forms a thin film around each grain (adhesive water) as well as rings surrounding each point of contact between the grains. These are called "pendular rings" or pendular water. The water phase is still continuous, the pressures are in principle transmitted, but the movement of water is very slow because the water film is so thin.

(b) The air phase is continuous but usually immobile. In this case, the evaporation inside the porous medium may become considerable as compared to the other modes of water movement. However, in order to leave the medium, the evaporated water must migrate by molecular diffusion toward the exterior, which is a very slow process. A migration

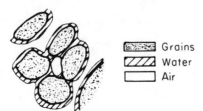

Grains
Water
Air

**Fig. 2.11.** Dry soil.

by density convection cells is also conceivable, but such a phenomenon has never been observed.

(4) Irreducible saturation. In order to get below the equilibrium saturation, we have already made use of phenomena other than circulation by gravity, i.e., evaporation and plant transpiration. If the moisture content continues to decrease, we are eventually left with only the adhesive water, which is sometimes called hygroscopic moisture. This irreducible saturation depends, in reality, on the drying methods that have been applied.

(a)   Irreducible saturation in a natural medium when the drying-out is caused by natural phenomena.

(b)   Irreducible saturation at 105°C. A soil sample is generally dried by heating it to 105°C. This temperature is chosen arbitrarily because beyond it there is a risk of decomposing certain minerals and extracting the water that is a constituent part of the solid phase. However, it is certain that a small fraction of the adhesive water is still present in the medium. Thus, a clay has to be heated to 900°C for all the water to be extracted. Indeed, the film of adhesive water creates a continuous layer that surrounds the grains, whatever the degree of saturation.

(c)   *Capillary pressure.*   Let us analyze the balance of pressures between the air phase and water phase in an unsaturated medium.

Between two fluids in contact with each other, or a fluid in contact with a solid, there is a free interfacial energy, created by the difference between the forces which attract the molecules toward the interior of each phase and those which attract them through the contact surface. The interfacial energy is characterized by the interfacial tension $\sigma_{ik}$, defined by the quantity of work needed to separate a surface of unit area of the substances $i$ and $k$. The tension $\sigma_{ik}$ is constant for two given substances, and varies only with the temperature. The interfacial tension $\sigma_i$ between a liquid and its own vapor is called vapor tension or surface tension.

For two fluids in contact with each other, Young's equation gives the connecting angle of the interface as in Fig. 2.12. Then $\theta$, measured from 0 to

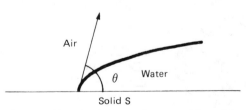

**Fig. 2.12.** Interface between air, water, and a solid.

$180°$ in the denser fluid (here, the water), is given at equilibrium by

$$\cos \theta = \frac{\sigma_{sa} - \sigma_{sw}}{\sigma_{aw}}$$

where the subscript sa stands for solid–air, sw stands for solid–water, and aw stands for air–water. There is no equilibrium possible if this ratio is larger than 1; in that case, one of the fluids (here, water) spreads indefinitely over the solid. If $\theta < 90°$, the fluid is said to be wetting. This is the case for the water here. If $\theta > 90°$, the fluid is said to be nonwetting. This is the case of the air here. The term $\sigma_{aw} \cos \theta$ is called the àdhesion tension.

In the fluids, on either side of the air–water interface the pressure is not the same. This difference in pressure is called the capillary pressure,

$$p_c = p_{air} - p_{water}$$

If $r$ is the mean radius of the interface curvature,

$$\frac{2}{r} = \frac{1}{r'} + \frac{1}{r''}$$

where $r'$ and $r''$ are the principal curvature radii (Fig. 2.13). Then the Laplace equation gives the capillary pressure,

$$p_c = \frac{2\sigma_{aw}}{r}$$

This pressure may be very high if the curvatures are small.

In a capillary tube, interfacial tension causes the water to rise and to form a meniscus above the level of the tank. The height of this rise is a function of the radius of the tube and measures the capillary pressure across the air–water interface in the tube.

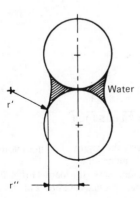

**Fig. 2.13.** Air-water interface curvature.

In porous media the shape of the interface is very complex, but there is also a capillary pressure, i.e., a difference in pressure between the air phase and the water phase. As the air phase, if it is continuous, is usually at atmospheric pressure, the water phase is at a negative pressure, which may reach several bars. One then speaks of positive suction or tension. For instance, the wilting point of certain plants is reached at a tension of the order of 15 bars if standard atmospheric pressure is the reference zero. As the air–water interfacial tension is of the order of 0.076 N/m at 20°C, this gives a mean curvature radius of the water menisci in the unsaturated medium of 0.1 μm, i.e., close to the dimension of the adhesive water layer.

To each value of the moisture content of a porous medium corresponds a certain distribution of the air and water phases. As the water phase is always continuous, the pressure at equilibrium must be uniform at a given elevation. As long as the air phase is also continuous, it stays at atmospheric pressure, and the capillary pressure must therefore be uniform at that elevation. On the average, the interfaces must thus take on a unique curvature radius. If the moisture content varies, this radius must change, and so must the capillary pressure. Hence, this capillary pressure is a function of the moisture content or the degree of saturation. Assuming that the air pressure is zero, it is usual to plot the pressure in the water versus the degree of saturation by defining the suction potential pF:

$$pF = \log\left(\frac{-p_{water}}{\rho g}\right)$$

where $\rho$ is the mass per unit volume of the water, $g$ is the acceleration due to gravity, and $p_{water}/\rho g$ is in centimeters. For example, we find curves such as in Fig. 2.14.

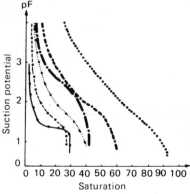

**Fig. 2.14.** Variation in the suction potential pF with the degree of saturation for different media. ——, Sands, grains of less than 500 μm; – – –, Ramona sands; – ··· –, Placentia clay loam; – · –, Hanford sandy loam; ···, Yolo clay loam; ··· Chino silty clay loam. [From Bear (1972). Reprinted by permission of the publisher from Dynamics of Fluids in Porous Media, by J. Bear. Copyright 1972 by Elsevier Science Publishing Co., Inc.].

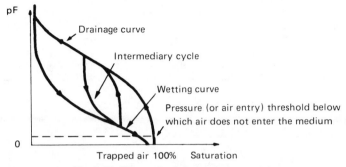

Fig. 2.15. Hysteresis of the suction potential.

However, this capillary pressure shows hysteresis with the saturation according to whether the soil is dried out or wetted. Indeed, the shape assumed by the interface at a given saturation is not the same if we soak a dry soil or drain a wet one: fluid "bubbles" remain imprisoned, the contact angles of the interfaces are not exactly the same, there are phenomena of dilation or compression, etc. Thus, we observe two suctions in Fig. 2.15. There is a whole series of intermediary cycles such as the one drawn between the two enclosing curves.

Finally, it must be said that if a sufficiently long time is allowed to elapse, in the end the trapped air is dissolved by the circulating water and the representative point moves frome one curve toward the other.

(d)  *Moisture-content profiles.*    Table 2.1 summarizes the main intervals that have been defined in the soil–water–air continuum. All these zones are also found on a soil profile such as the one in Chapter 1. This is illustrated in Fig. 2.16.

Above the level of the water table, there is first a zone with 100% saturation or nearly that, which is called the capillary fringe, where the water pressure is inferior to that of the atmosphere. This is the equivalent of the capillary rise in tubes. Indeed, there has to be a certain capillary pressure (threshold pressure) for air at atmospheric pressure and water to reach an equilibrium through the interface. However, there may be trapped air inside this zone (whence a saturation of less than 100%, e.g., 85–90%).

Above this zone, the capillary pressure increases and the saturation decreases until it reaches the equilibrium saturation and the profile is static.

On the ground surface we have shown a dried-out soil and a wet soil, both of which are in a transient state. In the case of the wet soil, the gravitational water is infiltrated and descends along the profile. In the dried-out soil, the drying-out of the ground surface causes an ascending circulation, which we shall study together with the circulation in unsaturated media in Chapter 9.

Next to the saturation profile in Fig. 2.16 we have shown the pressure profile. According to the laws of hydrostatics, a profile in equilibrium should

**Table 2.1**

Break Down of the Components of Porosity

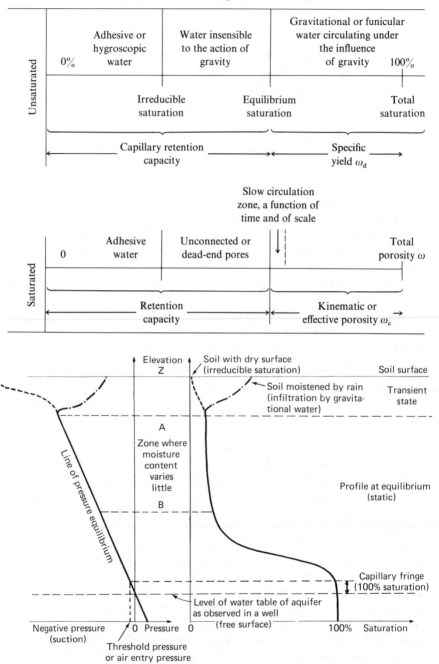

| Unsaturated | 0% | Adhesive or hygroscopic water | Water insensible to the action of gravity | Gravitational or funicular water circulating under the influence of gravity | 100% |
|---|---|---|---|---|---|
| | | Irreducible saturation | Equilibrium saturation | | Total saturation |
| | | Capillary retention capacity | | Specific yield $\omega_d$ | |

| Saturated | 0 | Adhesive water | Unconnected or dead-end pores | Slow circulation zone, a function of time and of scale | Total porosity $\omega$ |
|---|---|---|---|---|---|
| | | Retention capacity | | Kinematic or effective porosity $\omega_c$ | |

Elevation Z

Soil with dry surface (irreducible saturation) — Soil surface

Soil moistened by rain (infiltration by gravitational water) — Transient state

A — Zone where moisture content varies little

B

Profile at equilibrium (static)

Line of pressure equilibrium

Capillary fringe (100% saturation)

Level of water table of aquifer as observed in a well (free surface)

Negative pressure (suction)

0 Pressure 0

Threshold pressure or air entry pressure

100% Saturation

**Fig. 2.16.** Profiles of saturation and pressure in a soil.

exhibit a linear variation of the pressure with the elevation. By definition, the pressure is zero (i.e., equal to the atmospheric pressure) at the water table. Below it, the pressure grows linearly with the depth; above it, it decreases with the elevation and becomes a suction. This is evident if we consider that, as long as the water phase is continuous, two points at *hydrostatic equilibrium* at a vertical distance $\Delta Z$ from each other have a pressure difference of $\rho g \, \Delta Z$. All representative points situated to the left of the line of pressure equilibrium show that an ascending flow is occurring, and vice versa.

Note that the existence of the zone AB, where the saturation is approximately constant although the pressure varies, is related to the shape of the suction–moisture content curves shown in Fig. 2.15. On these graphs, the suction is marked on a logarithmic scale. Below a certain saturation, the profiles are almost vertical, i.e., a pressure variation by a factor of 10 only causes a very small variation of the saturation.

In practice, a medium is hardly ever in hydrostatic equilibrium and the real pressure profile nearly always deviates from the equilibrium line, but the orientation of this deviation actually gives the flow direction since inertial effects are negligible.

These high negative pressures (less than absolute zero) to which the water in an unsaturated medium may be subjected should not surprise us; they measure, in reality, a state of energy of the water in the soil, i.e., the quantity of energy needed to extract a molecule that is bound to the soil by electrostatic forces.

## 2.3. Porosity Measurements

### 2.3.1. *Direct Methods on Samples*

These methods are rather sophisticated and should be used in a specialized laboratory.

(1)  The total volume of the sample is measured, either by its dimensions (in particular the dimensions of a core sample of unconsolidated soil taken before the structure is destroyed), or by the volume of liquid it displaces after its surface has been made impermeable.

(2)  One can measure the volume of the solid phase by immersing it in a wetting liquid (saturation in vacuum, with boiling water or with $CO_2$ subsequently dissolved in water, etc.) and determining the buoyancy force by weight. Thus, we obtain the porosity of the interconnected voids. The sample has to be crushed, if one wants to find the porosity of all the voids including the unconnected ones.

(3)  One can also measure directly the volume of the connected pores by injecting mercury at high pressure into the rock while creating a vacuum in the

sample to expell the air it contains, by weighing the sample when it is dry and when it is saturated with water, etc.

### 2.3.2. Indirect Methods in Situ

(a) *Resistivity of the soil.* With the exception of clays, the minerals commonly found in the ground are insulators, and electricity circulates in the ground in the liquid phase. The resistivity is therefore dependent on the porosity. A formation factor $F$ is defined by

$$F = \frac{\text{resistivity of a rock}}{\text{resistivity of the water contained in the rock}}$$

Using $F$, geophysicists suggest the use of Archie's empirical formula for finding the total porosity $\omega$:

$$F = C/\omega^m \qquad C \simeq 1$$

where $m$ is the cementing factor, which varies from 1.3 for unconsolidated rocks to 2 for limestones. The formula may be corrected if there are known quantities of clay particles in the rock. The porosity obtained by measuring these two resistivities is close to the total porosity. These formulas are useful in interpreting electric logs in exploratory borings.

(b) *Neutron logging.* The ground is bombarded with fast neutrons, usually from sources containing americium, then one counts the number of slow neutrons produced by the deceleration of the fast neutrons on the hydrogen atoms, which are mainly present in the water phase.

In this way, we can determine the porosity of saturated media and especially the moisture content of unsaturated media. It is, however, preferable to evaluate the method on a sample of dry soil in order to deduct the fraction of the hydrogen atoms that are not related to the porosity. Water that is a constituent part of the minerals, clays, etc., will contain these atoms of hydrogen.

(c) *Density measurement (gamma–gamma method).* The ground is bombarded with gamma rays. We identify the part of the radiation which is not absorbed at a fixed distance from the source. This quantity is inversely proportionate to the mass per unit volume of the medium penetrated by the radiation. In turn, this mass per unit volume is linked to the porosity through the expression

$$\rho_r = \omega \rho_w + (1 - \omega)\rho_s$$

where $\rho_r$, $\rho_w$, and $\rho_s$ are, respectively, the mass per unit volume of the rock at hand, the water and the solid grains of which it is made up.

**Table 2.2**

Drainage of Homogeneous Sands

| | | Seepage: quantity of water collected (%) | | | | | |
|---|---|---|---|---|---|---|---|
| Grain size (mm) | Calculated total porosity (%) | First 0.5 hr | Next 0.5 hr | 9 Subsequent days | 10 Day to 2.5 yr | Total | Capillary retention after 2.5 yr |
| 0.475 | 38.86 | 10.68 | 4.88 | 8.72 | 2.60 | 26.88 | 6.87 |
| 0.083 | 39.73 | 1.26 | 0.90 | 11.29 | 2.01 | 15.46 | 18.87 |

(d)  *Sonic velocity.*   This quantity is linked to a number of parameters and especially to the porosity by the quantity of fluid contained in the rock. However, the method is seldom used.

### 2.3.3. *Some Porosity Values*

We have defined a certain number of physical quantities: total porosity $\omega$; specific yield or drainage porosity of an unsaturated soil $\omega_d$ and its complement, the capillary retention capacity; and kinematic porosity of a saturated medium $\omega_c$ and its complement, the saturated retention capacity. These concepts are not always easy to distinguish and evaluate. Table 2.2 is an example, using results obtained by King (as quoted by Geze, 1967) on the drainage of homogeneous well-sorted sands. The difference that exists between the calculated total porosity and the sum of specific yield and capillary retention stems from errors made in measuring and calculating the total porosity. Thus, we observe that the specific yield depends, in reality, on the length of time during which the rock is allowed to drain. If the object is to find out what quantity of water may be extracted from a rock by drainage, we must try to determine the specific yield or drainage porosity. If the object is to find out how much water flows through a saturated rock, for example for a calculation of flow velocity, we have to look for the kinematic porosity. Finally, if we are interested in the total quantity of water contained in a porous medium—for example, in problems concerning the compressibility of the fluid phase or the possible dilution of the ions in solution in the fluid phase— we must look for the total porosity.

It must be admitted that, in practice, one often speaks of porosity without specifying which one. Drainage and kinematic porosities as defined above are often lumped together under the term effective porosity. This is unfortunate.

The accompanying table gives a few orders of magnitude for interstitial porosity leaving aside fracture porosity.

| Medium | Total porosity |
| --- | --- |
| Unaltered granite and gneiss | 0.02–1.8% |
| Quartzites | 0.8% |
| Shales, slates, mica-schists | 0.5–7.5% |
| Limestones, primary dolomites | 0.5–12.5% |
| Secondary dolomites | 10–30% |
| Chalk | 8–37% |
| Sandstones | 3.5–38% |
| Volcanic tuff | 30–40% |
| Sands | 15–48% |
| Clays | 44–53% |
| Swelling clays, silt | Up to 90% |
| Tilled arable soils | 45–65% |

As a general rule, the smaller the grains in a rock, the greater the decrease in effective porosity and the increase in the retention capacity, as illustrated by Fig. 2.17. However, this must be used with caution for determining the porosity as a function of the grain size. (For instance, it hardly lends itself to the interpretation of King's experiments.)

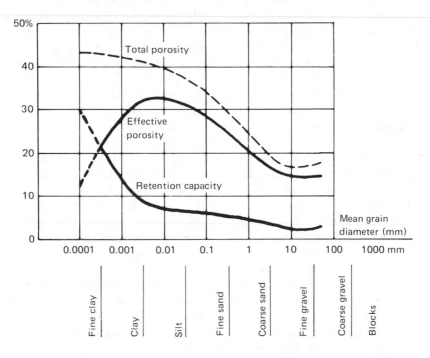

**Fig. 2.17.** Porosity components as a function of grain size. [After Castany (1967)].

## 2.4. Measurements of the Water Pressure in the Ground

One must distinguish between the pressure in the saturated part of the ground, where the pressure is positive, and the unsaturated part, where the pressure is negative.

### 2.4.1. *Measurement in the Saturated Medium*

(a)  *Piezometer.*  If the medium is fairly permeable, a hole is simply drilled in the ground and is fitted with a perforated tube if the sides of the hole are likely to collapse. The water level in the tube gives the elevation of the water table (or free surface), i.e., the point where the pressure is zero (not counting the atmospheric pressure). Under the free surface, the pressure increases linearly with depth if the system is hydrostatic.

(b)  *Pressure gauge.*  If the medium has low permeability (clay or clay–sand, for instance), a tube with a porous point (fritter metal) is inserted into the ground (e.g., by hammering). This is schematically illustrated in Fig. 2.18. Air is injected using a foot pump or bottle through a small plastic tube at the surface, and the pressure is monitored. The rubber membrane is opened, when the air pressure is equal to that of the water, which causes a return of air to the surface. This is visible if the return tube is immersed in a glass of water.

Electric pressure transducers can also be used.

### 2.4.2. *Measurement in the Unsaturated Medium*

To measure the suction in the unsaturated medium, one uses a porous cup made of ceramics, inserted vertically (or horizontally, from a well or trench). This is called a tensiometer (Fig. 2.19).

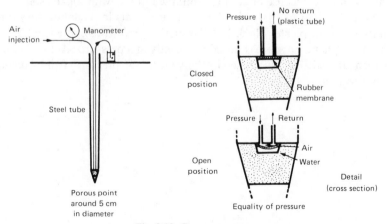

**Fig. 2.18.**  Pressure gauge.

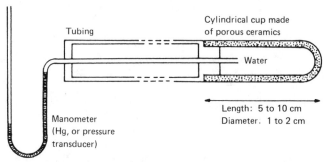

**Fig. 2.19.** Tensiometer.

Through the porous ceramic cup, the water inside attains a pressure equilibrium with the water in the soil (continuity of the water phase through the unsaturated soil and the cup, which is a porous medium like any other). Thus, the suction is measured with a manometer. However, this device is limited to a suction of around 800–900 millibars; beyond that, water starts to boil in the cup at the ordinary temperature, and the tensiometer "disconnects." Each tensiometer is also defined by its air entry pressure (or threshold pressure). As we have explained for the porous medium, the porous ceramic cup always remains 100% saturated, and no air can enter into the tensiometer if the suction is kept below this threshold value (generally between 1 and 10 bars for fine ceramics).

To get below 1 bar, we have to use indirect methods such as blocks of plaster fitted with electrodes and buried in the ground. The water they contain will then reach a pressure equilibrium with that of the soil. By quantitative analysis of the relationship of pressure–moisture-content–resistivity of the plaster block, we can then estimate the suction in the soil. However, this relation may vary with time because of solutes contained in the water of the soil.

Note that in order to make these measurements in the laboratory (on real soil samples or plaster blocks) of the suction–moisture-content relationship, the atmospheric pressure is raised artificially in a pressurized closed circuit to prevent the suction (difference in the water—air pressure) from causing the water to boil.

*Chapter 3*

# Basic Concepts in Hydraulics

## 3.1. General Equations of Fluid Mechanics

In this chapter we are mainly concerned with establishing the form of the continuity equation in porous media. This equation simply states that in a fixed closed volume, the variation per unit time of the fluid mass it contains is equal to the algebraic sum of the mass flux crossing the surface of the volume in question. This is, consequently, the basic principle of mass balance as expressed by Lavoisier: "nothing is lost, nothing is created."

To understand the development of this equation, the reader must have some notions of general fluid mechanics. If this is not the case, it is sufficient to read the beginning of Section 3.2.1, then Sections 3.2.2 and 3.3, before going on to the next chapter.

In fluid mechanics and thermodynamics, we know that solving any flow problem of a Newtonian* fluid means determining six unknowns:

The mass per unit volume $\rho$ (mass length$^{-3}$), the pressure $p$ (mass length$^{-1}$ time$^{-2}$), the temperature $\theta$, and $u_x$, $u_y$, $u_z$, the components of the velocity field **u**.

---

* A Newtonian fluid is an isotropic fluid with a pressure that only depends on the standard state variables $\rho$ and $\theta$, the viscosity tensor of which is a linear form of the velocity gradient with coefficients depending only on the standard state variable.

All these unknowns are functions of time $t$ and the point in space.

We shall use Eulerian coordinates (i.e., a fixed point of reference in the laboratory or medium) and try to express these six unknowns as functions of the space–time variables $x^i$ and $t$. To do this, we have:

(1)   The equation of continuity, which expresses the mass conservation:

$$\text{div}(\rho\mathbf{u}) + \frac{\partial\rho}{\partial t} = 0 \tag{3.1.1}$$

established in an elementary volume that is fixed in space.

This may also be written $\rho \, \text{div} \, \mathbf{u} + d\rho/dt = 0$ in Lagrangian coordinates, following the movement of the matter at its velocity $\mathbf{u}$.

(2)   The Navier–Stokes or dynamic equations, which express the basic principle of mechanics $f = m\gamma$ for viscous fluids, of which the viscosity coefficients are assumed constant:

$$\frac{\partial p}{\partial x^i} - \left(\zeta + \frac{\mu}{3}\right)\frac{\partial}{\partial x^i}(\text{div} \, \mathbf{u}) - \mu\nabla^2 u^i = \rho\left(F^i - \frac{du^i}{dt}\right) \tag{3.1.2}$$

where $\zeta$ is the coefficient of volume viscosity, negligible when compared with $\mu$ (mass length$^{-1}$ time$^{-1}$), $\mu$ the coefficient of dynamic viscosity (mass length$^{-1}$ time$^{-1}$)*, [with $v = \mu/\rho$, kinematic viscosity (length$^2$ time$^{-1}$)], $\nabla^2$ the Laplace differential operator $\sum_i \partial^2/\partial(x^i)^2$, $F_i$ the components of body forces acting at a distance per unit mass, e.g., gravity (length time$^{-2}$).

There are three Navier–Stokes equations, one for each direction $x^i$ in space. This, then, gives us four equations. In general, the two remaining equations are, on the one hand, the heat equation (conductive and convective heat transport by the fluid), and on the other, the equation of state of the fluid giving its mass per unit volume $\rho$ as a function of the pressure and the temperature. In a porous medium it is often possible to simplify the problem by observing that the high degree of division in the porous medium and its enormous heat capacity result in flows that are, in practice, mostly isothermal. The unknown, which is the temperature, then disappears and we only need one further equation.

(3)   The equation of state of the fluid, which we take as

$$\rho = \rho_0 e^{\beta(p - p_0)} \tag{3.1.3}$$

where $\beta$ is the compressibility coefficient of the fluid (mass$^{-1}$ length time$^2$).

The case where the temperature varies in the medium will be examined in Chapter 10.

We shall now examine how these laws may be transposed to the porous medium.

## 3.2. Continuity Equation in Porous Media

### 3.2.1. *Mean Filtration Velocity and Equations of Macroscopic Continuity*

Let us start by setting out our objective. Let $\mathbf{u}$ be the real fluid velocity in each of the pores of the porous medium (also called microscopic velocity). Let $\rho$ be the mass per unit volume at this scale and $\omega$ the point porosity ($\omega = 1$ in a pore, $\omega = 0$ in a grain). At this scale, the ordinary equation of continuity already mentioned holds for the interior of the pores.

We then define the macroscopic quantities or "averages" in the porous medium, which, for the time being, we shall call $\langle \mathbf{u} \rangle$, $\langle \rho \rangle$, and $\langle \omega \rangle$. These macroscopic quantities are defined either by spatial integration, as proposed by Marle (1967) [see Chapter 2, Section 2.1.d], through convolution by a weighting function $m$ or through a probabilistic definition, the mathematical expectation of $\mathbf{u}, \rho$, and $\omega$ at the considered point $x$, for all possible realizations of the medium.

We will then establish the equation of continuity in porous media, equivalent to Eq. (3.1.1):

$$\mathrm{div}[\langle \rho \rangle \langle \mathbf{u} \rangle] + \frac{\partial}{\partial t}[\langle \rho \rangle \langle \omega \rangle] = 0$$

where $\langle \ \rangle$ designates the average taken, and $\langle \mathbf{u} \rangle$ is the fictitious mean velocity, sometimes called filtration velocity.

We shall call it $\mathbf{U}$ later. Note the appearance of the term $\langle \omega \rangle$ in the second term.

It is important to completely understand the physical significance of the two terms of this equation. The equation shows that in a closed volume, the sum of the entering mass flux is equal to the variation of the mass contained in the volume. Although it is expressed at a point, it is always established for an elementary volume $D$ which is *fixed and completely rigid in space*. (In Chapter 5, we shall discuss the case of a volume that is *mobile in space*.)

If we use Ostrogradski's formula,* we find that the divergence of $\langle \rho \rangle \langle \mathbf{u} \rangle$ represents the mass flux of $\langle \rho \rangle \langle \mathbf{u} \rangle$ across the surface $\Sigma$ of $D$. However, one must keep in mind that $\langle \mathbf{u} \rangle$, which we shall define, is a fictitious mean velocity, i.e., the mean velocity of a fluid flowing through the entire space, pores plus

---

* Ostrogradski's formula is

$$\int_D \mathrm{div}\,\mathbf{V}\,dv = -\int_\Sigma \mathbf{V} \cdot \mathbf{n}\,d\sigma$$

where $D$ is the closed volume with outer surface area $\Sigma$, $\mathbf{n}$ is the outer normal on $\Sigma$, $\mathbf{V}$ is the continuous velocity in $D$ and over $\Sigma$, and $\partial V_i/\partial x_i$ is continuous in $D$ and over $\Sigma$.

grains, instead of only through the pores. Indeed, the term div $(\langle\rho\rangle\langle\mathbf{u}\rangle)$ means that we integrate $\langle\rho\rangle\langle\mathbf{u}\rangle$ over the *whole surface* $\Sigma$ of the volume $D$, and not only over the pores. This is why $\langle\mathbf{u}\rangle$ is called *filtration velocity*.

Finally, the fluid mass contained in $D$ is not $\int_D \rho\, dv$, but $\int_D \rho\omega\, dv$, as there is fluid only in the pores. It is therefore normal that the term $\langle\omega\rangle$ appears in the second term.

Let us now establish this equation rigorously. The readers who do not want to pursue the theory any further can skip to Section 3.2.2., but should look at the two definitions of the filtration velocity, compressible in Eq. (3.2.1.1) and incompressible in Eq. (3.2.1.2).

(a)  *Establishing the equation of continuity in porous media.*    This development follows that of Marle (1967). Let $\mathbf{u}$ be the local microscopic velocity inside the pores of a porous medium. To move to a larger scale, we shall use the notion of the representative elementary volume (REV), which we have defined in Section 2.1. Let us agree to extend the field of definition of $\mathbf{u}$ to the entire space with, of course, $\mathbf{u} = 0$ in the grains.

*Incompressible fluid and solid.*    The equation of continuity at the microscopic scale is reduced to

$$\mathrm{div}\,\mathbf{u} = 0$$

because $\rho$ is constant. Furthermore, the velocity $\mathbf{u}$ is continuous in the entire space, because $\mathbf{u}$ is zero at the walls in laminar flow and defined as zero in the grains.

To define the mean macroscopic velocity $\langle\mathbf{u}\rangle$ or filtration velocity, we shall integrate in space the local property weighted by a weighting function $m(\mathbf{x})$ such as

$$\int m(\mathbf{x})\,d\mathbf{x} = 1$$

where x stands for the coordinates in three dimensions and the integral may be extended either to a certain bounded domain $D$ if $m$ has a bounded support or to the whole space.

Hence, for example,

$$m(\mathbf{x}) = \begin{cases} 3/4\pi r^3 & \text{if} \quad |\mathbf{x}| \leq r \\ 0 & \text{if} \quad |\mathbf{x}| > r \end{cases}$$

where $m$ is the indicatrix of a sphere of radius $r$ centered at the origin, or again

$$m(\mathbf{x}) = \frac{1}{\sigma^3(2\pi)^{3/2}}\, e^{-|\mathbf{x}|^2/2\sigma^2} \qquad \forall\mathbf{x}$$

which is the normal probability distribution in three-dimensional space, $\sigma$ being the standard deviation.

If $a$ is a local magnitude, the mean $\langle a \rangle$ of $a$ at the point $\mathbf{x}$ in space is then defined by

$$\langle a \rangle(\mathbf{x}) = \int a(\mathbf{x} + \mathbf{x}')m(\mathbf{x}')\,d\mathbf{x}'$$

It is often advantageous to require $m$ to be continuous and continuously differentiable, so that $\langle a \rangle$ may have the same properties, even if $a$ does not; thus, for example,

$$m(\mathbf{x}) = \begin{cases} C \exp\left( -\dfrac{1}{r^2 - |\mathbf{x}|^2} \right) & \text{if} \quad |\mathbf{x}| \le r \\ 0 & \text{if} \quad |\mathbf{x}| > r \end{cases}$$

where $C$ is chosen in order that the integral of $m$ indeed be 1. However, there is a great deal of freedom in the choice of $m$.

As $\mathbf{u}$ is a vector, we can define a macroscopic velocity $\langle \mathbf{u} \rangle$ by taking the weighting by $m$ of each of the components $u_i$ of $\mathbf{u}$:

$$\langle u_i \rangle = \int u_i(\mathbf{x} + \mathbf{x}')m(\mathbf{x}')\,d\mathbf{x}' \qquad (3.2.1.1)$$

The equation of microscopic continuity may be written

$$\operatorname{div}\mathbf{u} = \frac{\partial u_1}{\partial x_1} + \frac{\partial u_2}{\partial x_2} + \frac{\partial u_3}{\partial x_3} = 0$$

We multiply by $m$ and integrate in space:

$$\sum_i \int \frac{\partial u_i}{\partial x_i}\bigg|_{\mathbf{x}+\mathbf{x}'} m(\mathbf{x}')\,d\mathbf{x}' = 0$$

Since $\mathbf{u}$ *is continuous*, and the domain of integration (or the entire space) is immobile, the differentiation and integration commute, i.e.,

$$\sum_i \frac{\partial}{\partial x_i} \int u_i(\mathbf{x} + \mathbf{x}')m(\mathbf{x}')\,d\mathbf{x}' = 0$$

i.e.,

$$\sum_i \frac{\partial \langle u_i \rangle}{\partial x_i} = 0$$

or

$$\operatorname{div}\langle \mathbf{u} \rangle = 0$$

Let us consider for a moment the physical significance of $\langle \mathbf{u} \rangle$, the filtration velocity. The integral which defines $\langle \mathbf{u} \rangle$ is extended to the entire space even if $\mathbf{u}$ is, in reality, zero in the grains of the porous medium. Therefore, $\langle \mathbf{u} \rangle$ is a fictitious mean velocity calculated as if the entire space were accessible to the flow (pore plus solid).

One must not confuse $\langle \mathbf{u} \rangle$ with the mean flow velocity inside the pores, which we shall define later.

*Case where the fluid is compressible and the flow steady* (*not a function of time*). This means that $\partial \rho / \partial t = 0$, i.e., that the equation of microscopic continuity is reduced to $\operatorname{div}(\rho \mathbf{u}) = 0$ and, furthermore, that the porous medium is immobile.

We shall start by defining a macroscopic mass per unit volume $\langle \rho \rangle$. As the microscopic mass per unit volume $\rho$ is defined only in the pores, we must similarly extend its definition to the entire space by agreeing that $\rho = 0$ in the grains. But remember that $\rho$ is now *discontinuous at the solid–liquid interface*.

Furthermore, if we were simply to define $\langle \rho \rangle$ by convolution of $\rho$ by the weighting function $m$ we would get a certain inconsistency, because the mean $\langle \rho \rangle$ would be very different from the local $\rho$, even in the case where $\rho$ is uniform in the pores. The problem stems from the fact that, in the convolution, $\rho$ would be weighted by the porosity as well.

Therefore, it is preferable to define the mean porosity first as

$$\langle \omega \rangle = \int \omega(\mathbf{x} + \mathbf{x}')m(\mathbf{x}')d\mathbf{x}' \qquad \text{where*} \qquad \omega = \begin{cases} 0 \text{ in a grain} \\ 1 \text{ in a pore} \end{cases}$$

Then, the macroscopic mass per unit volume is defined by

$$\langle \rho \rangle = \frac{1}{\langle \omega \rangle} \int \rho(\mathbf{x} + \mathbf{x}')m(\mathbf{x}')\,d\mathbf{x}'$$

(If $\rho = \text{const}$, then $\langle \rho \rangle = \rho$ with this definition.)

Finally, we could keep the same definition for the filtration velocity as in Eq. (3.2.1.1). However, this definition implies that the fluid is incompressible. Indeed, if it is not, there is no physical significance in adding (or taking the average of) the velocities. Mass is the only magnitude that can be added up, i.e., which satisfies an equation of continuity.

---

* Here we are talking about effective porosity—not in a kinematic sense, i.e., water that can circulate, but in the sense of compressibility: when we make $p$ vary, we want to identify that fraction of the medium which contains compressible water. All that is excluded in the end is the film of adhesive water bound to the solids, which is itself already greatly compressed, and which we shall assume to be part of the grain. In practice, we use the total porosity $\omega$.

Thus, we shall define the filtration velocity $\langle \mathbf{u} \rangle$ from the mass flux $\rho\mathbf{u}$ and the average mass per unit volume $\langle \rho \rangle$:

$$\langle u_i \rangle = \frac{1}{\langle \rho \rangle} \int \rho(\mathbf{x} + \mathbf{x}') u_i(\mathbf{x} + \mathbf{x}') m(\mathbf{x}')\, d\mathbf{x}' \qquad (3.2.1.2)$$

If $\rho$ is constant, this definition coincides with Eq. (3.2.1.1).

Although $\rho$ is not continuous in space, the product of $\rho\mathbf{u}$ is, as long as the porous medium is immobile. Therefore, we can also permute the signs of summation and differentiation, and write the equation of macroscopic continuity:

$$\int \operatorname{div}(\rho\mathbf{u}) m(\mathbf{x}')\, d\mathbf{x}' = \operatorname{div}\left[ \int \rho\mathbf{u}\, m(\mathbf{x}')\, d\mathbf{x}' \right] = 0$$

or from Eq. (3.2.1.2),

$$\operatorname{div}[\langle \rho \rangle \langle \mathbf{u} \rangle] = 0$$

*Case where the fluid is compressible, the flow a function of time, and the medium elastic.* We shall keep the same definition as above for $\langle \omega \rangle$, $\langle \rho \rangle$, and $\langle \mathbf{u} \rangle$, i.e.,

$$\langle \omega \rangle = \int \omega(\mathbf{x} + \mathbf{x}') m(\mathbf{x}')\, d\mathbf{x}' \qquad \text{with} \qquad \omega = \begin{cases} 0 \text{ in a grain} \\ 1 \text{ in a pore} \end{cases}$$

$$\langle \rho \rangle = \frac{1}{\langle \omega \rangle} \int \rho(\mathbf{x} + \mathbf{x}') m(\mathbf{x}')\, d\mathbf{x}' \qquad \text{with} \qquad \rho = \begin{cases} 0 \text{ in a grain} \\ \rho \text{ in a pore} \end{cases}$$

$$\langle \mathbf{u} \rangle = \frac{1}{\langle \rho \rangle} \int \rho(\mathbf{x} + \mathbf{x}') \mathbf{u}(\mathbf{x} + \mathbf{x}') m(\mathbf{x}')\, d\mathbf{x}' \qquad \text{with} \qquad \mathbf{u} = \begin{cases} 0 \text{ in a grain} \\ \mathbf{u} \text{ in a pore} \end{cases}$$

We shall use the complete microscopic equation of continuity and integrate it in space, with a weighting function $m$:

$$\int \left[ \operatorname{div}(\rho\mathbf{u}) + \frac{\partial \rho}{\partial t} \right]\Bigg|_{\mathbf{x}+\mathbf{x}'} m(\mathbf{x}')\, d\mathbf{x}' = 0$$

This integral is indeed zero, because, by definition, the term in brackets must be zero in the pores and the definition of $\rho$ and $\mathbf{u}$ in the grains result in their being zero in the grains as well. The fact that the spatial derivatives are not defined on the interface $\Sigma_1$ between the pores and the grain does not influence the calculation of the integral of the volume, because $\Sigma_1$ is a set of measure zero.

Although the final result is simple, the calculation is trickier, because this time the signs of differentiation and integration do not simply commute.

The problem is caused by the fact that if the medium is compressed the mass per unit volume of the water varies, but the porous medium itself subjected to these pressures becomes deformed. Therefore, the porosity varies and the liquid–solid boundary $\Sigma_1$ moves at a velocity that we shall call $\mathbf{u}_\sigma$. These velocities are, of course, very small and, more often than not, negligible. However, here we are endeavoring to rigorously establish the basic equations.

The consequence of this movement is that neither $\rho$ nor $\rho\mathbf{u}$ is continuous in space. One can then show that the summation and the differentiation only commute if differentiation is defined according to the theory of distributions and not in the usual sense (see Marle, 1967, and Schwartz, 1961). However, we will not use this approach here.

Let us now examine the term

$$\int \frac{\partial \rho}{\partial t}\bigg|_{\mathbf{x}+\mathbf{x}'} m(\mathbf{x}')\,d\mathbf{x}'$$

Without referring to distribution theory, we shall use Leibnitz' rule* for the derivative of an integral to evaluate our integral. From the definition of $\langle \rho \rangle$, we can write

$$\langle \rho \rangle \langle \omega \rangle = \int \rho(\mathbf{x}+\mathbf{x}')m(\mathbf{x}')\,d\mathbf{x}'$$

Let us assume that $m$ has bounded support, and let $D$ be the domain, centered in $x$, in which $m$ is not nil; the external surface of $D$ is called $\Sigma$. As $\rho$ is nil in the grains, we can even limit the integration to the domain $D_1$ occupied by the fluid and limited by the external surface $\Sigma$ and by the fluid–solid interface, which we call $\Sigma_1$. We now take the derivative of the above expression with respect to time:

$$\frac{\partial}{\partial t}[\langle \rho \rangle \langle \omega \rangle] = \frac{\partial}{\partial t}\left[\int_{\mathbf{x}+\mathbf{x}' \in D_1} \rho(\mathbf{x}+\mathbf{x}')m(\mathbf{x}')\,d\mathbf{x}'\right]$$

* Leibnitz' rule: If

$$f(x) = \int_{a(x)}^{b(x)} \psi(x,y)\,dy$$

then

$$\frac{\partial f}{\partial x} = \frac{db}{dx}\psi[x,b(x)] - \frac{da}{dx}\psi[x,a(x)] + \int_{a(x)}^{b(x)} \frac{\partial}{\partial x}\psi(x,y)\,dy$$

where $\psi(x,y)$ is continuous in $x$ and $y$, $\partial\psi/\partial x$ exists and is continuous, and $a$ and $b$ are differentiable.

As $D_1$ varies in time the porous medium is deformed continuously and Leibnitz' rule gives us two terms in the differentiation: the time-dependence of $\rho$, and the variation in time of the integration volume $D_1$.

The first one is simply

$$\int_{D_1} \frac{\partial}{\partial t} [\rho(\mathbf{x} + \mathbf{x}')] m(\mathbf{x}') \, dx'$$

which is exactly the one we wanted to estimate at the beginning of this paragraph.

The second term can be evaluated by noting that the volume swept by a surface element $d\sigma$ belonging to the fluid–solid interface $\Sigma_1$ during a time $dt$ is given by the scalar product

$$-\mathbf{u}_\sigma \cdot \mathbf{n} \, dt$$

where $\mathbf{u}_\sigma$ is the velocity of the interface and $\mathbf{n}$ is the normal line at this interface directed toward the fluid. The variation of the volume $D_1$ per unit time is therefore the integral of this term on the surface $\Sigma_1$ (the external surface $\Sigma$ of $D_1$ is immobile):

$$-\int_{\mathbf{x}+\mathbf{x}' \in \Sigma_1} \rho(\mathbf{x} + \mathbf{x}') \mathbf{u}_\sigma(\mathbf{x} + \mathbf{x}') \cdot \mathbf{n}(\mathbf{x} + \mathbf{x}') m(\mathbf{x}') \, dx'$$

Observe that this term appears only because $\rho$ is not continuous over $\Sigma_1$: if $\rho$ were equal to zero on $\Sigma_1$ here, the integral would disappear. Thus, we can write

$$\int \left. \frac{\partial \rho}{\partial t} \right|_{\mathbf{x}+\mathbf{x}'} m(\mathbf{x}') \, dx' = \frac{\partial}{\partial t} [\langle \rho \rangle \langle \omega \rangle] + \int_{\Sigma_1} \rho \mathbf{u}_\sigma \cdot \mathbf{n} m(\mathbf{x}') \, dx'$$

Now for the second term, $\int \operatorname{div}(\rho \mathbf{u}) \, m \, dx'$.

As the interface $\Sigma_1$ between fluid and solid moves, the velocity of the fluid at this interface is only zero as a relative velocity, in relation to the interface velocity*, i.e., for points on $\Sigma_i$.

$$\mathbf{u} - \mathbf{u}_\sigma = 0, \text{ or } \mathbf{u} = \mathbf{u}_\sigma.$$

* In the most general case, the relation of the mass balance that exists at the interface in a two-phase medium is described by

$$\rho_1(\mathbf{u}_1 \cdot \mathbf{n} - \mathbf{u}_\sigma \cdot \mathbf{n}) - \rho_2(\mathbf{u}_2 \cdot \mathbf{n} - \mathbf{u}_\sigma \cdot \mathbf{n}) = 0$$

where 1 and 2 designate the two phases, $\mathbf{n}$ is the normal line at the interface $\Sigma$ between 1 and 2, and $\mathbf{u}_\sigma$ is the velocity of $\Sigma$. This rule assumes that the interface is a single surface and ignores interface phenomena such as surface tension. It allows the exchange of matter at the interface (e.g., fusion of ice, chemical reaction). Here it is obvious that the velocity of the solid $\mathbf{u}_2$ over $\Sigma$ is equal to $\mathbf{u}_\sigma$; thus, similarly, $\mathbf{u}_1 = \mathbf{u}_\sigma$. See Slattery (1972).

Thus, $\mathbf{u}$ is discontinuous on each side of $\Sigma_1$ making it necessary to introduce one more term on $\Sigma_1$. We shall calculate a spatial derivative of $\langle\rho\rangle\langle u\rangle$. We integrate, as above, in a bounded volume $D$ and let $D_1$ be the domain occupied by the fluid in $D$; $\Sigma_1$ is the fluid–solid interface.

From the definition of the filtration velocity in Eq. (3.2.1.2);

$$\langle\rho\rangle\langle u_i\rangle = \int \rho u_i\Big|_{\mathbf{x}+\mathbf{x}'} m(\mathbf{x}')\,d\mathbf{x}'$$

We take the derivative:

$$\frac{\partial}{\partial x_i}[\langle\rho\rangle\langle u_i\rangle] = \frac{\partial}{\partial x_i}\int_{\mathbf{x}+\mathbf{x}'\in D} \rho u_i\Big|_{\mathbf{x}+\mathbf{x}'} m(\mathbf{x}')\,d\mathbf{x}'$$

$$= \frac{\partial}{\partial x_i}\int_{\mathbf{x}+\mathbf{x}'\in D_1} \rho u_i\Big|_{\mathbf{x}+\mathbf{x}'} m(\mathbf{x}')\,d\mathbf{x}' \quad \text{as} \quad \rho u_i = 0 \quad \text{in} \quad (D-D_1)$$

If we make the change of variable, $\mathbf{x}'' = \mathbf{x} + \mathbf{x}'$, we find

$$\frac{\partial}{\partial x_i}[\langle\rho\rangle\langle u_i\rangle] = \frac{\partial}{\partial x_i}\int_{\mathbf{x}''\in D_1} \rho u_i\Big|_{\mathbf{x}''} m(\mathbf{x}''-\mathbf{x})\,d\mathbf{x}''$$

However, now only $m$ is a function of $\mathbf{x}$. If we observe that

$$\frac{\partial m(\mathbf{x}''-\mathbf{x})}{\partial x_i} = -\frac{\partial m(\mathbf{x}''-\mathbf{x})}{\partial x_i''}$$

then

$$\frac{\partial}{\partial x_i}[\langle\rho\rangle\langle u_i\rangle] = -\int_{\mathbf{x}''\in D_1} \rho u_i\Big|_{\mathbf{x}''} \frac{\partial m(\mathbf{x}''-\mathbf{x})}{\partial x_i''}\,d\mathbf{x}''$$

$$= -\int_{\mathbf{x}''\in D_1} \left\{ \frac{\partial}{\partial x_i''}\left[\rho u_i\Big|_{\mathbf{x}''} m(\mathbf{x}''-\mathbf{x})\right] \right.$$
$$\left. -\frac{\partial}{\partial x_i''}\left[\rho u_i\Big|_{\mathbf{x}''}\right] m(\mathbf{x}''-\mathbf{x}) \right\}\,d\mathbf{x}''$$

Similarly, since the interface $\Sigma_1$ is a set of measure zero, the fact that the gradient is not defined on it is of no importance for the calculation of the integral. If we transform the first term with Ostrogradski's formula, then

$$\int_{D_1} \frac{\partial}{\partial x_i''}\left[\rho u_i\Big|_{\mathbf{x}''} m(\mathbf{x}''-\mathbf{x})\right]d\mathbf{x}'' = -\int_{\Sigma_1} \rho u_i\Big|_{\mathbf{x}''} m(\mathbf{x}''-\mathbf{x})n_i\,d\mathbf{x}''$$

where $n_i$ is the component in the direction $i$ of the normal line to $\Sigma_1$ oriented from the solid toward the fluid. Note that the integral is limited to $\Sigma_1$ and not to the external surface of $D_1$, because if $D$ is sufficiently large, then $m$ is nil there.

We can now go back to the initial variable $x$ through the change of variables $\mathbf{x}'' = \mathbf{x} + \mathbf{x}'$:

$$\frac{\partial}{\partial x_i}[\langle \rho \rangle \langle u_i \rangle] = \int_D \frac{\partial}{\partial x_i}\left[\rho u_i\Big|_{\mathbf{x}+\mathbf{x}'}\right]m(\mathbf{x}')\,d\mathbf{x}' + \int_{\Sigma_1} \rho u_i\Big|_{\mathbf{x}+\mathbf{x}'} m(\mathbf{x}')n_i\,d\mathbf{x}'$$

i.e., finally,

$$\int_D \operatorname{div}(\rho\mathbf{u})\Big|_{\mathbf{x}+\mathbf{x}'} m(\mathbf{x}')\,d\mathbf{x}' = \operatorname{div}[\langle \rho \rangle \langle \mathbf{u} \rangle] - \int_{\Sigma_1} \rho\mathbf{u}\cdot\mathbf{n}\Big|_{\mathbf{x}+\mathbf{x}'} m(\mathbf{x}')\,d\mathbf{x}'$$

The fact that $D$ is bounded does not influence the demonstration, which remains general.

If we regroup the two terms of the equation of continuity, we get

$$\operatorname{div}[\langle \rho \rangle \langle \mathbf{u} \rangle] + \frac{\partial}{\partial t}[\langle \rho \rangle \langle \omega \rangle] + \int_{\Sigma_1} \rho(\mathbf{u}_\sigma - \mathbf{u})\cdot\mathbf{n}\,m(\mathbf{x}')\,d\mathbf{x}' = 0$$

but on $\Sigma_1$ we have shown that $\mathbf{u} = \mathbf{u}_\sigma$; there remains only

$$\operatorname{div}[\langle \rho \rangle \langle \mathbf{u} \rangle] + \frac{\partial}{\partial t}[\langle \rho \rangle \langle \omega \rangle] = 0 \qquad (3.2.1.3)$$

Observe that we could also define $\langle \omega \rangle$, $\langle \rho \rangle$, and $\langle u \rangle$ in the sense of random functions as the mathematical expectations, at point $\mathbf{x}$, of all the values assumed by the infinite set of possible realizations of the medium. In that case, the differentiation operator must be understood as in the theory of distributions in order that it can be commuted with the expectation operator:

$$\frac{\partial}{\partial \mathbf{x}}[E(a(\mathbf{x},t))] = E\left[\frac{\partial a}{\partial \mathbf{x}}(\mathbf{x},t)\right]$$

### 3.2.2. Simplification of the Notation; Source Term

In order to avoid cumbersome expressions, we shall now dispense with the sign $\langle\,\rangle$ for $\rho$ and $\omega$ and denote the filtration velocity $\mathbf{U} = \langle \mathbf{u} \rangle$, while remembering that these magnitudes have been defined, in a porous medium, by the operation of taking averages, on which we have commented abundantly.

However, we shall add one more term to the equation of continuity. Indeed, this equation expresses the balance of matter inside a closed volume. However, in hydrogeology, one often has to add a source or sink term, which accounts for the withdrawal (or recharge) of water that may be made in the medium (e.g., bore holes).

We shall define the source term $q$, which will represent the volumetric flow rate of fluid withdrawn (or added, if it is negative) per unit volume at each point. The withdrawn mass flow rate is therefore $\rho q$, with $q$ defined on the macroscopic scale. This term is added to the equation of continuity, which is then

$$\text{div}(\rho \mathbf{U}) + \frac{\partial}{\partial t}(\rho \omega) + \rho q = 0 \qquad (3.2.2.1)$$

### 3.2.3. Mean Microscopic Velocity

From the filtration velocity $\mathbf{U}$, we can define a "mean microscopic velocity" of the fluid simply by saying that $\mathbf{u}$ is nil in the grains. Let $\Sigma$ be a section of the porous medium and $\omega_{cs}$ the kinematic surface porosity over $\Sigma$.

$$\omega_{cs} = \frac{\text{surface area of effective pores}}{\text{total surface area of the section}}$$

The mean microscopic velocity is defined by

$$\mathbf{u} = \frac{\mathbf{U}}{\omega_{cs}}$$

However, this velocity does not have a great physical significance as opposed to $\mathbf{U}$, which, by definition, satisfies the equation of continuity.

In practice, it is generally assumed that the porous medium is isotropic in so far as the distribution of the porosity over a section is concerned, and we admit that $\omega_{cs} = \omega_c$, although generally $\omega_{cs} < \omega_c$, in reality. The mean microscopic velocity (or mean actual velocity in the pores) is then

$$\mathbf{u} = \frac{\mathbf{U}}{\omega_c}$$

where $\omega_c$ is the kinematic porosity.

## 3.3. Hydraulic and Piezometric Head

In courses in hydraulics, the hydraulic head at a point $M$ of an incompressible fluid subjected only to gravity is defined by the relationship

$$h = \frac{u^2}{2g} + \frac{p}{\rho g} + z$$

where $u$ is the real velocity of the fluid at the point $M$, the elevation of which is $z$ (measured positively upwards). Each term can be interpreted in terms of energy.

Furthermore, we know (Bernouilli's theorem) that the head can only decrease in the direction of the flow and that, if the fluid is immobile, its head is constant in space.

In porous media, the real velocities are always very slow, and we are justified in omitting the term for the dynamic head $u^2/2g$, which reduces the head to the static or piezometric head:

$$h = \frac{p}{\rho g} + z \qquad (3.3.1)$$

Thus, the piezometric head merges with the hydraulic head, the value of which is, of course, dependent on the origin chosen on the axis $z$. Hydraulic heads are generally expressed in relation to the mean sea level in the same way as topographic elevations.

If we want to measure this head at a point A of a saturated porous medium, it is necessary to bore a hole and sink an open-ended tube. After stabilization, the elevation $z_B$ reached by the water in the tube is equal to the head $h$ at the point of the lower opening of the tube (Fig. 3.1). This kind of apparatus is called a piezometer. The elevation $z_B$ is equal to the head in the piezometer at point B, which is the same as the one at point A, because the fluid is immobile in the tube of the piezometer.

$$h_A = \frac{p_A}{\rho g} + z_A = \frac{p_B + \rho g(z_B - z_A)}{\rho g} + z_A = \frac{p_B}{\rho g} + z_B = h_B$$

As one always chooses the atmospheric pressure as the zero reference pressure, indeed, $h_A = h_B = z_B$.

If the fluid were immobile in a water table aquifer that is directly recharged by rainfall, the hydraulic head would be the same at all points of the porous medium. Consequently, the head $z_B$ in the piezometer would define the "free

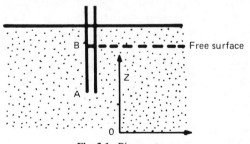

**Fig. 3.1.** Piezometer.

surface" of the aquifer, i.e., the boundary (where the water pressure is zero) that separates the saturated porous medium from the unsaturated one.

If the aquifer flows horizontally in the saturated medium, the head varies in the horizontal direction. However, the hydraulic head remains the same in the vertical direction, and the elevation of the free surface is given by the one measured by the piezometer independently of its depth. This is no longer the case, if the flow is not horizontal; the hydraulic head then also varies with the depth of the piezometer and the free surface is defined by the head obtained, when the piezometer begins to enter the saturated medium.

In practice, the piezometer is often perforated along its entire length (punctures or slots), and the "mean hydraulic head" in the aquifer is measured in this way.

In order to account for the compressibility of the fluid, the hydraulic head is sometimes defined by

$$h = z + \int_{p_0}^{p} \frac{dp}{\rho(p)g} \tag{3.3.2}$$

where $p_0$ is pressure at the origin of the axis $z$ and $p$ is pressure at the point of elevation $z$ (see Remson *et al.*, 1971; Hubbert, 1940). We shall not use this formulation.

## 3.4. Simplification and Integration of the Navier–Stokes Equations for Schematic Porous Media

The Navier–Stokes equations are not in practice applicable as such in porous media, because we do not know precisely what happens to pressures and velocities in the pores on the microscopic scale. Therefore, one must find a macroscopic law, which may be used on the scale of the elementary domain of the porous medium, linking pressure, velocity, and external forces. This is an experimental law, Darcy's law, which we shall study in Chapter 4.

We shall, however, simplify the Navier–Stokes equations by choosing the case of slow (laminar) steady flow of an incompressible fluid. Once simplified, they will be applied to two simple geometric cases: flow between two parallel plates placed close together, and flow through a cylindrical tube as in the example borrowed from Houpeurt (1974). We then obtain a macroscopic law that can be compared with Darcy's experimental law.

Our purpose is not to prove Darcy's law, which is a phenomenological law and has to be admitted, but to rely on theoretical reasoning as a basis for the generalization of Darcy's law from the elementary experiment.

However, it is worth mentioning that there are works by Matheron (1967) and Marle (1967) that deal with the justification of Darcy's law by integration of Navier's equations in a real medium. In particular, Matheron shows that Darcy's law is the result of the linearity of the Navier–Stokes equations, not of their form.

*Simplifications.*   For steady flow, we can write

$$\partial u^i/\partial t = 0$$

and, by using the ordinary equation of continuity;

$$\mathrm{div}(\rho \mathbf{u}) = -\partial \rho/\partial t = 0$$

If, moreover, the fluid is incompressible, this equation reduces to

$$\mathrm{div}\, \mathbf{u} = 0$$

Then the Navier–Stokes equations reduce to

$$\frac{\partial p}{\partial x^i} - \mu \nabla^2 u^i - \rho F^i = 0 \qquad (3.4.1)$$

We shall integrate them in three simple cases.

(a)   *Parallel isothermal and steady movement of an incompressible viscous fluid in a fracture of width e without any influence of external forces.*   The fracture is assumed to extend indefinitely in the horizontal $(x-y)$ plane and its opening $e$ is oriented along $z$.

The parallel flow runs in the direction $x$. The velocity has only one component, $u_x$. It is evident that the velocity $u_x$ then depends neither on $x$ nor on $y$, but only on $z$:

$$u_y = u_z = 0, \qquad \frac{\partial u_x}{\partial x} = \frac{\partial u_x}{\partial y} = 0$$

Then, without any influence of external forces $(F^i = 0)$, the Navier–Stokes equations of Eq. (3.4.1) reduce to

$$\frac{\partial p}{\partial x} = \mu \frac{\partial^2 u_x}{\partial z^2} \qquad (3.4.2.)$$

$$\frac{\partial p}{\partial y} = 0 \qquad (3.4.3.)$$

$$\frac{\partial p}{\partial z} = 0 \qquad (3.4.4.)$$

We isolate, in our mind, a length $L$ of the fracture along $x$, a width $b$ along $y$ (Fig. 3.2), and as $p$ is independent of all but $x$, we can determine the boundary

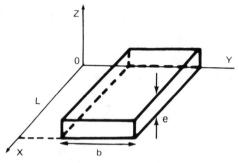

**Fig. 3.2.** Fracture geometry.

conditions:

$$p = p_1 \quad \text{for} \quad x = 0$$

$$p = p_2 \quad \text{for} \quad x = L, \quad \text{with} \quad p_2 < p_1$$

If we include gravity as an external force, the last equation of Navier–Stokes, Eq. (3.4.4), would be

$$\frac{\partial p}{\partial z} = -\rho g$$

and $p$ would also be a function of $z$. We ignore the gravity here in order to simplify the analysis [see Section (c)].

The first of the Navier–Stokes equations Eq. (3.4.2), depends only on the independent variables $x$ on the left-hand side and $z$ on the right-hand side. The only means of ensuring the equality of these two quantities is to make each of them, on its own side, equal to the same constant $C$.

Consequently, Eq. (3.4.2) is replaced by

$$\frac{\partial p}{\partial x} = \frac{dp}{dx} = C \quad \text{and} \quad \mu \frac{\partial^2 u_x}{\partial z^2} = \mu \frac{d^2 u_x}{dz^2} = C$$

The integration of these two equations leads to

$$\left. \begin{aligned} p &= p_1 + \frac{p_2 - p_1}{L} x \\ u_x &= \frac{1}{\mu} C \frac{z^2}{2} + C'z + C'' \end{aligned} \right\} \quad C', C'' \text{ constants}$$

For $z = 0$ and $z = e$, one should have $u = 0$; thus we get

$$u_x = \frac{1}{2\mu} \frac{p_2 - p_1}{L} (z^2 - ez) \quad \text{(parabolic velocity profile)}$$

We calculate the flow $q$ across the fracture for the width $b$:

$$q = \int_0^e b u_x \, dz$$

$$q = be \frac{e^2}{12\mu} \frac{p_1 - p_2}{L}$$

If there are $n$ parallel fractures over a depth of $l$ of an otherwise impermeable rock, its porosity is then

$$\omega = \frac{ne}{l}$$

The total section of the medium $A$ is $bl$ and gives the flow $Q = nq$:

$$Q = A \frac{\omega e^2}{12} \frac{1}{\mu} \frac{p_1 - p_2}{L} \tag{3.4.5}$$

Thus, the Navier–Stokes equations lead to the following conclusions: the flow $Q$ is proportional to the total section $A$ of the rock and to the pressure gradient $(p_1 - p_2)/L$ and inversely proportional to the viscosity $\mu$. The coefficient of proportionality for the medium in question is here $\omega e^2/12$.

(b)   *Flow through a circular tube of radius r (Poiseuille's formula).*   If we use radial symmetry and introduce polar coordinates with the flow direction $x$ as their axis, the first of the Navier–Stokes equations, Eq. (3.4.2), given for the case of the fracture, becomes

$$\frac{dp}{dx} = \mu \frac{d^2 u}{dr^2} + \frac{1}{r} \frac{du}{dr}$$

The integration leads to

$$q = \frac{\pi r^4}{8\mu} \frac{p_1 - p_2}{L}$$

Let us consider a porous medium composed of an impermeable matrix pierced by $n$ circular ducts of radius $r$, all parallel to each other. If $A$ is the total surface area of the medium, perpendicular to the direction of the ducts, it has a porosity of

$$\omega = n\pi r^2 / A$$

Then the total flow through the porous medium is $Q = nq$, i.e.,

$$Q = A \frac{\omega r^2}{8} \frac{1}{\mu} \frac{p_1 - p_2}{L} \tag{3.4.6}$$

and this expression is similar in all points to the expression of Eq. (3.4.5) for the flow in the fractured medium, but the proportionality coefficient here is $\omega r^2/8$ instead of $\omega e^2/12$.

These two calculations suggest—but do not prove—that the flow $Q$ of an incompressible fluid with viscosity $\mu$ through a cross-section of area $A$ of a porous medium under a pressure gradient $dp/dx$ has the form

$$Q = -A\frac{k}{\mu}\frac{dp}{dx}$$

where $k$ is the coefficient of proportionality of the porous medium in question. We shall see that this is indeed the result found experimentally by Darcy.

(c) *Introduction of external forces.* If we want to include external forces in the Navier–Stokes equations, for example the gravity, we orient the parallel fracture vertically along the $y$–$z$ plane. Then, the Navier–Stokes equations are

$$\frac{\partial p}{\partial x} = 0 \qquad \frac{\partial p}{\partial y} = 0 \qquad \frac{\partial p}{\partial z} = \mu\frac{\partial^2 u_z}{\partial z^2} - \rho g$$

which, when similarly integrated, lead to

$$p = p_1 + \frac{p_2 - p_1}{L}z$$

$$u_z = \frac{1}{2\mu}\left(\frac{p_2 - p_1}{L} + \rho g\right)(x_2 - ex) \tag{3.4.7}$$

The conclusion is that the force of gravity $\rho g$ plays the same role as the pressure gradient $dp/dz$, to which it should be added. The flow through the fractured medium becomes

$$Q = A\frac{\omega e^2}{12}\frac{1}{\mu}\left(\frac{p_1 - p_2}{L} - \rho g\right)$$

If we calculate the filtration velocity, defined above as that of a fluid, which might flow through the entire cross section $A$ of the fractured medium, we get

$$U = \frac{Q}{A} = \frac{\omega e^2}{12}\frac{1}{\mu}\left(\frac{p_1 - p_2}{L} - \rho g\right)$$

and generalizing for all directions in space,

$$U = -\frac{\omega e^2}{12}\frac{1}{\mu}(\mathbf{grad}\,p + \rho g\,\mathbf{grad}\,z) \tag{3.4.8}$$

where **grad** $z$ is a vector of coordinates $(0, 0, 1)$ and the axis $z$ is vertical and oriented upward. The minus sign is due to the fact that the fluids flow from high pressure toward low pressure or from above downward.

As the fluid is assumed incompressible, we can write

$$\mathbf{grad}\, p + \rho g\, \mathbf{grad}\, z = \rho g [\mathbf{grad}(p/\rho g + z)]$$

$$= \rho g\, \mathbf{grad}\, h$$

where $h = p/\rho g + z$ is the hydraulic head, which we have defined in Eq. (3.3.1), i.e.,

$$\mathbf{U} = -\frac{\omega e^2}{12}\frac{\rho g}{\mu}\mathbf{grad}\, h$$

The role of the pressure is taken over by that of the hydraulic head $h$ if the fluid is incompressible.

However, we must remember that in the Navier–Stokes equations as such, we can associate pressure gradients and external forces:

$$\frac{\partial p}{\partial x^i} - \rho F^i$$

That is, if the forces $F^i$ derive from a potential such as gravity, then $\mathbf{grad}\, p + \rho g\, \mathbf{grad}\, z$, but the definition of a unique potential $p/\rho g + z$ assumes that the fluid is incompressible, which is not always the case in a porous medium. We shall use the two forms alternately.

# Chapter 4

# Darcy's Law

## 4.1. Darcy's Experiment, Hydraulic Conductivity, Permeability, and Transmissivity

Henri Darcy, while studying the fountains in the city of Dijon, France, around 1856, established empirically that the flux of water through a sandy formation (Fig. 4.1) may be calculated by

$$Q = KA\,\Delta h/L \qquad (4.1.1)$$

where $A$ is the area of the cross-section of the sandy formation, $\Delta h$ the difference in hydraulic head in the water between the top and the bottom of the sandy formation, $K$ a constant that depends on the porous medium, called hydraulic conductivity in hydrogeology, or sometimes coefficient of permeability, and $L$ the thickness of the sandy formation.

By dividing both sides by $A$, we obtain the fictitious velocity $U$ of the fluid at the outlet of the formation, bearing in mind that this definition of the velocity $U$ considers the entire section to be open to the flow. This is what we have called the *filtration velocity*:

$$U = Q/A$$

Furthermore, if the difference in hydraulic head per unit length of porous medium traveled by the flow is denoted $i = \Delta h/L$, also called *hydraulic*

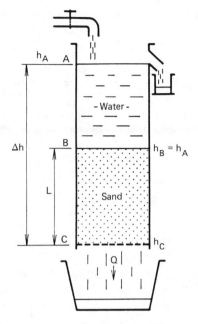

**Fig. 4.1.** Darcy's experiment.

*gradient*, we get

$$U = Ki \tag{4.1.2}$$

which is the simplest expression of Darcy's law.

(a) *Intrinsic permeability.* If we take Eq. (3.4.8) using dimensional analysis and experimental verification, we find that the constant $K$ actually varies inversely with the dynamic viscosity $\mu$ of the fluid. Moreover, we know from the calculations based on the Navier–Stokes equations that the real reasons for fluid displacement in a porous medium are, on the one hand, the pressure gradients and, on the other, the external gravity forces as in Eq. (3.4.8). Consequently, Darcy's law should be expressed in the generalized form

$$\mathbf{U} = -\frac{k}{\mu}(\mathbf{grad}\, p + \rho g\, \mathbf{grad}\, z) \tag{4.1.3}$$

which we admit for steady and unsteady flow of compressible fluids. Note that since $\mathbf{U}$ is a macroscopic magnitude, this is also true for $\mu$, $\rho$, and $p$, regarding the averages (given in angle brackets) defined in Chapter 3.*

---

* In particular, one can show that Darcy's law applies to the gradient of average pressure, **grad** $\langle p \rangle$, and not to the average gradient of pressure, $\langle \mathbf{grad}\, p \rangle$. See Marle (1967).

The intrinsic or specific permeability $k$ relates to the porous medium regardless of the characteristics of the fluid. It is only defined on the macroscopic scale. Its dimension, from Eq. (4.1.3), is that of a surface area,

$$[k] = \frac{[Q][\mu]}{[A][pL^{-1}]} = \frac{(\text{length}^3\text{time}^{-1})(\text{mass length}^{-1}\text{time}^{-1})}{(\text{length}^2)(\text{mass length}^{-2}\text{time}^{-2})} = (\text{length}^2)$$

However, it is often expressed in darcys. One darcy is equal to $0.987 \times 10^{-12}$ m$^2$, and is defined by a medium for which a flow of 1 cm$^3$/s is obtained through a section of 1 cm$^2$, for a fluid of viscosity 1 cP, and a pressure gradient of 1 atm/cm (760 mm Hg/cm).

In practice, the petroleum industry uses the millidarcy (md $10^{-3}$ darcy) because the most common permeabilities usually lie between one and a few thousands of millidarcies.

(b)    *The hydrogeologist's hydraulic conductivity.*    The relation between the intrinsic permeability $k$ and the hydraulic conductivity $K$ used by hydrogeologists is established by treating the flow as a function of the hydraulic head gradient $\Delta h/L = -\mathbf{grad}\, h$.

If we assume that the fluid is incompressible, we can write Eq. (4.1.3) as follows:

$$\mathbf{U} = -\frac{k}{\mu}\mathbf{grad}(p + \rho gz)$$

or yet, remembering that the hydraulic head $h$ is defined by $h = p/\rho g + z$, and taking $\rho g$ out of the gradient,

$$\mathbf{U} = -\frac{k\rho g}{\mu}\mathbf{grad}\, h \qquad (4.1.4)$$

When Eq. (4.1.2) is compared to Eq. (4.1.4), we see that

$$K = k\rho g/\mu$$

Note that the two forms of Darcy's law,

$$\mathbf{U} = -\frac{k}{\mu}[\mathbf{grad}\, p + \rho g\,\mathbf{grad}\, z] = -K\,\mathbf{grad}\, h$$

are strictly equivalent even for compressible fluids, if the definition of the hydraulic head is taken to be

$$h = z + \int_0^p \frac{dp}{\rho g}$$

which we have already mentioned above. However, we shall not use it here.

The dimension of $K$ is that of a velocity:

$$[K] = \frac{(\text{length}^2)(\text{mass length}^{-3})(\text{length time}^{-2})}{(\text{mass length}^{-1}\text{time}^{-1})} = (\text{length time}^{-1})$$

It is usually expressed in meters per second (see the conversion factor in Appendix 2 for U.S. non-SI units). The hydraulic conductivity of aquifer layers range from $10^{-9}$ to $10^{-2}$ m/s.

The hydraulic conductivity depends not only on the fluid, which is not very disturbing since we are always dealing with water, but also on its viscosity, and the viscosity varies a great deal according to the temperature. The following figure gives the variations in the viscosity of the water with temperature, compared to the viscosity measured at 20°C, which is equal to $1.002 \times 10^{-3}$ Pa s (or 1.002 cP) (Fig. 4.2).

In spite of the hypothesis of an isothermal porous medium, which we have formulated, we must be careful when dealing with very superficial aquifers where the climatic variations between summer and winter result in considerable variations in hydraulic conductivity: it is reduced by 40% if the water temperature drops from 25 to 5°C. This will be discussed later in relation to geothermal problems.

In order to compare the intrinsic permeability and the hydraulic conductivity, it is useful to keep the following relation in mind: for water at 20°C, 1 millidarcy gives

$$0.987\frac{10^{-15} \times 10^3 \times 9.81}{1.002 \times 10^{-3}} = 0.966 \; 10^{-8} \text{ m/s}$$

Thus, 1 millidarcy is close to $10^{-8}$ m/s for water at 20°C. (For air at 15°C and normal pressure, $\mu = 1.8 \times 10^{-5}$ Pa s and $\rho = 1.25$ kg m$^{-3}$.)

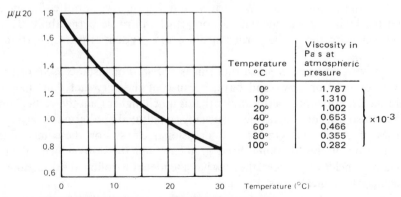

Fig. 4.2. Water viscosity as a function of temperature.

(c) *Permeability and porosity.* From the analogy of flow in a fractured medium or through a circular tube, attempts have been made at linking permeability to the porosity or to the dimension of the pores.

$$\text{Darcy's law:} \quad \mathbf{U} = -\frac{k\rho g}{\mu}\,\mathbf{grad}\,h$$

$$\text{Fractured medium:} \quad \mathbf{U} = -\frac{\omega e^2 \rho g}{12\mu}\,\mathbf{grad}\,h$$

$$\text{Tubular medium:} \quad \mathbf{U} = -\frac{\omega r^2 \rho g}{8\mu}\,\mathbf{grad}\,h$$

One could therefore consider linking $k$ to $\omega e^2/12$ or $\omega r^2/8$ but, unfortunately, all attempts to do this have yielded mediocre results. The best known empirical formulas are that of Koseny–Carman,

$$k = \frac{\omega^3}{5S_0^2(1-\omega)^2}$$

where $S_0$ is the surface area exposed to the fluid per unit volume of the solid (and not porous) medium and $\omega$ is the total porosity; that of Hazen,

$$\log k = 2\log d_{10} - 3$$

where $d_{10}$ is the "effective diameter" of the grains in the soil (see Section 2.1.e), and $k$ is in cm$^2$ and $d_{10}$ in cm; and that of Bretjinski (for sands), with $K$ in m/day,

$$\omega = 0.117\,(K)^{1/7}$$

(d) *Permeability tensor.* The experiment with Darcy's permeameter is made by observing a flow in one direction. When we went from $U = Ki$ to $\mathbf{U} = -K\,\mathbf{grad}\,h$, we already admitted that it was possible to generalize the law to three-dimensional space. Moreover, in doing this, we admitted implicitly that the hydraulic conductivity $K$, or yet, the intrinsic permeability $k$, are *isotropic* properties of the porous medium, independent of the orientation in space.

However, we know *a priori* that this is not so. For instance, sedimentary layers of sand or clay–sand have, because of the very fact that they are stratified, a horizontal permeability that is much higher than the vertical one. This is also true for alluvial media, usually constituted by alternating layers or lenses of sands and gravels and occasional clays. For these media the orientation of the hydraulic head gradients and the flow velocity do not usually coincide any longer: the flow has a tendency to follow the directions of the highest permeabilities (Fig. 4.3).

This leads us to consider the permeability as a tensorial property, which is simply the mathematical translation of this observation. To do this, one

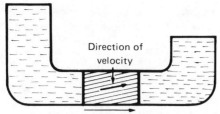

Direction of the hydraulic head gradient

**Fig. 4.3.** Evidence of the anisotropy of a layered medium.

defines a permeability tensor **k**, which we take to be a second order*
symmetrical[†] tensor (i.e., **k** is a matrix of nine coefficients, symmetrical with
respect to the diagonal):

$$\mathbf{k} = \begin{vmatrix} k_{xx} & k_{xy} & k_{xz} \\ k_{yx} & k_{yy} & k_{yz} \\ k_{zx} & k_{zy} & k_{zz} \end{vmatrix} \quad \text{with} \quad \begin{matrix} k_{xy} = k_{yx} \\ k_{xz} = k_{zx} \\ k_{yz} = k_{zy} \end{matrix}$$

Some authors have tried to prove this on the basis of models representing the
porous medium (models of capillary tubes, fractures, etc.). These demon-
strations justify the generalization by analogy but do not prove it. However,
Matheron (1967) has established the symmetry of the permeability tensor
through integration of the Navier–Stokes equations.

Thus, we write

$$\mathbf{U} = -\mathbf{K} \operatorname{grad} h \tag{4.1.5}$$

$$\mathbf{U} = -\frac{\mathbf{k}}{\mu} [\operatorname{grad} p + \rho g \operatorname{grad} z] \tag{4.1.6}$$

---

* A tensor of the second order is defined by the rule of transformation of the tensor components
in a rotation of the cartesian coordinate system: if in one coordinate $(x_1, x_2, x_3)$, the components
of the tensor are $K_{ij}$, then the components $K_{ij}'$ in a coordinate $(x_1', x_2', x_3')$ are

$$K_{ij}' = \sum_l \sum_m \cos \alpha_{li} \cos_{mj} K_{lm}$$

where $\alpha_{li}$ is the angle of the axis $Ox_l$ with the axis $Ox_i'$. We can easily establish that this is indeed
how the components of the permeability tensor are transformed by writing a flow balance
equation.

† One can show macroscopically that the symmetry of this tensor is a sufficient condition, at
least for describing the observations. In a stratified medium, it is obvious that the directions
parallel and perpendicular to the stratification are special directions of the flow, for which the
hydraulic head gradient and the flow velocity again coincide, i.e., that the components of the
tensor are reduced to its diagonal component. We know that a symmetric matrix is a sufficient
condition for its eigenvalues to be distinct and its eigenvectors orthogonal. However, to prove
that this condition is necessary, one has to make use of the first and second principle of
thermodynamics.

Develop, for example, this last relationship by calculating the three components of the velocity $\mathbf{U}$ in the most general manner:

$$U_x = -\frac{k_{xx}}{\mu}\frac{\partial p}{\partial x} - \frac{k_{xy}}{\mu}\frac{\partial p}{\partial y} - \frac{k_{xz}}{\mu}\left(\frac{\partial p}{\partial z} + \rho g\right)$$

$$U_y = -\frac{k_{xy}}{\mu}\frac{\partial p}{\partial x} - \frac{k_{yy}}{\mu}\frac{\partial p}{\partial y} - \frac{k_{yz}}{\mu}\left(\frac{\partial p}{\partial z} + \rho g\right) \qquad (4.1.7)$$

$$U_z = -\frac{k_{xz}}{\mu}\frac{\partial p}{\partial x} - \frac{k_{yz}}{\mu}\frac{\partial p}{\partial y} - \frac{k_{zz}}{\mu}\left(\frac{\partial p}{\partial z} + \rho g\right)$$

Indeed, it is clear that if $\mathbf{k}$ is defined as a tensor, it is possible for a gradient in a given direction $x$ to generate components of the flow in the perpendicular directions $y$ and $z$, which tallies with the experiment. This relationship is written with six different permeability coefficients and takes the symmetry into account.

This rather cumbersome expression may be simplified by using a new set of orthogonal axes $X$, $Y$, and $Z$, deduced from the former by a rotation such that the permeability tensor is reduced to its diagonal components. Mathematically, $X$, $Y$, and $Z$ are the directions of the eigenvectors of the matrix $\mathbf{k}$. Physically, $X$, $Y$, and $Z$ are the directions in which the flow is actually parallel to the hydraulic head gradient (in practice, one direction at right angles to the stratification and two directions parallel to it). These directions are called the principal axes of anisotropy of the medium. In these axes, the tensor $\mathbf{k}$ is reduced to three diagonal components

$$\mathbf{k} = \begin{vmatrix} k_{xx} & 0 & 0 \\ 0 & k_{yy} & 0 \\ 0 & 0 & k_{zz} \end{vmatrix} \qquad (4.1.8)$$

and the Eq. (4.1.7) becomes

$$U_x = -\frac{k_{xx}}{\mu}\frac{\partial p}{\partial x}$$

$$U_y = -\frac{k_{yy}}{\mu}\frac{\partial p}{\partial y} \qquad (4.1.9)$$

$$U_z = -\frac{k_{zz}}{\mu}\left(\frac{\partial p}{\partial z} + \rho g\right) \qquad \text{(if } z \text{ is still the vertical direction)}$$

In practice, there are two distinct permeabilities in sedimentary media with more or less horizontal stratification: a vertical permeability $k_{zz}$ and a horizontal permeability $k_{xx} = k_{yy}$. The anisotropy ratio $k_{zz}/k_{xx}$ generally ranges between 1 and 100.

As the hydraulic conductivity $K$ is equal to the intrinsic permeability $k$, except for one scalar factor, the anisotropy concept already developed for **k** applies to **K** as well. In the rest of the analysis, we shall always assume that the cartesian coordinates are the principal axes of the permeability tensor, while $z$ remains the vertical axis. [Otherwise the term $\rho g$ **grad** $z$ in Eq. (4.1.6) will be distributed on the three equations in $X, Y, Z$ of Eq. (4.1.9), making the writing cumbersome. This difficulty disappears if the fluid is incompressible because the hydraulic head $h$ may then be used.]

Note that if the anisotropy of the medium is uniform (the same at all points in space) we can turn it into an equivalent isotropic medium by anamorphosis on the coordinates (see Section 7.1.6).*

(e) *The fractured medium.* At present, there are two methods for tackling flow in a fractured medium: modeling of the flow, accounting for the fractures one by one, or modeling with an equivalent continuous medium. In an elementary fracture, the laws governing the flow are (summarizing Louis, 1974) for laminar flow $V = K_f J_f$ and for turbulent flow $V = K_f' J_f^\alpha$, where $V$ is the mean velocity of the flow in the fracture, i.e., a velocity assumed uniform over the total aperture of the fracture and producing the same flow[†] as the real one; $K_f$ the hydraulic conductivity of the fracture (length time$^{-1}$); $K_f'$ the turbulent conductivity of the fracture (length time$^{-1}$); $J_f$ the right-angle projection of the hydraulic head gradient on the fracture plane; and $\alpha$ the degree of nonlinearity of the flow ($0.5 \leq \alpha \leq 1$).

The transition from laminar to turbulent flow is governed by the values of the Reynolds number $R_e$ on the one hand and of the relative roughness $R_r$ on the other.

The Reynolds number (dimensionless) is defined, for a cylindrical pipe, by

$$R_e = \frac{V d \rho}{\mu} \tag{4.1.10}$$

where $V$ is the mean velocity of the fluid, $d$ the diameter of the pipe, $\mu/\rho$ the kinematic viscosity.

In classical hydraulics, the flow regime is laminar for $R_e < 2000$ and turbulent for $R_e > 2000$.

---

* It is also possible to define directional hydraulic conductivities either in the direction of the flow (ratio between **U** and the component of the head gradient along **U**) or in the direction of the gradient (ratio between the component of **U** along the gradient and the head gradient itself). See Bear (1972).

† We have shown in Section 3.4 that the real profile of the velocity, in laminar flow, is parabolic if the fracture is smooth.

For a plane fracture, the diameter of the pipe is replaced by the "hydraulic diameter" defined by

$$D_h = 4S/p \tag{4.1.11}$$

where $S$ is the cross-section area of the flow in the fracture and $p$ is the outside perimeter of this cross-section area of the flow. For a very long fracture, $D_h$ is equal to twice its aperture.

The relative roughness (dimensionless) is defined by

$$R_r = \varepsilon/D_h \tag{4.1.12}$$

where $\varepsilon$ is the mean height of the irregularities in the fractures and $D_h$ is the hydraulic diameter of Eq. (4.1.11).

Depending on the values of $R_e$ and $R_r$, Louis (1974) defines empirically five flow regimes and their domains of validity, which are represented on Fig. 4.4.

The laws of the steady-state flow in each regime depend on the aperture $e$, the kinematic viscosity $\mu/\rho$, the relative roughness $R_r$, and the hydraulic head gradient in the fracture plane $J_f$:

Type 1; smooth laminar:
$$V = -\left(\frac{\rho g e^2}{12\mu}\right)J_f \tag{4.1.13}$$

Type 2; smooth turbulent:
$$V = -\left[\frac{g}{0.079}\left(\frac{2\rho e^5}{\mu}\right)^{1/4}J_f\right]^{4/7} \tag{4.1.14}$$

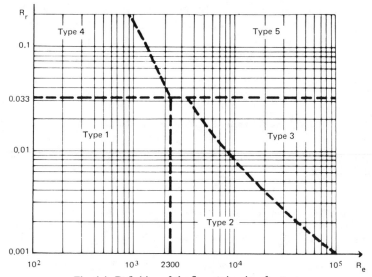

**Fig. 4.4.** Definition of the flow regime in a fracture.

Type 3; rough turbulent:
$$V = -\left(4\sqrt{eg}\ln\frac{3.7}{R_r}\right)\sqrt{J_f} \qquad (4.1.15)$$

Type 4; rough laminar:
$$V = -\left[\frac{\rho g e^2}{12\mu(1 + 8.8R_r^{1.5})}\right]J_f \qquad (4.1.16)$$

Type 5; very rough turbulent:
$$V = -\left(4\sqrt{eg}\ln\frac{1.9}{R_r}\right)\sqrt{J_f} \qquad (4.1.17)$$

(In these expressions, ln is the natural logarithm.)

Finally, if the fracture is not completely open (the two edges touch in places), we have to multiply the right-hand side of Eqs. (4.1.13)–(4.1.17) by the "degree of separation of the fracture" $F$:

$$F = \frac{\text{open fracture surface area}}{\text{total fracture surface area}} \qquad (4.1.18)$$

For a system of parallel and continuous fractures, in laminar flow, the equivalent hydraulic conductivity of the medium can be calculated from

$$K = \frac{e}{b}K_f + K_m \qquad \text{(length time}^{-1}) \qquad (4.1.19)$$

where $b$ is the mean distance between fractures, $K_f$ the hydraulic conductivity of fractures, Eq. (4.1.13) or (4.1.16), and $K_m$ the hydraulic conductivity of the rock matrix (length time$^{-1}$), if not zero.

The term $K$ is a directional permeability, i.e., defined for a hydraulic gradient parallel to the fracture plane.

If the fracture system is discontinuous (the fractures are of finite length and unconnected), the largest part of the transfer happens in the matrix and the fractures work as "short cuts." For the equivalent directional conductivity, Louis proposes

$$K = K_m\left[1 + \frac{1}{2}\left(\frac{l}{L-l} - \frac{l}{L}\right)\right] \qquad (4.1.20)$$

where $l$ is the mean extension of the fractures and $L$ is the mean distance between two unconnected fractures.

These conductivities are the directional conductivities of the equivalent continuous medium. In the case of continuous fractures, the continuous directional conductivity of the equivalent medium is therefore dependent on the cube of the fracture aperture:

$$K = e^3\frac{Fg\rho}{12\mu bC}$$

where $F$ = degree of fracture separation, Eq. (4.1.18), $C = 1$, regime of type 1, and $C = 1 + 8.8R_r^{1.5}$, regime of type 4.

Maini and Hocking (1977) give the equivalence between the hydraulic conductivity in a fractured medium and that of a porous medium in Fig. 4.5. For example, the flow through a 100-m-thick cross-section of a porous medium with a hydraulic conductivity of $10^{-7}$ m/s could also come from *one single fracture* with an opening not wider than 0.2 mm in a fractured medium with an impervious rock matrix! This shows the immense importance for the flow of one single fracture that is not even wide. Figure 4.5 gives the relation between the aperture of the single studied fracture, the hydraulic conductivities of the equivalent medium, and the thickness of the section of the continuous medium equivalent to this fracture.

Consequently, there are two possible ways of modeling the flow in a medium constituted by several conducting fractures. The first is the method of the continuous medium: each family of fractures defines a directional conductivity, thus constituting a hydraulic conductivity tensor. As we know the intensity and the direction of these conductivities, we can calculate the principal axes of anisotropy of the tensor and the conductivities in these directions.

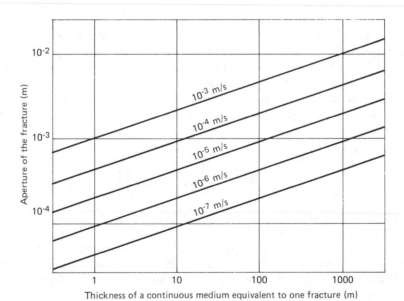

**Fig. 4.5.** Comparison between the hydraulic conductivity of the porous medium and the fractured medium versus the aperture. [From Maini and Hocking (1977). Reproduced with permission from the Geological Society of America.]

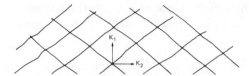

**Fig. 4.6.** Principal axes of anisotropy of a fractured medium.

For example, in two dimensions, two fracture systems with the same directional conductivity give the principal axes of anisotropy shown in Fig. 4.6. Maini and Hocking (1977) give the following expressions for calculating the directions of anisotropy and the principal hydraulic conductivities of the equivalent medium:

$$\psi_i = \frac{1}{2} \arctan\left(\frac{\sin 2\theta}{\cos 2\theta \, K_a/K_b}\right)$$

$$K_i = \frac{K_a K_b \sin^2 \theta}{K_a \sin^2 \psi_i + K_b \sin(\theta - \psi_i)}$$

where $K_a$ and $K_b$ are the equivalent directional hydraulic conductivities of the fracture networks a and b, as shown in Fig. 4.7.

In three dimensions, Feuga (1981) gives the following expressions for determining the hydraulic conductivity tensor of a fractured medium with several fracture directions:

$$\mathbf{K} = \frac{1}{l} \sum_{i=1}^{N} e_i k_i \mathbf{R}_i$$

where $l$ is the arbitrary dimension of the side of a square block of the fractured medium, large enough to statistically sample all the families of fractures, $N$ the

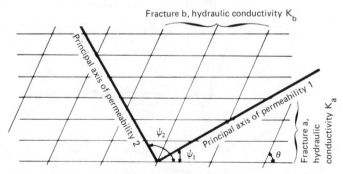

**Fig. 4.7.** Orientation of the principal axes of anisotropy in a fractured medium in two dimensions. [From Maini and Hocking (1977). Reproduced with permission from the Geological Society of America.]

number of fractures in the block of side $l$, $e_i$ the aperture of each individual fracture, and $k_i$ the hydraulic conductivity of each individual fracture

$$\mathbf{R}_i = \begin{bmatrix} 1 - \cos^2 d_i \sin^2 p_i & \frac{1}{2}\sin 2 d_i \sin^2 p_i & -\frac{1}{2}\sin 2 p_i \cos d_i \\ \frac{1}{2}\sin 2 d_i \sin^2 p_i & 1 - \sin^2 d_i \sin^2 p_i & \frac{1}{2}\sin 2 p_i \sin d_i \\ -\frac{1}{2}\sin 2 p_i \cos d_i & \frac{1}{2}\sin 2 p_i \sin d_i & \sin^2 p_i \end{bmatrix}$$

In the matrix $\mathbf{R}_i$, the direction $d_i$ and the dip $p_i$ of each fracture are defined as in Fig. 4.8.

Once the tensor $\mathbf{K}$ has been determined, the principal axes of anisotropy and the diagonal components of $\mathbf{K}$ in these directions can be determined by calculating the eigenvalues and the eigenvectors of the matrix $\mathbf{K}$.

This method of the continuous medium approximation is valid for a certain scale of observation: the flow velocities or the hydraulic heads in each fracture are not described with precision, but a mean value of these magnitudes is taken over all the fractures.

The definition of the hydraulic conductivities of each family of fractures may be approached in two ways: either (1) by measuring (or estimating) the mean geometric properties of the fractures (aperture, distance from each other, roughness, etc.) and using the expressions given above, or (2) through *in situ* tests by injecting water and measuring the hydraulic conductivities $K_f$ of the elementary fractures directly.

The drawback of both methods is that they assume the fractures to be infinite and to have the same properties everywhere. Their results must be taken with caution. The directions of the principal axes of the conductivity tensor are probably more accurate than the value of the conductivities; these

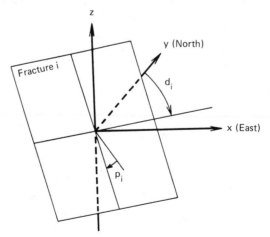

**Fig. 4.8.** Direction and dip of a fracture in three dimensions.

are better defined by large-scale flow and pressure head measurements (e.g., pumping tests; see Section 8.2).

Research is at present being done on systems of fractures of finite length using the theory of percolation: see Clerc et al. (1983), Hammersley and Welsh (1980), Kirkpatrick (1973), Long et al. (1982), Shante and Kirkpatrick (1971), Wilke et al. (1985), Engelman et al. (1983), and Rouleau and Gale (1985).

The second method of modeling the flow in a fractured medium, the method of the "discontinuous medium," takes into account either the elementary fractures of the system or equivalent fractures representing several elementary fractures of the same family. This contrasts with the first method.

The model is composed of "nodes," where the fractures cross each other, joined by planes, where the fluids flow according to the directional laws given above. The hydraulic head is calculated at the nodes, and the velocities are calculated in the planes. Louis (1974) showed that in laminar flow a potential $\Gamma$ may be defined by

$$\Gamma = K_f(p/\rho g + z)$$

The velocity is then given by gradients of this potential.

Although this method enables us to represent the flows with more precision on the small scale, it requires precise knowledge of the position in space and the properties of each of the fractures, taken one by one or grouped into families.

*Remark: unsteady flow in fractured media.*    So far, we have assumed that the water flow is steady, i.e., does not vary with time. If we introduce transient flows, one of the fundamental properties of fractured media appears: double porosity.

Indeed, in the general case, the fractured medium may be looked upon as two coexisting systems of voids: the apertures of the fractures and the porosity between the grains of the blocks of rock separating the fractures. The definition of the equivalent permeability of the medium, given in Eq. (4.1.19), really points to this double system, since it adds the hydraulic conductivity $K_f$ of the fractures to the conductivity $K_m$ of the blocks.

In the steady state, this double porosity and the double conductivity are accounted for by the notion of equivalent hydraulic conductivity. However, it is easily understood that in a transient state the transmission of the pressure variations is much faster in the fractures than in the matrix of the blocks if $K_f \gg K_m$. It therefore becomes necessary to define, in a representative elementary volume, two different pressures, one in the fractures and the other in the matrix, as well as a term for the exchange of mass between the intergranular porosity and the porosity of the fracture.

These problems have been studied by, among others, Warren and Root (1963), Barenblatt et al. (1960), Braester (1972), and Lefebvre du Prey and Weill (1974). Barrenblatt suggests a law of movement as follows:

$$\text{div}\left[\frac{k_f}{\mu}\,\mathbf{grad}\,p_f + \eta\beta_0\frac{\partial}{\partial t}\,\mathbf{grad}\,p_f\right] = \beta_0\frac{\partial p_f}{\partial t}$$

where $k_f$ is the intrinsic permeability of the fracture ($k_f = K_f\mu/\rho g$), $p_f$ is the pressure in the fractures, $\eta = k_f/\alpha$ is a characteristic parameter of the degree of fracturing, where $\alpha$ is the intensity of transfer between the blocks and the fractures, and $\beta_0$ is the usual coefficient of elastic compressibility of the complex water plus porous medium (see Chapter 5).

This leads us in fact to adopt a special darcian law for fractured media, which is dependent on time and is written

$$\mathbf{U} = -\frac{k_f}{\mu}\,\mathbf{grad}\,p_f - \eta\beta_0\frac{\partial}{\partial t}\,\mathbf{grad}\,p_f$$

(f)  *Transmissivity.*  If the aquifer is a layer of thickness $e$, as in Fig. 4.9, and we want to calculate the flow $Q$ in the direction $x$ through the layer over a unit length in the direction perpendicular to the figure, we get

$$Q = \int_0^e \mathbf{U}\cdot\mathbf{n}\,dz = \int_0^e U_x\,dz$$

where $\mathbf{n}$ is the normal line to the axis $Oz$ and $U_x$ is the velocity component in the direction $x$.

Assuming that $z$ is a principal direction of anisotropy [i.e., that the two other directions are in the same plane as the layer $(x, y)$] then at all points $M$ of $Oz$,

$$\mathbf{U} = -\mathbf{K}_M\,\mathbf{grad}\,h$$

where $\mathbf{K}_M$ is the hydraulic conductivity tensor in the plane $x$–$y$ and $\mathbf{grad}\,h$ is the hydraulic head gradient in this plane. If we further assume that this gradient is constant on the transverse line $Oz$, then

$$Q = -\mathbf{grad}\,h\int_0^e \mathbf{K}_M\,dz$$

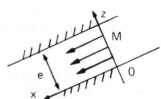

**Fig. 4.9.** Aquifer layer.

This integral has been named the transmissivity;

$$\mathbf{T} = \int_0^e \mathbf{K}\, dz$$

If $K$ is isotropic and constant along $Oz$,

$$T = Ke$$

where $T$ is expressed in meters squared per second and is very often used in the case of groundwater aquifers, whether they are horizontal or not.

## 4.2. Limitations on the Validity of Darcy's Law

The various generalizations of Darcy's elementary experimental law are in fact validated by practice: we find that the calculations made with the help of this generalized law tally with what we observe. However, at the extremes, toward the weak as well as toward the strong hydraulic gradients, there are distortions of the law, which, in truth, are not encountered very often.

(a) *Where the hydraulic gradients have low values.*    In compact clays, the most general law of variation for the low values of the gradient is given by Fig. 4.10 (Jacquin, 1965a, b): below a value $i_0$, the permeability is zero; between $i_0$ and $i_1$, the relation is not linear; and the proportionality corresponding to Darcy's law only applies for $i > i_1$ and is expressed by a formula such as $U = K(i - i_2)$.

The values of $i_0$, $i_1$, and $i_2$ vary a great deal according to the type of clay and its structure; the mineral content of the water also plays a part. As an example, montmorillonite often has reported $i_2$ values on the order of several tens. However, these concepts are still controversial.

(b) *Where the hydraulic gradients have high values.*    When the hydraulic gradient is increased, we observe *experimentally* that there is no longer any

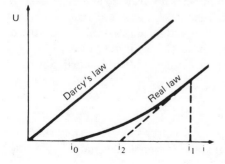

**Fig. 4.10.** Darcy's law for small hydraulic gradients.

proportionality between the gradient and the filtration velocity:

$$\text{grad } h = \alpha U + \beta U^2$$

where $\alpha U$ is the loss due to the viscuous friction against the walls of the matrix and $\beta U^2$ the loss due to the inertia of the fluid (dissipation of kinetic energy in the pores, where the flow lines converge and then diverge again rapidly; these losses are similar to those in bends or narrowing sections of a pipe).

The borderline hydraulic gradient, beyond which Darcy's linear law is no longer valid, depends largely on the medium. In order to make this borderline gradient an intrinsic property of the medium, we sometimes define a "Reynolds number in porous media" (dimensionless) by

$$R_e = U\rho\sqrt{k}/\mu$$

or

$$R'_e = Ud\rho/\mu$$

where $U$ is the filtration velocity (length time$^{-1}$), $\sqrt{k}$ the square root of the intrinsic permeability (length), $\rho/\mu$ the kinematic viscosity (length$^2$ time$^{-1}$), and $d$ the mean diameter of the grains (length) or effective diameter $d_{10}$ (see Section 2.1.e).

Note that the exact definition of the Reynolds number in a circular pipe is $ud\rho/\mu$ ($u$ is mean velocity of the fluid in a pipe of diameter $d$). In view of the difference between the definitions, one must not try to compare these numbers to each other.

In practice, we admit that Darcy's law is valid if the Reynolds number in a porous medium (taking the mean diameter of the grains) is below a limit somewhere between 1 and 10. In this case, the flow is purely laminar inside the pores (Chauveteau and Thirriot, 1967). From 10 to 100 there is the beginning of transient flow, where the forces of inertia are no longer negligible and Darcy's law no longer holds. Beyond 100, the state of flow is turbulent inside the pores and Darcy's law applies even less.

In practice, with the exceptions of karstic systems and the immediate vicinity of wells, the critical Reynolds number is not reached and the flow stays laminar. The result is that, even in the vicinity of a well, the quadratic terms appear only in a limited zone, usually within the gravel pack (gravel introduced as a filter around the well), and is of little importance.

Sichardt's empirical formula for the borderline gradient is worth noting:

$$i = 1/15\sqrt{K}$$

where $K$ is expressed in m/s.

(c) *Darcy's law in the transient state.* Darcy's law is established for steady-state flow (independent of time) both experimentally and theoretically.

We have already said, in Section 4.1.e, that in fractured media the phenomenon of double porosity causes a transient term to appear, which is new to Darcy's law. It is also possible to prove theoretically that, in a porous medium, an additional term appears in Darcy's law in the transient state.

We shall return for a moment to the Navier–Stokes equations. We have seen, in Section 3.4, how in artificial media in a steady state we move from the case where the external forces are nil to that where they exist by simply adding the term $\rho F_i$ to the pressure gradient $\partial p / \partial x_i$. Going back to the complete Navier–Stokes equations Eq. (3.1.2), we see that the transient terms $\rho \, \partial u_i / \partial t$ have the same role in the equations as the external forces. In the steady state if we write

$$ \mathbf{U} = -\frac{k}{\mu}(\mathbf{grad}\, p + \rho g \, \mathbf{grad}\, z) $$

then in the transient state we write

$$ \mathbf{U} = -\frac{k}{\mu}\left( \mathbf{grad}\, p + \rho g \, \mathbf{grad}\, z - \frac{\rho}{\omega}\frac{\partial \mathbf{U}}{\partial t}\right) $$

The factor $1/\omega$ for the transient term originates in the integration, in the REV, of the microscopic transient term* $\rho \, \partial u / \partial t$. However, this additional term is in practice *always disregarded*, because, as $\mathbf{U}$ is small in porous media, $\partial \mathbf{U}/\partial t$ is negligible versus the other terms, except maybe during a time of the order of a second, when the flow gets underway in a porous medium.

## 4.3. Permeability Measurements on Samples

(a)  *Medium with high hydraulic conductivity.*   If the hydraulic conductivity of the medium is not too low, we can use a difference in hydraulic head generated solely by gravity.

*Constant-head permeameter.*   We return to Darcy's experiment (Fig. 4.11). If $A$ is the cross-sectional area of the sample of porous medium, Darcy's law takes the form

$$ Q = -KA \, \mathrm{grad}\, h \qquad \text{i.e.} \qquad K = \frac{\mathcal{Q}L}{(h_1 + L - h_2)A} $$

*Falling-head permeameter.*   If the hydraulic conductivity is less than $10^{-5}$ m/s the constant-head permeameter must be replaced by the falling

---

* The theoretical demonstration is made by first rewriting the Navier–Stokes equations as a partial derivative of time instead of a total derivative.

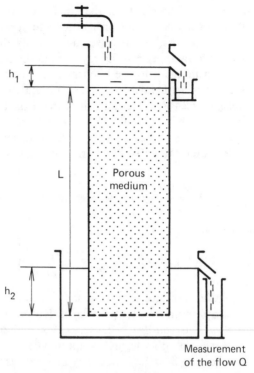

**Fig. 4.11.** Constant-head permeameter.

head permeameter (Fig. 4.12), where a larger head gradient is created through a long pipe with a small section $a$. If $Q$ is the flow through the sample of cross-sectional area $A$, we can write

$$Q = KAh/L \qquad \text{(Darcy's law)}$$

$$Q = -a\,dh/dt$$

and thus

$$\frac{dh}{h} = -\frac{A}{a}\frac{dt}{L}$$

$$\ln\frac{h}{h_0} = -\frac{AK}{aL}(t - t_0)$$

If we trace $\ln h$ on a graph versus time, we obtain a straight line, the slope of which is proportional to $K$.

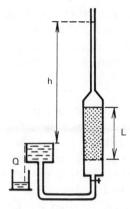

Fig. 4.12. Falling-head permeameter.

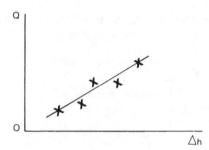

Fig. 4.13. Flow versus head losses in a porous medium.

(b) *Medium with low hydraulic conductivity.* If we want to measure lower hydraulic conductivities, we apply larger pressure differences with the help of pumps and measure the pressures upstream and downstream for different values of the flow $Q$. The slope of the line that gives $Q$ versus $\Delta h$ (see Fig. 4.13) makes it possible to calculate the permeability. Quite often, the permeability to a gas is measured, since it is easier to obtain. The knowledge of $\rho$ and $\mu$ enables us to pass from permeability to hydraulic conductivity.

The various measurements carried out in the laboratory do not reflect the *in situ* hydraulic conductivity, which may be quite different.* In order to measure the latter, the reaction of the terrain to pumping or injection is used, depending on whether we are dealing with a permeable or impermeable terrain. This reaction is examined in detail in Sections 8.2 and 8.6.

Hydraulic conductivity could also conceivably be determined indirectly in an aquifer by measuring the mean pore velocity of water in the medium, using tracers, as $\mathbf{u}^* = \mathbf{U}/\omega$. If the kinematic porosity and the head gradient are

---

* If the rock is not compact, the sample is often modified by the sampling technique; moreover, the permeability usually varies a great deal in space, and one sample may not be representative.

known, $K$ can thus be obtained. Hanshaw and Back (1974) used natural $^{14}C$ as a tracer for this purpose in a limestone aquifer in Florida. However, the mechanisms of diffusion and dispersion in the medium (see Chapter 10) make the determination of the mean velocity very imprecise [see also Pearson *et al.* (1983)].

(c)  *Hydraulic conductivity values.*   The permeability of a rock is, of course, due to its effective porosity, i.e., to the existence of interconnected voids.

In the same way as we have defined an interstitial porosity and a fracture porosity, it would be possible to define two types of permeability (interstitial and fracture), which were formerly described as small-scale permeability and large-scale permeability, because the REV used to define them was not the same. In practice, it is difficult to distinguish between the two types of permeability that may coexist in the field.

*For unconsolidated detrital rocks with interstices*, the hydraulic conductivity depends on the size of the grains, as in the accompanying table.

| Medium | $K$ (approximate) (m/s) |
|---|---|
| Coarse gravels | $10^{-1}$–$10^{-2}$ |
| Sands and gravels | $10^{-2}$–$10^{-5}$ |
| Fine sands, silts, loess | $10^{-5}$–$10^{-9}$ |
| Clay, shale, glacial till | $10^{-9}$–$10^{-13}$ |

The limit separating permeable rocks from impermeable ones is arbitrarily set at $10^{-9}$ m/s. The clays are impermeable in spite of their great total porosity, because their small pores give them a very low effective porosity.

*For hard rocks*, hydraulic conductivity depends on the permeability of the matrix and that of the fractures. The following table of ranges is given for unfractured rocks.

| Medium | $K$ (m/s) |
|---|---|
| Dolomitic limestones | $10^{-3}$–$10^{-5}$ |
| Weathered chalk | $10^{-3}$–$10^{-5}$ |
| Unweathered chalk | $10^{-6}$–$10^{-9}$ |
| Limestone | $10^{-5}$–$10^{-9}$ |
| Sandstone | $10^{-4}$–$10^{-10}$ |
| Granite, gneiss, compact basalt | $10^{-9}$–$10^{-13}$ |

*For fractured rocks*, the hydraulic conductivity depends very much on the density and aperture of the joints. However, fractures may either seal with time or, on the contrary, increase in aperture. In limestones, $CO_2$ is dissolved by water in the atmosphere and in the superficial soil; $H_2CO_3$ then dissolves limestone deep in the aquifer, thus enlarging the fissures. This may evolve into a karstic system, where some of the fractures may locally become very large and form an underground system of chambers, tunnels, pipes, and siphons, through which most of the water flows. The concept of hydraulic conductivity no longer applies in such cases. However, all limestone aquifers are not necessarily pure karstic systems: the dissolution of limestone may create a network of open fractures with hydraulic conductivities in the range $10^{-3}$ to $10^{-1}$ m/s.

In crystalline rocks, on the other hand, fractures are very often sealed (partially or totally) by deposits of calcite, silica, or clay. Fractured crystalline rocks have hydraulic conductivities in the range $10^{-4}$ to $10^{-8}$ m/s.

Fractured basalt can be highly permeable. In some circumstances, the cooling of a basalt layer creates a dense network of vertical joints, which divide the layer into contiguous pillars or prisms of basalt (diameter in the order of 0.5 m with around six facets). The hydraulic conductivity may reach $10^{-1}$ m/s.

In fractured rocks, the hydraulic conductivity generally decreases with depth due to the increase in the mechanical stress, causing the fractures to close. In crystalline rocks, Snow (1968) and Carlsson and Olsson (1977) have suggested the following empirical laws:

$$K(z) = (K_s)(10^{-z/l})$$

$$K(z) = (K_s)(z^{-2.5})$$

$$K(z) = (K_s)(z^{-1.6})$$

where $l$ is in the range of 100–500 m, $z$ is in meters with origin at the surface, positive downward, and $K_s$ is hydraulic conductivity at the surface.

These laws may apply in the average, i.e., for a large number of measurements of hydraulic conductivity as a function of depth in boreholes (see also Section 8.6). Occasionally, in a given borehole, an open fracture or a crushed zone with high hydraulic conductivity may be encountered, even at great depth.

In a given network of fractures, high fluid pressure may locally increase the aperture of the fractures and thus the conductivity (e.g., near an injecting well); see Gale (1975) or Witherspoon *et al.* (1973). At even higher pressures, an injected fluid may create a new fracture in the rock. This is known as hydraulic fracturing, and is often used for increasing the permeability of an oil reservoir. [See Cornet (1979, 1980), Fairhurst and Cornet (1981), and Cornet and Valette (1984).]

## 4.4. Probabilistic Approach to Permeability and Spatial Variability

We have seen in Section 2.1.d that a probabilistic definition of a property like porosity can be given in porous media. However, the definition of the permeability as a random function requires a change of scale, which was proposed by Matheron in 1967 referring to the works of Schwydler (1962). As a matter of fact, point permeability cannot be used in the same way as point porosity, because, on the microscopic scale, Darcy's law implied by the notion of permeability does not apply to the flow: it is the Navier–Stokes law that governs the relationship between the hydraulic head and the velocity.

Matheron (1967) has shown that Darcy's law is simply a consequence of the linearity of Navier's equations, not of their form. It is, however, the spatial integration of Navier's equation in the very complex geometry of a porous medium that leads to Darcy's law and the definition of permeability. We can thus, conceptually at least, link permeability to the geometric description of a porous medium. Such a geometric description of a medium (e.g., size and shape of pores) can be made stochastically, exactly as we have done in Section 2.1.d for porosity. For instance, we have seen in Section 3.4 that the permeability of simple geometrical media (fractures, tubes) depends on the aperture of the fractures or the diameter of the tubes. These can be given a stochastic definition at a point in space (probability distribution function, expected value, spatial covariance, etc.). In a more complex medium, the number of descriptors of the geometry increases, but conceptually, each of them can be given a stochastic definition on the microscopic scale.

As a consequence, permeability on the macroscopic scale, depending on stochastic microscopic quantities, can be regarded as a stochastic property and can be defined conceptually as a random function. This will have a probability distribution function, expected values, spatial covariance, etc.

Quite a number of authors have studied the pdf of permeability, hydraulic conductivity or transmissivity (see Section 4.1.f) in a given aquifer. Their analysis is biased most of the time because they assume that the measurements taken at different locations are statistically independent, whereas, in reality, permeability usually displays a strong spatial correlation. Nevertheless, following Law (1944), Walton and Neill (1963), Krumbein (1936), Farengolts and Kolyada (1969), Ilyin *et al.* (1971), Jetel (1974), and Rousselot (1976), we can admit that permeability usually has a log-normal probability distribution function, whatever the nature of the rock. The variance of this spatial variability of permeability is quite high: if $Y = \ln k$, $\sigma_Y^2$ is generally in the range between 1 and 2 but can reach 10 in some cases.

The spatial correlation of transmissivity has also been studied, e.g., by Delhomme (1974, 1978a,b, 1979). He found that, in general, the stationarity hypothesis did not hold, and that stationarity on the first increments (called the intrinsic hypothesis) should be used. Instead of the covariance $C(h)$ (Section 2.1.1), one must then use the variogram $\gamma(h)$, which we will define in Chapter 11. The spatial correlation is important over distances that can be short (e.g., 10 m) or very long (up to 100 km), depending on the type of aquifer. There is, however, very often a strong erratic component (spatially uncorrelated) in the transmissivity, which may cause two wells not very far apart to have quite different transmissivities.

This spatial variability of permeability (or hydraulic conductivity, or transmissivity) leads us to the question of how to compose local permeability values in order to obtain an average permeability. In a deterministic approach, it is easy to show that the composition of uniform "blocks," placed side by side in space, gives:

(1)    A law of harmonic composition, if the blocks are in series (Fig. 4.14):

$$\frac{\sum l_i}{K_{\text{mean}}} = \sum \frac{l_i}{K_i}$$

(2)    A law of arithmetic composition, if the blocks are in parallel (Fig. 4.15):

$$K_{\text{mean}} \sum e_i = \sum e_i K_i$$

Here, we recognize the same law as that of the composition of resistances derived from Ohm's law in electricity.

In a probabilistic approach, where the permeability may vary in all directions of space, Matheron (1967) has obtained the following results:

(1)    If the flow is uniform (parallel flow lines), whatever the spatial correlation of the permeability and whatever the number of dimensions of the

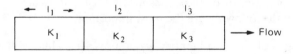

**Fig. 4.14.** Blocks in series.

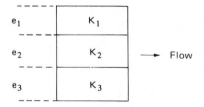

**Fig. 4.15.** Blocks in parallel.

space, the average permeability always ranges between the harmonic* mean and the arithmetic* mean of the local permeabilities.

(2)  If the probability distribution function of the permeability is log-normal and if the flow is two-dimensional, the average permeability is exactly equal to the geometric* mean of the local permeabilities in uniform flow.

(3)  If the flow is not uniform (converging radial, for example), there is no law of composition, constant in time, that makes it possible to define a mean darcian permeability. This problem is quite worrying from the conceptual viewpoint in so far as it is precisely through pumping tests in wells that the permeability (or transmissivity) of an aquifer is measured *in situ* (see Chapter 8). On this point, research continues.

Gelhar (1976), Bakr *et al.* (1978), and Gutjahr *et al.* (1978) also give linearized approximations of the average permeability in uniform flow, for a normal probability distribution function of permeability:

$$1\text{-D:} \quad k_M = k_G(1 - \sigma_Y^2/2)$$

$$2\text{-D:} \quad k_M = k_G$$

$$3\text{-D:} \quad k_M = k_G(1 + \sigma_Y^2/6)$$

where $k_M$ is average permeability, $k_G$ is geometric mean permeability, and $\sigma_Y^2$ is the variance of $Y = \ln k$.

## 4.5. Movement of Water due to the Influence of Other Forces

The hydraulic head gradient is the main force influencing the movement of water in the ground. It is, however, not the only one. Indeed, experiments show that the flow of water through porous media is caused by other gradients as well, of which the following are the most important:

(1)  Gradient of electric potential: water moves from high voltage towards low voltage. This principle has been used for electrokinetic drainage of soils with weak permeability; see Terzaghi and Peck (1967), Casagrande (1952), and Rocheman, in Filliat (1981).

---

\* Harmonic mean:     $1/K_M = E(1/K)$

  Arithmetic mean:    $K_M = E(K)$

  Geometric mean:     $\ln K_M = E(\ln K)$

(2)   Gradient of chemical concentration: water moves from zones with high concentrations towards those with low concentrations. This effect is also part of the osmotic effect, which generates a selective filtration of the ions in solution.

(3)   Thermal gradient: flow from zones of high temperature to zones of low temperatures. This phenomenon is important in the formation of ice lenses in the soil (Harlan, 1973).

We can then write a generalized darcian law as follows:

$$U = -K_1 \operatorname{grad} h - K_2 \operatorname{grad} E - K_3 \operatorname{grad} C - K_4 \operatorname{grad} \theta$$

The coefficients $K_i$ may be scalar or tensorial.* Similarly, the other flows in porous media (electricity, solutes, heat) are linked to the same gradients by other series of coefficients:

$$i = -K_1' \operatorname{grad} h - K_2' \operatorname{grad} E - K_3' \operatorname{grad} C - \cdots$$

A hydraulic head gradient therefore causes flow of electricity, of solutes, of heat, etc.

In thermodynamics, we therefore need to study all the flows and gradients simultaneously according to what are called coupled transport processes.

**Table 4.1**

Coupled-Process Terminology

| Flow | Gradients | | | |
| | Hydraulic head | Electric potential | Temperature | Concentration |
| --- | --- | --- | --- | --- |
| Fluid | Darcy | Electro-osmosis, Casagrande | Thermal osmosis | Chemical osmosis |
| Electricity | Rouss | Ohm | Seebeck or Thompson | Sedimentation current |
| Heat | Thermal filtration | Peltier | Fourier | Dufour |
| Solutes | Ultra-filtration | Electrophoresis | Soret | Fick |

* Casagrande has found that the "electro-osmotic permeability" $K_2$ does not vary a great deal for disturbed or loose soils and is of the order of $5 \times 10^{-9}$ m$^2$ V$^{-1}$ s$^{-1}$ [Rocheman, in Filliat (1981)].

Refer to the works of Onsager (1931) or Casimir (1945) quoted by Bear (1972) on the subject of the thermodynamics of irreversible processes. The coefficients $\mathbf{K}$ are called "phenomenological coefficients" and must be measured experimentally. In certain cases, we find relations of symmetry and non-negativity in the matrix of these coefficients. In practice, however, the nondiagonal coefficients (i.e., those that are different from the coefficient of the hydraulic head for the velocity, from that of the electric potential for the current, from that of the temperature for the heat flow) are relatively small and negligible versus the diagonal terms.

Table 4.1 briefly reviews the main names given to the mechanisms of coupling. The word "law" is used for the diagonal terms, and "effect" for the nondiagonal ones (Fourier's law, Darcy's law; Soret's effect, Dufour's effect).

# Chapter 5

# Integration of the Elementary Equations, the Diffusion Equation, and Consolidation

The three equations for the circulation of a fluid in a porous medium, established in the two preceding chapters, are significant only for elementary volumes of a porous medium. The first is the continuity (or mass balance) equation:

$$\text{div}(\rho \mathbf{U}) + \frac{\partial}{\partial t}(\rho \omega) + \rho q = 0 \qquad (3.2.3)$$

where $\rho$ is mass per unit volume of fluid (mass length$^{-3}$), $\mathbf{U}$ is filtration velocity of the fluid (length time$^{-1}$) (as if the whole section were accessible to the flow), $\omega$ is the total porosity of the porous medium[†] (dimensionless), and $q$ is volumetric flow rate of fluid per unit volume of rock withdrawn (or added if it is negative) in the porous medium (time$^{-1}$), to which is added a term for the displacement of the fluid–solid interface if the medium is deformed.

The second equation is Darcy's law:

$$\mathbf{U} = -\frac{\mathbf{k}}{\mu}(\mathbf{grad}\ p + \rho g\, \mathbf{grad}\, z) \qquad (4.1.6)$$

where $\mathbf{k}$ is the intrinsic permeability tensor (length$^2$), $\mu$ is dynamic viscosity of the fluid (mass length$^{-1}$ time$^{-1}$), $p$ is fluid pressure (mass length$^{-1}$ time$^{-2}$), $g$ is acceleration due to gravity (length time$^{-2}$), $z$ is the vertical axis directed

---

[†] See footnote to Section 3.2.1, p. 44.

upward, and $\mathbf{grad}\, z$ is a vector with components $(0, 0, 1)$. This law can be simplified for incompressible fluids as follows:

$$\mathbf{U} = -\frac{\mathbf{k}\rho g}{\mu}\,\mathbf{grad}\, h = -\mathbf{K}\,\mathbf{grad}\, h \qquad (4.1.5)$$

where $h$ is the hydraulic head or piezometric head (length),

$$h = \frac{p}{\rho g} + z \qquad (3.3.1)$$

and $\mathbf{K}$ is the permeability tensor (length time$^{-1}$). The law is also expressed by Eq. (4.1.5) for compressible fluids if we agree to define the hydraulic head as

$$h = z + \int_{p_0}^{p} \frac{dp}{\rho(p)g} \qquad (3.3.2)$$

where $p_0$ is pressure at the origin of the axis $z$.

The third is the isothermal equation of state of the fluid,

$$\rho = \rho_0 e^{\beta(p - p_0)} \qquad (3.1.3)$$

where $\beta$ is the coefficient of fluid compressibility (mass$^{-1}$ length time$^{-2}$).

We shall combine these laws in what is called the diffusion equation, the integration of which allows us to calculate the evolution of the fluid in porous media, retaining only one unknown: the pressure $p$ or the hydraulic head $h$, from which we can deduce the other four unknowns, $\rho$ and the velocity $\mathbf{U}$ (three components). This equation is equivalent to what is called "the heat equation" in thermal problems,

$$\nabla^2 \theta = \frac{\rho C}{\lambda}\frac{\partial \theta}{\partial t}$$

where $\theta$ is the temperature, $\rho C$ the heat capacity, $\lambda$ the conductivity, and $\nabla^2$ the Laplacian operator.

It is easier to establish this equation separately in two special cases according to the hypotheses concerning the behavior of the porous medium before establishing its more general form. We shall look at (1) the unconfined aquifer (incompressible water, incompressible medium), (2) the theory of consolidation (incompressible water, compressible porous medium), and (3) the general case (compressible water and porous medium).

## 5.1. Diffusion Equation in Unconfined Aquifers

A water table aquifer is a porous medium that is only saturated up to a certain elevation and overlaid by a dry or unsaturated porous medium. The aquifer is generally limited at the bottom by impermeable bedrock.

In this case, we can disregard the compressibility of the water ($\rho$ constant), as well as that of the porous medium ($\omega$ constant). All variations in hydraulic head cause a movement of the free surface, which increases or decreases the amount of stored water by saturating or draining the porous medium; in the continuity equation, one must consider an elementary volume that includes a section of mobile free surface. Consequently, we take a vertical prism, of thickness, $e$, which penetrates the aquifer between the impermeable bedrock and the free surface.

We now assume that in this water table aquifer, all the velocities are horizontal and parallel to each other along the same vertical line. This hypothesis, called Dupuit's hypothesis, is quite well borne out in reality at some distance from the outlets or from the water divide.

We assume that the permeability tensor allows the vertical axis to be one of its principal directions. Then, according to Darcy's law, if there is no vertical velocity component, there is no vertical hydraulic gradient ($\partial h/\partial z = 0$). We then take the hydraulic head $h(x, y)$ as the unknown, thus making it a two-dimensional problem, since $h$ is independent of $z$; $h$ then represents the hydraulic head at any point on the vertical axis and is, in particular, equal to the elevation of the free surface of the aquifer (Fig. 5.1).

We choose the axes $x$ and $y$ along the two principal directions of anisotropy in the plane. Here, we reestablish the three terms of the continuity equation for the prism $dx$, $dy$, $(h - \sigma)$.

*Mass flux per unit time entering the two faces perpendicular to $Oz$.*

$$F_x = \rho \, dy \left[ \int_{\sigma(x,y)}^{h(x,y)} U_x(x, y, z) \, dz - \int_{\sigma(x+dx,y)}^{h(x+dx,y)} U_x(x + dx, y, z) \, dz \right]$$

$U_x$ is the component of the filtration velocity along $x$. This yields

$$F_x = -\rho \, dy \frac{\partial}{\partial x} \left[ \int_{\sigma}^{h} U_x \, dz \right] dx$$

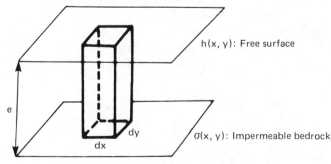

**Fig. 5.1.** Elementary prism in an unconfined aquifer.

Darcy's law allows us to calculate $U_x$

$$U_x = -K_{xx}(x, y, z)\frac{\partial h}{\partial x}$$

If we substitute, we notice that $\partial h/\partial x$ does not depend on $z$. If the term corresponding to the flux entering through the face perpendicular to $Oy$ is added, we get

$$F = +\rho\, dx\, dy \left\{ \frac{\partial}{\partial x}\left[ \int_\sigma^h K_{xx}\, dz\, \frac{\partial h}{\partial x} \right] + \frac{\partial}{\partial y}\left[ \int_\sigma^h K_{yy}\, dz\, \frac{\partial h}{\partial y} \right] \right\}$$

It is assumed that no flux enters or exits through the upper and lower faces (see below).

*Variation in the elemental mass.*    The water mass that can be moved by gravity (specific yield, or drainage porosity, $\omega_d$) contained in the element is $\rho\omega_d(h - \sigma)\, dx\, dy$, and its variation per unit time is

$$\rho\omega_d \frac{\partial h}{\partial t}\, dx\, dy$$

The variation of the elevation $h$ in the free surface indeed causes the specific yield $\omega_d$ to come into play and not the total porosity $\omega$.

*The volumetric flow rate of fluid withdrawn from the element.*    This is found by integration; $q$ is positive if withdrawn and negative if injected.

$$\int_\sigma^h q\, dz\, dx\, dy = Q\, dx\, dy$$

where $Q$ is now the flow rate per unit surface area withdrawn from the aquifer. The mass flux is then $\rho Q\, dx\, dy$. This term for the flux per unit surface area makes it possible to take into account the exchanges between the aquifer and its surroundings (withdrawal, infiltration, etc.), assuming that they take place over the whole thickness of the aquifer. This hypothesis means that the vertical component of the velocity of the fluid is negligible compared to the horizontal one: it is again the Dupuit hypothesis.

*Balance.*    When we write the mass balance, adding together these three quantities and dividing by $\rho$, which is constant, and by $dx\, dy$, which is the elementary area of the aquifer, we get

$$\frac{\partial}{\partial x}\left[ \int_\sigma^h K_{xx}\, dz\, \frac{\partial h}{\partial x} \right] + \frac{\partial}{\partial y}\left[ \int_\sigma^h K_{yy}\, dz\, \frac{\partial h}{\partial y} \right] = \omega_d \frac{\partial h}{\partial t} + Q \qquad (5.1.1)$$

This is the diffusion equation in a water table aquifer. It is nonlinear in $h$.

If $K_{xx}$ and $K_{yy}$ are constant along the entire vertical axis, we can make the integral on $z$ disappear

$$\frac{\partial}{\partial x}\left[K_{xx}(h-\sigma)\frac{\partial h}{\partial x}\right]+\frac{\partial}{\partial y}\left[K_{yy}(h-\sigma)\frac{\partial h}{\partial y}\right]=\omega_{\mathrm d}\frac{\partial h}{\partial t}+Q \qquad (5.1.2)$$

It is still nonlinear in $h$. However, it can, in some cases, be linearized by considering the quantities

$$T_{xx}=\int_{\sigma}^{h}K_{xx}\,dz \qquad\text{and}\qquad T_{yy}=\int_{\sigma}^{h}K_{yy}\,dz$$

which have already been defined as the anisotropic transmissivities of the aquifer (integral of the permeability over the thickness of the aquifer). This transmissivity may sometimes be assumed to vary little with the hydraulic head—i.e., the variations of $h$ are negligible compared to $(h-\sigma)$, for example, less than 10%. Alternatively, the vertical distribution of $K$ can be assumed to be such that the variations in $h$ do not cause a variation in $T$ of more than 10% (this is the case when the permeability is higher at depth than on the surface, e.g., a deep layer of gravel overlayed with fine sands). It then becomes

$$\frac{\partial}{\partial x}\left(T_{xx}\frac{\partial h}{\partial x}\right)+\frac{\partial}{\partial y}\left(T_{yy}\frac{\partial h}{\partial y}\right)=\omega_{\mathrm d}\frac{\partial h}{\partial t}+Q \qquad (5.1.3)$$

Finally, if the transmissivity is isotropic and constant in the entire aquifer,

$$\nabla^2 h=\frac{\partial^2 h}{\partial x^2}+\frac{\partial^2 h}{\partial y^2}=\frac{\omega_{\mathrm d}}{T}\frac{\partial h}{\partial t}+\frac{Q}{T} \qquad (5.1.4)$$

which is a partial differential equation of the second order and parabolic type, similar to the heat equation. The symbol $\nabla^2$ is the Laplace operator, already defined for two dimensions.

As will be seen later, the expressions Eqs. (5.1.3) and (5.1.4) are very often used in practice.

Yet another solution may be suggested in the case where the bedrock $\sigma$ is horizontal. If $\sigma=0$ is chosen as the reference plane for the elevation $z$, $h-\sigma=h$ is the thickness of the aquifer and Eq. (5.1.2) becomes

$$\frac{\partial}{\partial x}\left[K_{xx}h\frac{\partial h}{\partial x}\right]+\frac{\partial}{\partial y}\left[K_{yy}h\frac{\partial h}{\partial y}\right]=\omega_{\mathrm d}\frac{\partial h}{\partial t}+Q$$

If we can assume that $K_{xx}=K_{yy}=K$ is constant in space (isotropic and uniform medium), we find:

$$\nabla^2 h^2=\frac{2\omega_{\mathrm d}}{K}\frac{\partial h}{\partial t}+\frac{2Q}{K}$$

i.e., an equation in $h^2$. In steady state $(\partial h/\partial t = 0)$, the equation is linear in $h^2$. It will be used for studying the flow around a well.

### 5.2. Terzaghi's Theory of Consolidation. Effect of Interstitial Water on Porous Media

First, we shall examine the interactions between solid and liquid as developed by Schneebeli (1966). This paragraph mainly concerns civil engineering in relatively shallow layers. The medium is assumed to be made up of grains without cohesion (sand, silt, clay).

(a) *Effective stress and fluid pressure.* The porous medium is assumed to be saturated and to contain only grains (solid phase) and a liquid phase filling all the interstices.

What is the effect of an external load acting on such a medium? Terzaghi's experiment can be described as in Fig. 5.2 (Terzaghi and Peck, 1967).

In case b (external load = column of water), the pressure at the surface of the porous medium is $\rho l$. It does not cause any compaction. In case c (external load = lead pellets), the same pressure on the porous medium causes a compaction $\delta e$.

*Conclusion.* Only the loads applied directly to the solid skeleton have mechanical effects on the porous medium. The effect of a load of water is simply that the pressure increases in the liquid that fills the pores, and since the solid grains are virtually incompressible in the range of pressure of interest here, there is no apparent effect on the medium.

*Definition.* Terzaghi uses the term "effective stress" $\bar{\sigma}$ to describe the stress that is transmitted directly from grain to grain as in the case of the lead pellets. It is the only one that influences the solid phase, as opposed to the pressure $p$ of the fluid filling the interstices. The total stress $\sigma$ applied to the liquid–solid complex is thus composed of effective stress and fluid pressure. We get

$$\sigma = \bar{\sigma} + p$$

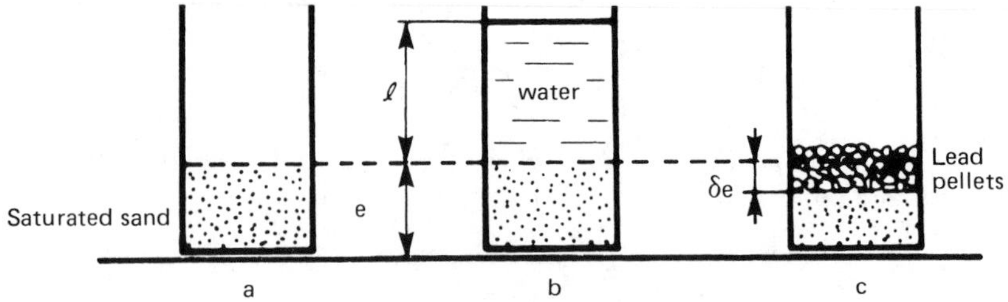

**Fig. 5.2.** Terzaghi's experiment of compaction.

This is the equation on which this section is based. In the most general case, $\sigma$ and $\bar{\sigma}$ are tensors with three normal stresses and three tangential stresses.

*Hypotheses.* For Section 5.2, the following are assumed: (1) the liquid is incompressible, i.e., $\rho$ is constant; (2) the solid grains of the medium are incompressible; and (3) the porous medium is compressible by reduction of the porosity $\omega$.

(b) *Buoyancy.* Take a column of dry soil (Fig. 5.3) and let us conceptually divide it into two parts with a section at elevation $z_0$. Let $l$ be the height of the column above $z_0$. The lower part of the column is subjected to a stress corresponding to the weight of the upper part (overburden). By definition, this is an effective stress, since it is transmitted by the grains.

$$\bar{\sigma}_z = \rho_d gl = \rho_s(1 - \omega)gl$$

where $\bar{\sigma}_z$ is the effective stress in the vertical direction, $\rho_d$ the mass per unit volume of the dry soil, $\rho_s$ the mass per unit volume of the solid grains in the soil, and $\omega$ the total porosity.

Note that in soil mechanics one usually works with specific weight $\gamma = \rho g$, but we shall keep the usual notation of mass per unit volume (sometimes called mass density or density). Here, the total stress is equal to the effective stress $\sigma_z = \bar{\sigma}_z$.

If the column is now saturated with immobile water up to the top, the total stress at the section $z_0$ of the column becomes (weight of the soil + weight of the water):

$$\sigma_z = \rho_s(1 - \omega)gl + \rho\omega gl = \rho_w gl$$

where $\rho_w = \rho_s(1 - \omega) + \rho\omega$ = mass per unit volume of saturated soil, and $\rho$ is the mass per unit volume of water.

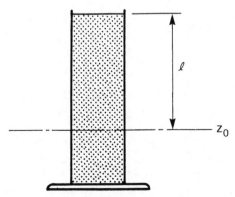

**Fig. 5.3.** Stress in a soil column.

According to its definition, the effective stress is

$$\bar{\sigma}_z = \sigma_z - p = \rho_w gl - \rho gl = (\rho_w - \rho)gl$$

From the mechanical point of view, everything happens as if the mass per unit volume of the soil were now smaller than $\rho_d$ and given by

$$\rho_a = \rho_w - \rho = (1 - \omega)(\rho_s - \rho)$$

where $\rho_a$ is the buoyant mass per unit volume of the saturated soil.

The apparent decrease in the mass per unit volume of soil is in reality only the effect of the buoyancy of the water (Archimedes' force).

(c) *Seepage force.* Let us consider an elementary volume $dx\, dz \times l$ of a saturated porous medium, where the interstitial water is now moving at a filtration velocity **U** in the plane $x$–$z$.

Three types of forces act on the system: (1) forces due to the fluid pressure, (2) forces of gravity, and (3) forces transmitted by grain-to-grain contact due to the effective stress.

*Fluid pressure.* A normal force $p\, dz$ acts on the face AD and a normal force $[p + (\partial p/\partial x)dx]dz$ on the face BC (Fig. 5.4). The sum of the two forces directed along $Ox$ is

$$-\frac{\partial p}{\partial x}\, dx\, dz$$

Similarly, the sum on AB and CD is

$$-\frac{\partial p}{\partial z}\, dx\, dz$$

i.e., the sum of the pressure forces is $-\mathbf{grad}\; p$ per unit volume.

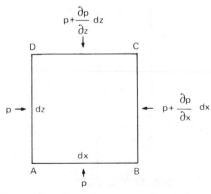

**Fig. 5.4.** Elementary volume of porous medium and pressure forces.

*Gravity.* The sum of the forces of gravity on the element (weight of the solid and water) gives

$$-\rho_w g \, \mathbf{grad} \, z$$

*Sum of fluid pressure and gravity.* If we introduce the hydraulic head instead of the pressure, namely $h = p/\rho g + z$, we find

$$-\mathbf{grad} \, p = -\rho g \, \mathbf{grad} \, h + \rho g \, \mathbf{grad} \, z$$

whence the sum of the fluid pressure and gravity on the element:

$$\mathbf{R} = -\rho g \, \mathbf{grad} \, h + \rho g \, \mathbf{grad} \, z - \rho_w g \, \mathbf{grad} \, z$$

$$= -\rho g \, \mathbf{grad} \, h - \rho_a g \, \mathbf{grad} \, z$$

The second term $-\rho_a g \, \mathbf{grad} \, z$ is again the buoyant weight of the saturated soil. The first term $-\rho g \, \mathbf{grad} \, h$ is called the *seepage force*. It is a volumetric force working in the opposite direction to the hydraulic gradient, i.e., in the direction of the filtration velocity $\mathbf{U}$ if the medium is isotropic (for anisotropic media, the seepage force is simply in the opposite direction of the hydraulic gradient, not in the direction of the velocity $\mathbf{U}$. It is very important to notice that the seepage force is *independent of the magnitude of the hydraulic conductivity or of the velocity*: it depends only on the magnitude of the hydraulic gradient. Thus the seepage force can be identical in a medium of very low hydraulic conductivity, where the velocity of the flow is almost negligible, and in a coarse medium, where the velocity is very high. This must be kept in mind when dealing with civil engineering problems.

*The variation in effective stress* balances these two forces in order to arrive at an equilibrium in the element. In conclusion, the flow of water gives rise to variations in the effective stress affecting the solid phase which sometimes have to be taken into account in civil engineering.

*Example: Quicksands.* The following experiment (Fig. 5.5) is carried out on ascending flow in a column of sand. The flow is uniform and the hydraulic gradient is $\mathbf{grad} \, h = H/l$ directed upwards. The sum $R$ of the seepage force and the buoyant weight is the volume force:

$$(\rho_a - \rho \, \mathbf{grad} \, h)\mathbf{g}$$

If we gradually increase the hydraulic head $H$, there comes a moment when this volume force vanishes. The sand appears to be freed from the influence of gravity: it becomes "unstable" and "quick." A heavy object placed on the column sinks into it. If $H$ is increased even more, the entire sand column rises up. The critical gradient at which all volume forces disappear in this particular case of upward vertical flows is

$$\mathbf{grad} \, h = \frac{\rho_a}{\rho}$$

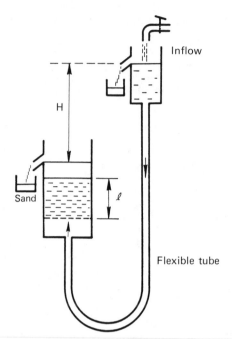

**Fig. 5.5.** Experiment for obtaining quicksands.

This is fundamental in soil mechanics. Take a dike of homogeneous soil without any sealing curtain, i.e., through which a small amount of water can leak. At first sight, it might seem as if the upstream face of the dike were subjected to the uplift pressure of the water in the dam. This is completely erroneous. As a matter of fact, the pressure acting on an element of the upstream facing is a *fluid pressure*, not an effective stress, which is therefore not transmitted by the solid grains. The force of the water is not transmitted on the upstream facing of the dike but is decomposed into a system of volume forces (seepage force) affecting the whole of the saturated volume. The solidity of the dike depends essentially on the nature of the seepage flow through the dike, which therefore has to be calculated.

When the seepage forces in a porous medium are able to initiate a movement of the constituent grains (e.g., at the outer wall of a dike) *piping* is said to occur.

(d) *Theory of consolidation according to Terzaghi.* When certain saturated low-permeability soils are loaded (e.g., buildings are constructed on them), there is at first only a slight compaction. However, eventually, sometimes after a long period of time, compaction may attain a considerable degree. This phenomenon of compaction in the course of time is called *consolidation*. It occurs especially in clay soils.

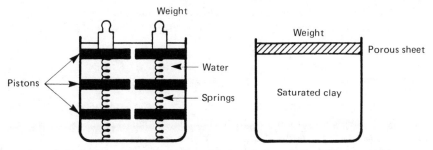

**Fig. 5.6.** Terzaghi's analogy for the consolidation of clay: cylinder with pierced pistons.

Terzaghi has shown that the phenomenon of consolidation is caused by the slow outflow of the interstitial water contained in the soil as shown by the analogy of the pierced pistons (Fig. 5.6). If there is no water in the container to the left when the overload is applied, this overload is entirely absorbed by the springs, which are shortened: the compaction is immediate and elastic. However, if the container is filled with water and the holes in the pistons are very small, the latter will not immediately move downward. The overload will first cause a pressure in the water (without compaction if the water is taken to be incompressible). The water will then gradually escape from the system and leave the springs to react to the overload by contracting.

Similarly, the compaction of saturated clays is obtained by expelling water, which has to be drained by means of a porous sheet. This can be demonstrated with an oedometer, which is an apparatus for measuring the consolidation of clay by draining it while it is under compressive stress, as shown in the right-hand side of Fig. 5.6.

The theory of consolidation assumes that:

(1) The outflow of the interstitial water obeys Darcy's law.

(2) The permeability $K$ of the soil does not vary during the consolidation process (which is only an approximation of reality).

(3) The water and the solid elements in the soil are incompressible; compression then means decrease in porosity.

(4) The compressibility of the soil (decrease in porosity) is "elastic," i.e., there is a linear relation between the effective compression stress and decrease in soil volume. This is also an approximation of reality (see Section 5.3).

The mechanism of consolidation assumes that an external overload applied to the soil is absorbed in part by the solid phase (increase in effective stress) and in part by the interstitial water (increase in fluid pressure). As a result of this increase in pressure, a transient flow is started, the water is drained, and the effective stress gradually increases, causing compaction.

We shall try to establish the *state equation of the soil*. During consolidation, the external loads remain constant as well as the resulting total stress.

$$\sigma = \bar{\sigma} + p = \text{constant}$$

Thus

$$d\bar{\sigma} + dp = 0 \tag{5.2.1}$$

At the beginning of the consolidation process, the excess load is entirely absorbed by $p$ but is gradually transformed into increased effective stress until the pressure reaches a hydrostatic equilibrium (no outflow).

According to hypothesis (4), the relative variation in volume of a soil element should become

$$-dV/V = \alpha\, d\bar{\sigma} \tag{5.2.2}$$

with $\alpha$ the compressibility coefficient of the soil (mass$^{-1}$ length time$^2$) and $\bar{\sigma}$ the effective stress.

According to hypothesis (3), the variation in volume of the element is altogether due to the variation of its porosity. If $V$ is the total volume of the soil element, $V_P$ is the volume of the pores and $V_S$ is the volume of the solid phase:

$$V = V_S + V_P \qquad \text{and} \qquad dV = dV_P$$

According to this assumption, when we calculate $dV/V$ as a function of the total porosity $\omega$ we get

$$\omega = \frac{V_P}{V_S + V_P}$$

$$d\omega = \frac{V_S + V_P - V_P}{V^2}\, dV_P = \frac{1 - \omega}{V}\, dV_P = (1 - \omega)\frac{dV}{V}$$

That is, taking into account Eqs. (5.2.1) and (5.2.2),

$$d\omega = (1 - \omega)\alpha\, dp$$

Further, if we consider the local derivatives of these magnitudes (in an Eulerian coordinates system), we have

$$\frac{\partial \omega}{\partial t} = (1 - \omega)\alpha \frac{\partial p}{\partial t} \tag{5.2.3}$$

which we use to describe the behavior of the porous medium.

The compaction is given directly by Eq. (5.2.2) if the variation of effective stress is known. The latter can be calculated by Eq. (5.2.1) if the evolution of the pressure is known. Therefore, the transient evolution of the pressure in the soil must be calculated.

We choose the pressure as the principal unknown and write the consolidation equation using:

(1)  the equation of continuity:

$$\text{div}(\rho \mathbf{U}) + \frac{\partial}{\partial t}(\rho \omega) + \rho q = 0 \tag{3.2.3}$$

(2)  Darcy's law:

$$\mathbf{U} = -\frac{\mathbf{K}}{\rho g}(\text{grad } p + \rho g \, \text{grad } z) \tag{4.1.6}$$

(3)  the state equation of water:

$$\rho = \text{constant} \qquad \text{(incompressible fluid)} \tag{5.2.4}$$

(4)  the equation of state of the porous medium:

$$\frac{\partial \omega}{\partial t} = (1 - \omega)\alpha \frac{\partial p}{\partial t} \tag{5.2.3}$$

These equations are easily combined to yield

Eq. (3.2.3) + Eq. (5.2.4) $\rightarrow \text{div } \mathbf{U} + \dfrac{\partial \omega}{\partial t} + q = 0$

same + Eq. (5.2.3) $\rightarrow -\text{div } \mathbf{U} = (1 - \omega)\alpha \dfrac{\partial p}{\partial t} + q$

same + Eq. (4.1.6) $\rightarrow \text{div}(\mathbf{K} \, \text{grad } p) = \rho g(1 - \omega)\alpha \dfrac{\partial p}{\partial t} + \rho g q \quad$ (5.2.5)

because when $\rho g$ is constant, $\text{div}(\text{grad } z) = 0$.

This is the consolidation equation. Remember that $q$ represents the withdrawn (or added, if it is negative) flow rate per unit volume in the porous medium. Here it is usually zero, unless drains (e.g., well points) are set into the soil to accelerate the consolidation.

If the permeability $K$ is isotropic and constant, the equation is simplified as follows:

$$\nabla^2 p = \frac{(1 - \omega)\alpha \rho g}{K} \frac{\partial p}{\partial t} \tag{5.2.6}$$

where $\nabla^2$ is the Laplace operator and the flow $q$ is assumed to be zero.

The coefficient $C_v = (1 - \omega)\alpha \rho g/K$ is called the consolidation coefficient (length$^{-2}$ time). One sometimes disregards the term $(1 - \omega)$ if it is close to 1.

Freeze and Cherry (1979) give the following ranges of values for soil compressibility ($\alpha$) in $m^2/N$ or $Pa^{-1}$:

| Clay | $10^{-6}-10^{-8}$ |
|---|---|
| Sand | $10^{-7}-10^{-9}$ |
| Gravel | $10^{-8}-10^{-10}$ |
| Jointed rock | $10^{-8}-10^{-10}$ |
| Sound rock | $10^{-9}-10^{-11}$ |

Having calculated the evolution of the pressure $p$, we know that of the effective stress $\bar{\sigma}$ because $\bar{\sigma} + p = $ constant. The compactions are deduced from

$$\alpha \, \Delta\bar{\sigma} = -\Delta V/V = -\Delta l/l$$

where $l$ is the thickness of the consolidating layer if the compaction occurs only in the vertical direction.

Remember that in clay media the compression is, as a rule, elastic only in the very first approximation. In particular, the compaction is almost irreversible (for the same $|\Delta\bar{\sigma}|$, expansion would be $\sim \frac{1}{10}$ of compaction). Clay subjected to successive cycles of compression shows a change in the slope of its compaction when the stress reaches or exceeds the maximum stress which it has previously undergone. This is called consolidation stress (Fig. 5.7).

The same behavior is found for clay sands, but with a smaller compressibility. Clean sands and gravel, which do not contain interbedded clay layers, tend indeed to have an elastic behavior, and compressibility is almost reversible. Jointed or sound rocks generally follow the elastic hypothesis, but their compressibility is very small.

In hydrogeology, consolidation is generally referred to as subsidence. The problem is not that of an additional external load on a soil but of a decrease of

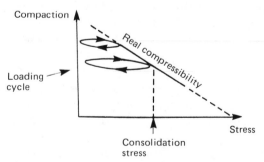

Fig. 5.7. Compaction of clay showing the consolidation stress.

the pressure in the aquifer caused by heavy withdrawals. Because of the relation $d\bar{\sigma} + dp = 0$, the result is identical to compaction: the effective stress $\bar{\sigma}$ increases, the aquifer is compacted, and the land surface subsides. Inside the aquifer layer, subsidence is instantaneous (i.e., a variation of pressure $-\Delta p$ instantly gives a variation of effective stress $+\Delta\bar{\sigma}$, which instantly causes a compaction $\Delta l/l = -\alpha\,\Delta\bar{\sigma}$: the delay in the consolidation due to an external load is caused by the time necessary to drain the water, as already explained in the analogy with the pierced pistons). However, the layers of material above and below the aquifer that is pumped may be slowly drained by this pressure variation $\Delta p$ (see Section 8.3 on leakage). As they drain they also compact, and this delay in drainage causes a delay in the additional compaction of the system.

Subsidence due to heavy withdrawals can be very important: several tens of centimeters in Venice, several meters in Mexico City! When it is high, it means that $\alpha$ is large, that the aquifer or its overlaying and underlying beds are rich in clay, and the subsidence is almost irreversible, even if the withdrawals are stopped and the pressure recovered. This is what is observed in Venice at present [see Gambolati and Freeze (1973) and Gambolati *et al.* (1974).]

Finally, it must be pointed out that in a few rare cases, where there are unconnected pores between which the pressures are not transmitted, the relation $\sigma = \bar{\sigma} + p$ does not hold: an increase in the total stress $\sigma$ may be sustained almost immediately by the effective stress $\bar{\sigma}$.

(e) *Effective stress in an unsaturated medium.*    In a saturated medium with constant total stress, there is a linear relation of slope $-1$ between $p$ and $\bar{\sigma}$. However, in an unsaturated one, when the pressure descends below the atmospheric pressure, this relation becomes more complex. It is described in Fig. 5.8 (Freeze and Cherry, 1979).

Indeed, we observe empirically that as a first approximation the unsaturated soil only sustains the total stress through the effective stress (curve 1). The pressure plays no role.

In reality, the actual behavior is closer to curve 2, which is dependent on the structure of the soil and its history of saturation and drainage.

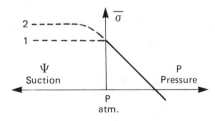

**Fig. 5.8.** Effective stress in an unsaturated medium.

## 5.3. General Diffusion Equation: Confined Aquifers

The complete theory is not very easy to establish, which is the reason why we have left it to the end. It must be assumed that the fluid is compressible as well as both the pores and solid grains of the porous medium, but rigorously, if the porous medium is compressible, we must take into account its displacement in the equation of continuity: in the fixed elementary volume in Euler coordinates in which the mass balance is written, there will be a flux of solid grains as well as of fluid.

The porous medium is assumed to be totally saturated with fluid, since the complete equation including the three compressibilities only applies to deep, confined aquifers, i.e. aquifers trapped between two impermeable layers (Fig. 5.9). We shall use the following relationships.

(a)   *The equation of continuity for the fluid in an elementary volume fixed in space.*

$$\text{div}(\rho \mathbf{U}) + \frac{\partial}{\partial t}(\rho \omega) + \rho q = 0 \tag{3.2.3}$$

(b)   *The equation of continuity for the flux of solid grains in the same elementary volume in space.*   If we define the same mean quantities for the solid as for the fluid,

$$\langle \rho_s \rangle = \frac{1}{1 - \langle \omega \rangle} \int \rho_s m \, d\mathbf{x}'$$

where $\rho_s$ is the mass per unit volume of the soil, equal to 0 in the pores and to that of the solid in the grains; and

$$\langle \mathbf{u}_s \rangle = \frac{1}{\langle \rho_s \rangle} \int \rho_s \mathbf{u}_s m \, d\mathbf{x}'$$

where $\mathbf{u}_s$ is the real velocity of the solid, equal to 0 in the pores and to that of each point of the solids in the grains.

It can be shown by exactly the same reasoning as for the fluid (Section 3.2.1) that the equation of continuity for the solid is

$$\text{div}(\langle \rho_s \rangle \langle \mathbf{u}_s \rangle) + \frac{\partial}{\partial t}[(1 - \langle \omega \rangle)\langle \rho_s \rangle] = 0$$

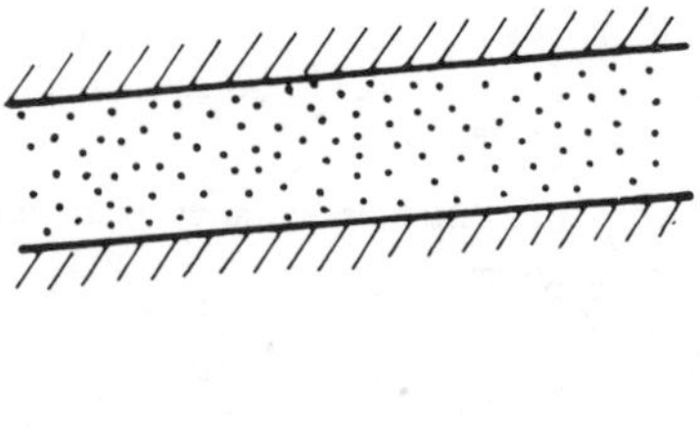

Fig. 5.9. Confined aquifer layer.

If the angles for averages are omitted in order to simplify the notation, the equation becomes

$$\text{div}(\rho_S U_S) + \frac{\partial}{\partial t}[(1 - \omega)\rho_S] = 0 \qquad (5.3.1)$$

where $U_S = \langle u_S \rangle$ is the fictitious displacement velocity of the solid as if the whole section were open to the flow of solid.

In the same way as for the fluid, a "mean microscopic velocity" is defined for the solid by saying that $u_S$ is zero in the pores:

$$u_S^* = \frac{U_S}{1 - \omega}$$

(c)  *Darcy's law.*  In its classical form, this law applies in effect to the real mean velocity of the liquid $(u^* = U/\omega)$ in relation to that of the solid $[u_S^* = U_S/(1 - \omega)]$ and not in relation to stationary space[†]. Therefore, it is necessary to geometrically add the velocities $u^*$ and $u_S^*$ to obtain an exact expression of Darcy's law: it is $(u^* - u_S^*)$, which is proportionate to the pressure gradient and to the gravity, or alternatively $(U - \omega u_S^*)$ [see Biot (1955, 1956), Cooper (1966), Remson *et al.* (1971)].

From the general expression, Eq. (4.1.6), of Darcy's law, we obtain

$$U - \omega u_S^* = -\frac{k}{\mu}(\text{grad } p + \rho g \text{ grad } z) \qquad (5.3.2)$$

(d)  *Combining the equation of continuity* (3.2.3) *with Eqs.* (5.3.1) *and* (5.3.2).  We assume that all the magnitudes $\rho$, $\rho_S$, $p$, $\omega$, $U$, and $U_S$ are Euler functions, i.e., defined in relation to a fixed point in space. We get

$$\text{Eq. (3.2.3) + Eq. (5.3.2)} \rightarrow \text{div}\left[\rho\frac{k}{\mu}(\text{grad } p + \rho g \text{ grad } z)\right]$$

$$= \text{div}(\rho\omega u_S^*) + \frac{\partial}{\partial t}(\rho\omega) + \rho q$$

but

$$\text{div}(\rho\omega u_S^*) = \rho\omega \text{ div } u_S^* + u_S^* \text{ grad}(\rho\omega)$$

and

$$u_S^* \text{ grad}(\rho\omega) + \frac{\partial}{\partial t}(\rho\omega) = \frac{d}{dt}(\rho\omega)$$

---

[†] Rigorously, the kinematic porosity $\omega_c$ should be used here and not the total porosity $\omega$.

which is the *material derivative*[†] of $\rho\omega$ following the mean displacement of the solid at the velocity $\mathbf{u}_s^*$. Therefore, we look for the variations of $\omega$ and $\rho$ inside the elementary domain formed by the solid during its deformation, i.e., while it contains a constant quantity of solid.

Furthermore, if we substitute $(1 - \omega)\mathbf{u}_s^*$ for $\mathbf{U}_s$ in Eq. (5.3.1), we get

$$(1 - \omega)\rho_s \operatorname{div} \mathbf{u}_s^* + \mathbf{u}_s^* \operatorname{grad}[(1 - \omega)\rho_s] + \frac{\partial}{\partial t}[(1 - \omega)\rho_s] = 0$$

Similarly,

$$(1 - \omega)\rho_s \operatorname{div} \mathbf{u}_s^* + \frac{d}{dt}[(1 - \omega)\rho_s] = 0$$

Finally, by combining these,

$$\operatorname{div}\left[\frac{\rho k}{\mu}(\operatorname{grad} p + \rho g \operatorname{grad} z)\right] = \omega \frac{d\rho}{dt} + \frac{\rho}{1 - \omega}\frac{d\omega}{dt} - \frac{\rho\omega}{\rho_s}\frac{d\rho_s}{dt} + \rho q \quad (5.3.3)$$

(e)  *The state equations of the liquid and the solid.*  We choose the pressure $p$ as our only unknown. Thus, we have to estimate $d\rho/dt$, $d\omega/dt$, $d\rho_s/dt$ in an element of the deforming porous medium, which is mobile but contains *a constant quantity of solid.*

*For the liquid,* we know the result: it is the equation of isothermal compressibility, Eq. (3.1.3):

$$\rho = \rho_0 e^{\beta_l(p - p_0)}$$

or alternatively:

$$\frac{d\rho}{dt} = \rho\beta_l \frac{dp}{dt} \tag{5.3.4}$$

where $\beta_l$ may easily be measured. For water, $\beta_l = 5 \times 10^{-10}$ Pa$^{-1}$.

*For the solid,* things become more complicated. The calculations fill several pages, and an army of coefficients have to be defined in order to find a solution. Let us begin, keeping in mind that our purpose is to express $d\omega/dt$ and $d\rho_s/dt$ as a function of $dp/dt$[‡].

---

[†] The material or substantial derivative: it is the variation in the unit time interval of a property (here $\rho\omega$) at a point which moves with the solid grain in a Lagrangian coordinate system. It is denoted as a total derivative $d/dt$. The material derivative of a quantity "$a$", $da/dt$ in a Lagrangian coordinate system with a velocity $\mathbf{u}$, is related to the ordinary derivative in an Eulerian coordinate system, $\partial a/\partial t$, by

$$\frac{da}{dt} = \frac{\partial a}{\partial t} + \mathbf{u} \operatorname{grad} a$$

[‡] The reader may be interested only in the results of the following laborious calculations. In this case, he should simply look at the definition of the coefficients of compressibility in Eq. (5.3.6), note the resulting values of $d\omega/dt$ and $d\rho_s/dt$, and proceed directly to Subsection f.

Unlike what happens in the theory of consolidation, presented above, the volumes of pores and solids depend not only on the effective stress $\bar{\sigma}$ but also on the pressure $p$. We note that $V$ is the total volume of the element of mobile porous medium, $V_S$ the volume of the solid, and $V_P$ the volume of the pores ($V = V_S + V_P$).

We define:

(1) *The compressibility coefficient of the solid grains*

$$\frac{dV_S}{V_S} = -\frac{d\rho_S}{\rho_S} = -\beta_S \, dp \qquad (5.3.5)$$

such that the product $\rho_S V_S =$ mass of solids is a constant in the element of the mobile porous medium, and $\beta_S$ is measurable on pure minerals or with a triaxial cell. For quartz, $\beta_S \simeq 2 \times 10^{-11}$ Pa$^{-1}$.

(2) *Compressibility of the porous matrix*: The theory of elasticity of continuous media, applicable also to porous media, expresses a linear relationship between the deformation tensor and the tensor of effective stress increment. We usually choose the case of a medium which is isotropic as far as the mechanical properties are concerned, i.e., defined by only two coefficients, Young's modulus $E$ and Poisson's ratio $v$. This hypothesis is not imperative, however.

If $\Delta\bar{\sigma}_i$ are the three increments in normal effective stress in the three principal directions $(i, j, k)$ of the tensor of stress increments, and if $\varepsilon_i$ are the relative deformations in these directions ($\varepsilon_i = \Delta l_i / l_i$, $l$ length element), the theory of elasticity provides that

$$-\varepsilon_i = \frac{1}{E}\Delta\bar{\sigma}_i - \frac{v}{E}(\Delta\bar{\sigma}_j + \Delta\bar{\sigma}_k)$$

The volumetric expansion is the sum of the three relative deformations

$$-\frac{\Delta V}{V} = -\sum \varepsilon_i = \frac{1}{E}\left(\sum \Delta\bar{\sigma}_i\right) - \frac{2v}{E}\left(\sum \Delta\bar{\sigma}_i\right)$$

If $\Delta\bar{\sigma}$ is the mean stress increment, $\Delta\bar{\sigma} = \frac{1}{3}\sum \Delta\bar{\sigma}_i$, then

$$\frac{\Delta V}{V} = -\frac{3(1 - 2v)}{E}\Delta\bar{\sigma}$$

The minus sign means that the volume $V$ decreases if the effective stress $\bar{\sigma}$ increases (compression).

Thus it is shown that the possible anisotropy of the stress increment is unimportant; it is the average increase in effective stress that is significant. When we speak of increase in stress, this will always mean average stress in an isotropic medium.

Thus, there is a linear relation between $dV/V$ and $d\bar{\sigma}$. It is the one we have written from Eq. (5.2.2) as

$$\frac{dV}{V} = -\alpha\, d\bar{\sigma} \qquad \text{whence} \qquad \alpha = \frac{3(1 - 2\nu)}{E}$$

in the theory of consolidation. We must realize that, in this theory, we assume that the total stress $\sigma$ is a constant: the coefficient $\alpha$ is defined for a *particular transformation* of the state of stress in the rock, whence $d\bar{\sigma} + dp = 0$. It may be found by direct measurements on a sample with a triaxial cell.

We will assume that $V$, $V_S$, and $V_P$ are functions of both $\bar{\sigma}$ and $p$ and that these functions are linear, as in the theory of linear elasticity described already. Actually, $V$, $V_S$, and $V_P$ are usually given as functions of $\sigma$ and $p$, total stress and fluid pressure, which simplifies the calculations. We can always come back to $\bar{\sigma}$ instead of $\sigma$ by using $\sigma = \bar{\sigma} + p$. We write

$$\frac{dV}{V} = -C\, d\sigma + \alpha\, dp$$

$$\frac{dV_P}{V_P} = -C_P\, d\sigma + \alpha_P\, dp \qquad\qquad (5.3.6)$$

$$\frac{dV_S}{V_S} = -C_S\, d\sigma + \alpha_S\, dp$$

These six coefficients of compressibility are positive. The coefficient $\alpha$ of Eq. (5.3.6) is really the same as that of Eq. (5.2.2), because if we transform the state of stress with $d\sigma = 0$, i.e., $dp = -d\bar{\sigma}$, the first part of Eq. (5.3.6) gives $dV/V = -\alpha\, d\bar{\sigma}$.

These coefficients are not independent; it will be shown that the following relations may be established between them:

$$C = \alpha + \beta_S$$

$$C_P = \frac{\alpha}{\omega}$$

$$\alpha_P = \frac{\alpha}{\omega} - \beta_S \qquad\qquad (5.3.7)$$

$$\alpha_S = \frac{\omega}{1 - \omega}\beta_S$$

$$C_S = \frac{\beta_S}{1 - \omega}$$

*Proof:*  The equations (5.3.6) are general. During this demonstration, we disregard the real conditions of stress variations encountered in hydrogeology and draw up the five relations (5.3.7) between the six coefficients by imagining that the porous medium is subjected to three special transformations. The relations, which are made to appear between the coefficients in this way, will be general as, by definition, these coefficients are constant.

*First transformation*:   Assuming that $d\sigma = 0$ (constant total stress),

$$\frac{dV}{V} = \alpha\, dp, \qquad \frac{dV_P}{V_P} = \alpha_P\, dp, \qquad \frac{dV_S}{V_S} = \alpha_S\, dp$$

Moreover, we have

$$V = V_P + V_S$$

and their differentials

$$dV = dV_P + dV_S$$

or

$$\frac{dV}{V} = \frac{V_P}{V}\frac{dV_P}{V_P} + \frac{V_S}{V}\frac{dV_S}{V_S} = \omega\frac{dV_P}{V_P} + (1-\omega)\frac{dV_S}{V_S}$$

Thus, the first relation is written

$$\alpha = \omega\alpha_P + (1-\omega)\alpha_S$$

*Second transformation*:   Assume that $d\bar{\sigma} = 0$ (constant effective stress). The arrangement of the grains in the porous matrix is in reality only dependent on the effective stress: if it is increased, the medium is compacted and vice versa. Therefore, if $d\bar{\sigma} = 0$, the arrangement is unchanged. The variation in volume of the porous medium, which may occur, can only be caused by the expansion or contraction of the grains themselves, and the medium will be similarly deformed. Consequently, the porosity $\omega$ of the medium should not vary. From $\omega = V_P/V$, we deduce

$$d\omega = 0 = \frac{V\,dV_P - V_P\,dV}{V^2}$$

whence

$$\frac{dV_P}{V_P} = \frac{dV}{V} = \frac{dV - dV_P}{V - V_P} = \frac{dV_S}{V_S}$$

keeping in mind that

$$V = V_P + V_S$$

If $d\bar{\sigma} = 0$, we have $d\sigma = dp$, and therefore

$$\frac{dV_S}{V_S} = (\alpha_S - C_S)\,dp \quad \text{and} \quad \frac{dV_P}{V_P} = (\alpha_P - C_P)\,dp \quad \text{and} \quad \frac{dV}{V} = (\alpha - C)\,dp$$

However, we have already defined the compressibility coefficient for the solid:

$$\frac{dV_S}{V_S} = -\beta_S \, dp$$

defined on an isolated mineral, i.e., in fact, when $d\bar{\sigma} = 0$. The following three relations are then deduced:

$$\beta_S = C_S - \alpha_S = C_P - \alpha_P = C - \alpha$$

*Third transformation*:    Geertsma (1957) suggests the use of the Maxwell–Betti theorem (also called Betti and Rayleigh), which states:
Given two imposed elementary hydrostatic loads $d\sigma$ and $dp$, the action of the forces due to the first load in the displacement due to the second is equal to the action of the forces due to the second in the displacement due to the first.

$$\left[ \left( \frac{dV}{\partial p} \right)_{\bar{\sigma}} dp \right] d\sigma = - \left[ \left( \frac{\partial V_P}{\partial \sigma} \right)_p d\sigma \right] dp$$

The parentheses and subscripts remind us that the derivatives are taken with either $\bar{\sigma}$ or $p$ constant.

In this expression, the action of the total stress concerns indeed the entire volume $V$, (giving the product $\Delta V d\sigma$), whereas the pressure only acts on the volume of the pores in which it occurs, giving the product $\Delta V_p \, dp$.

This theorem is a direct consequence of the linearity of the compression equation, which we have admitted (elasticity). It is proved from the calculation of the potential elastic energy:

$$2V = \sum \bar{\sigma}_i \varepsilon_i + p \sum \varepsilon_i$$

which is a quadratic form of the deformations $\varepsilon_i$. See textbooks on the mechanics of continuous media for details.

$$\frac{\partial V}{\partial p} = -\frac{\partial V_P}{\partial \sigma} \qquad \text{i.e.} \qquad \alpha V = C_P V_P \qquad \text{or} \qquad \alpha = \omega C_P$$

Thus, we have established the five relations (5.3.7) by reordering.

We now return to our unknowns $d\omega/dt$ and $d\rho_S/dt$. In order to study the elementary volume of the mobile porous medium, we make the assumption that the total stress $\sigma$ does not vary, which is generally well borne out in the field, where deep confined aquifers are being exploited: the total stress due to the weight of overlying material does not vary.

(1) First we consider $d\omega/dt$. We can write

$$V_S = (1 - \omega)V$$

whence by differentiation

$$dV_S = (1 - \omega)\,dV - V\,d\omega$$

or

$$\frac{V_S}{V}\frac{dV_S}{V_S} = (1 - \omega)\frac{dV}{V} - d\omega$$

i.e.

$$d\omega = (1 - \omega)\left(\frac{dV}{V} - \frac{dV_S}{V_S}\right)$$

According to Eq. (5.3.6), when we take $d\sigma = 0$, we get

$$d\omega = (1 - \omega)(\alpha - \alpha_S)\,dp$$

We can take the material derivative of $\omega$ and $p$ in a Lagrangian coordinate system moving with the deforming solid. We finally get

$$\frac{d\omega}{dt} = (1 - \omega)(\alpha - \alpha_S)\frac{dp}{dt}$$

(2) Next we consider $d\rho_S/dt$. The mass balance of the solid in the elementary mobile volume is written as

$$d(\rho_S V_S) = 0 \qquad \text{i.e.} \qquad \frac{d\rho_S}{\rho_S} + \frac{dV_S}{V_S} = 0$$

Similarly, according to Eq. (5.3.6) with $d\sigma = 0$, and using the material derivatives of $\rho_S$ and $p$ in the same Lagrangian system moving with the deforming solid,

$$\frac{d\rho_S}{dt} = -\rho_S \alpha_S \frac{dp}{dt}$$

We now have all the state equations of the porous medium.

(f) *Synthesis: diffusion equation and simplifications.* When we introduce these three state equations into Eq. (5.3.3) while taking into account the value of $\alpha_S$ obtained from Eq. (5.3.7), we get

$$\operatorname{div}\left[\frac{\mathbf{k}\rho}{\mu}(\operatorname{grad} p + \rho g \operatorname{grad} z)\right] = \rho\omega\left[\beta_l - \beta_s + \frac{\alpha}{\omega}\right]\frac{dp}{dt} + \rho q$$

and multiplying by $g$ yields

$$\operatorname{div}[\mathbf{K}(\operatorname{grad} p + \rho g \operatorname{grad} z)] = \rho\omega g\left[\beta_l - \beta_s + \frac{\alpha}{\omega}\right]\frac{dp}{dt} + \rho g q \qquad (5.3.8)$$

The coefficient     $$S_s = \rho\omega g\left(\beta_l - \beta_s + \frac{\alpha}{\omega}\right)$$

is called the *specific storage coefficient* of the aquifer (its dimension is length$^{-1}$).
In Section 5.2.d, we have given the ranges of values of $\alpha$, between $10^{-6}$ and
$10^{-11}$ Pa$^{-1}$, depending on the aquifer type; $\beta_l \simeq 5 \times 10^{-10}$ Pa$^{-1}$; and finally,
$\beta_s \simeq 2 \times 10^{-11}$ Pa$^{-1}$ for quartz and most minerals: in practice, $\beta_s$ is often
neglected. Values of $S_s$ thus range between $10^{-2}$ m$^{-1}$ (highly compressible
clays) to $10^{-7}$ m$^{-1}$ (low porosity hard rocks); in the former case, $\beta_l$ is negligible
with respect to $\alpha$, in the latter case, $\beta_l$ can play a major role.

*Simplifications.*     Although theoretically correct, Eq. (5.3.8) is impractical. It
is usual to make the following simplifications, which are not theoretically very
"elegant."

First, the hydraulic head $h$ is substituted for the pressure $p$ in Eq. (5.3.8) by
making the assumption, acceptable in reality, that $\rho$ is variable in time
(compressibility) but less so in space. We can then remove the term $\rho g$ from the
divergence operator:

$$\operatorname{div}[\mathbf{K}(\operatorname{grad} p + \rho g \operatorname{grad} z)] \simeq \rho g \operatorname{div}\left[\mathbf{K}\operatorname{grad}\left(\frac{p}{\rho g} + z\right)\right] = \rho g \operatorname{div}(\mathbf{K}\operatorname{grad} h)$$

Moreover, as the velocity of the solid $\mathbf{u}_s^*$ is very small, the term $\mathbf{u}_s^* \operatorname{grad} p$ is
disregarded when compared with $\partial p/\partial t$ in the definition of the material
derivative, and we can write

$$\rho g \operatorname{div}(\mathbf{K}\operatorname{grad} h) = S_s \frac{\partial p}{\partial t} + \rho g q$$

The same result may be obtained by keeping the equation with a material
derivative because, since the measurement instruments (i.e., piezometers) are
connected to the solid, it is actually $dp/dt$ and not $\partial p/\partial t$ that can be measured.
Furthermore, when the expression $p = \rho g(h - z)$ is differentiated, we get

$$\frac{\partial p}{\partial t} = \rho g \frac{\partial h}{\partial t} + g(h - z)\frac{\partial \rho}{\partial t}$$

That is, when we take Eq. (5.3.4) into account,

$$\frac{\partial \rho}{\partial t} = \rho\beta_l \frac{\partial p}{\partial t}$$

$$\rho g \frac{\partial h}{\partial t} = \frac{\partial p}{\partial t}[1 + \rho g(z - h)\beta_l]$$

where $\rho g(z - h)\beta_l$ may be disregarded in comparison with 1. Indeed, for
$g = 10$ m/s$^2$, $\beta_1 = 5 \times 10^{-10}$, and $\rho = 10^3$ kg/m$^3$, this term is less than $10^{-2}$

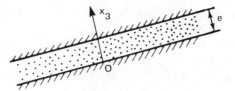

**Fig. 5.10.** Confined aquifer.

as long as $(h - z)$ is less than 2000 m[†]. Hence

$$\rho g \frac{\partial h}{\partial t} \simeq \frac{\partial p}{\partial t}$$

If we substitute in Eq. (5.3.8) and simplify, dividing by $\rho g$, we get

$$\text{div}(\mathbf{K} \, \text{grad} \, h) = S_s \frac{\partial h}{\partial t} + q \tag{5.3.9}$$

This is the diffusion equation in use for confined aquifers.

(g) *Integration of the diffusion equation taking the confining beds into account.* Take a confined aquifer with two confining beds (Fig. 5.10). We shall neither assume these beds to be necessarily horizontal nor completely impermeable. We shall try to reduce Eq. (5.3.9) to two dimensions, assuming that the flow is parallel to the confining beds. In order to do so, we integrate without approximation the diffusion equation along the $x_3$ axis perpendicular to the confining beds. The following assumptions must then be made.

(1) The beds are plane and parallel; the thickness $e$ of the aquifer is constant.

(2) One of the principal directions of anisotropy is orthogonal to the confining beds ($x_3$ on the figure); we shall use the two other principal directions of anisotropy, $x_1$ and $x_2$, which are parallel to the plane of the confining beds, as the coordinate system in two dimensions.

(3) We assume that the *hydraulic head gradient* in the plane $x_1 x_2$ does not depend on $x_3$:

$$\frac{\partial^2 h}{\partial x_1 \, \partial x_3} = \frac{\partial^2 h}{\partial x_2 \, \partial x_3} = 0$$

---

[†] The term $z$ is the elevation of a point in the aquifer; $h$ is the hydraulic head at that point. A difference of 2000 m would be found, for instance, for an aquifer 2000 m deep with a head close to the ground surface (e.g., artesian). This situation is not uncommon. But even $(h - z) = 4000$ m would only give an error of 2%.

(4)    Finally, we assume that the variation of the hydraulic head per unit time, $\partial h/\partial t$, is not a function of $x_3$. In other words, the head may vary with $x_3$ between the top and bottom of the aquifer, but at every moment, the *gradient* and the *variation of the hydraulic head* are the same at all points in the aquifer on the same transverse line $0x_3$. With these assumptions, the integration is simple:

$$\int_0^e \left\{ \frac{\partial}{\partial x_1}\left[ K_1 \frac{\partial h}{\partial x_1} \right] + \frac{\partial}{\partial x_2}\left[ K_2 \frac{\partial h}{\partial x_2} \right] + \frac{\partial}{\partial x_3}\left[ K_3 \frac{\partial h}{\partial x_3} \right] \right\} dx_3$$

$$= \int_0^e S_s \frac{\partial h}{\partial t} dx_3 + \int_0^e q \, dx_3$$

For the left-hand side, we know that (Leibnitz' rule):

$$\frac{\partial}{\partial u}\int_{a(u)}^{b(u)} F(u,v)dv = \int_{a(u)}^{b(u)} \frac{\partial}{\partial u} F(u,v)\, dv + \frac{\partial b}{\partial u} F[u,b(u)] - \frac{\partial a}{\partial u} F[u,a(u)]$$

Here, according to assumption (1), the third and the fourth terms vanish. Therefore, we can write

Left-hand side:

$$\frac{\partial}{\partial x_1}\left( \int_0^e K_1 \frac{\partial h}{\partial x_1} dx_3 \right) + \frac{\partial}{\partial x_2}\left( \int_0^e K_2 \frac{\partial h}{\partial x_2} dx_3 \right) + \int_0^e \frac{\partial}{\partial x_3}\left( K_3 \frac{\partial h}{\partial x_3} \right) dx_3$$

According to assumption (3), we can take out $\partial h/\partial x_1$ and $\partial h/\partial x_2$ from the first two integrals. Hence, the transmissivity of the aquifer, defined in Section 4.1.f, is shown:

$$T_1 = \int_0^e K_1 \, dx_3 \qquad T_2 = \int_0^e K_2 \, dx_3$$

The third integral can be integrated immediately and gives

$$\left( K_3 \frac{\partial h}{\partial x_3} \right)_{x_3 = e} - \left( K_3 \frac{\partial h}{\partial x_3} \right)_{x_3 = 0}$$

According to Darcy's law, this may be interpreted in terms of flux: it is the flow per unit surface area entering the aquifer, through its upper and lower limits respectively. If the orthogonal line to the confining beds is directed inward, these terms are $F = -(K \, \partial h/\partial n)$ at the interface with the confining beds.

These fluxes exchanged between the confined aquifer and its confining beds are called *leakage fluxes*. They will be denoted by $F_t$ and $F_b$ (top and bottom), and appear on the right-hand side as positive terms, if they are incoming.

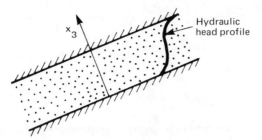

**Fig. 5.11.** Hydraulic head profile in a confined aquifer.

For the right-hand side, $\partial h/\partial t$ comes out of the integral, according to assumption (4), and we can define a new quantity:

$$S = \int_0^e S_s \, dx_3 = \rho \omega g e \left( \beta_l - \beta_s + \frac{\alpha}{\omega} \right)$$

As $S_s$ is in length$^{-1}$, this new quantity $S$ is dimensionless. It is called the *storage coefficient* of the aquifer and varies roughly between $5 \times 10^{-2}$ and $10^{-5}$.

The integration of the source term has already been defined for the unconfined aquifer in Section 5.1.c:

$$\int_0^e q \, dx_3 = Q$$

where $Q$ is now the withdrawn flow rate per unit surface area of the aquifer. Eventually we get (see Fig. 5.11)[†]

$$\frac{\partial}{\partial x_1} \left( T_1 \frac{\partial h}{\partial x_1} \right) + \frac{\partial}{\partial x_2} \left( T_2 \frac{\partial h}{\partial x_2} \right) = S \frac{\partial h}{\partial t} - (F_t + F_b) + Q$$

If the leakage flux is nil (totally impervious confining beds; see also

---

[†] If $Q$, $F_t$, and $F_b$ are given, we can try to integrate this equation and calculate $h$. For example, a solution of the following form may be found:

$$h = h(x_1, x_2, t)$$

The general solution as a function of $x_1, x_2, x_3$ according to assumptions (3) and (4) becomes

$$h = h(x_1, x_2, t) + f(x_3)$$

where $f(x_3)$ is a function independent of $x_1, x_2$, and $t$. In other words, the hydraulic head profile at one point as a function of $x_3$ is not defined, but it is the same at all points and all dates: if it is known at one point, it is known everywhere. As a rule, this profile (e.g., as shown in Fig. 5.11) is not considered important, and the assumption is made that $h$ varies little with $x_3$.

Section 8.3), we write

$$\text{div}(\mathbf{T} \, \textbf{grad} \, h) = S \frac{\partial h}{\partial t} + Q \qquad (5.3.10)$$

This is an equation that we shall use continuously.

Finally, if $T$ is isotropic and constant in space, Eq. (5.3.10) becomes

$$\nabla^2 h = \frac{\partial^2 h}{\partial x^2} + \frac{\partial^2 h}{\partial y^2} = \frac{S}{T} \frac{\partial h}{\partial t} + \frac{Q}{T} \qquad (5.3.11)$$

The ratio $T/S$ is called the aquifer diffusivity. Equations (5.3.10) and (5.3.11) are identical to those of the unconfined aquifer, Eqs. (5.1.3) and (5.1.4), but here $S$ replaces the specific yield $\omega_d$. However, we must remember that even though the two equations for the unconfined–confined aquifer are identical, the mechanisms coming into play (movement of the free surface on the one hand, and compressibility of water, grains, and soil, on the other) are different as are the ways of establishing the equations and the approximations used.

## 5.4. Highly Compressible Soils

Gambolati (1973), while studying the compaction in Venice due to the pumping at Mestre, has established a slightly different expression for the storage coefficient of highly compressible soils (e.g., clay, mud).

The expression assumes that the grains are incompressible ($\rho_s = $ constant) but that the compressibility coefficient of the porous medium, $\alpha$, is important. Furthermore, the analysis is limited to a one-dimensional vertical flow, defining the linear dilation instead of the volumetric one:

$$\varepsilon_z = \frac{\Delta l}{l} = -\alpha \, \Delta \bar{\sigma} = +\alpha \, \Delta p \qquad \text{if} \qquad \Delta \sigma = 0$$

Moreover, this dilation may be explained as the differential in Lagrangian coordinates of the position vector of the point under examination (Fig. 5.12).

$$\varepsilon_z = \frac{\Delta r}{\Delta \zeta}$$

linked to the velocity $\mathbf{u}_s^*$ of the deforming solid.

Finally, the analysis expresses the velocity $\mathbf{u}_s^*$ by

$$\mathbf{u}_s^* = \frac{\partial \mathbf{r}}{\partial t}$$

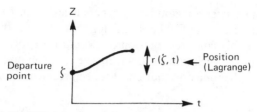

Fig. 5.12. Movement of an "average point" with time.

which implies that $\mathbf{r}$ designates an "average point" of the porous medium, since $\mathbf{u}_s^*$ is an average velocity.

Then, the analysis combines, as shown above, the equations of continuity of the liquid and of the moving solid. However, instead of giving the result in Eq. (5.3.3) as $d\rho/dt$ and $d\omega/dt$ (with $d\rho_s/dt = 0$), the analysis chooses $d\rho/dt$ and div $\mathbf{u}_s^*$ as the unknowns.

As the compaction is one-dimensional along $z$, the component in $z$, $\partial u_{zs}^*/\partial z$, is expressed in Eulerian coordinates from

$$\frac{\partial r_{zs}}{\partial \zeta} = \varepsilon_z = \alpha \, \Delta\rho \quad \text{(in Lagrangian coordinates)}$$

$$\frac{\partial r_{zs}}{\partial t} = u_{zs}^*$$

(Euler) $\qquad z = \zeta + r(\xi, t) \qquad$ (Lagrange)

Finally,

$$\frac{\partial u_{zs}^*}{\partial z} = \frac{\alpha}{1 + \alpha(p - p_0)} \frac{dp}{dt}$$

In the storage coefficient, $\alpha$ must therefore be replaced by $\alpha/[1 + \alpha(p - p_0)]$. It becomes important to take this term into account if $\alpha(p - p_0) > 0.5$, which thus represents a compaction $\varepsilon_z$ of more than 5%. Consequently, this effect is negligible except for special cases of large subsidence.

From the theoretical point of view, this difference in the results is caused by the change from the system of Eq. (5.3.6) with the three compressibilities of the total volume, the solid, and the pores to the linear relations in terms of material derivatives $d\omega/dt$ and $d\rho_s/dt$, functions of $dp/dt$ that are not wholly satisfactory. However, Gambolati's more rigorous result cannot be transposed to three dimensions.

Gambolati also shows that the variation in the hydraulic conductivity with $\rho$ (in the term grad $\rho g$), which we disregarded in order to arrive at Eq. (5.3.9), is actually negligible, when the aquifer thickness is less than 10,000 m and the pressure variation less than 500 bars, which greatly exceeds the usual ranges.

However, in highly compressible soils, it would probably be necessary to take into account the variation of $K$ during the compaction (when the pores close), which is a phenomenon that has not been studied a great deal.

## 5.5. Other Diffusion Equations

We have discussed the three most important cases. However, there are other cases where different equations are used:

(1)    Movement of the water in the unsaturated zone: see Section 9.1.2.
(2)    Exact equations of the movement of the free surface: see Section 6.3.d, and Schneebeli (1966) and Bear (1972).
(3)    Multiphase flow of immiscible fluids: see Section 9.1.1.
(4)    Flow of miscible fluids of different density: see Chapter 10.

In Chapter 7, solutions of the diffusion equation are given for the steady state ($\partial h/\partial t = 0$, the hydraulic head does not vary with time), and in Chapter 8 they are given for the transient state ($\partial h/\partial t \neq 0$, the hydraulic head varies with time).

# Chapter 6

# Aquifer Systems

We shall examine briefly the main aquifer types encountered in nature, their reserves, and, finally, their most common boundary conditions.

To begin at the beginning, what is an aquifer? It is a layer, formation, or group of formations of permeable rocks, saturated with water and with a degree of permeability that allows economically profitable amounts of water to be withdrawn.

In practice, an aquifer is an abstraction; it is a more or less isolated "layer" of rock saturated with water, limited in space at the top, at the bottom and on the sides, rather like a thin layer of mist in a forest.

An aquifer is by no means equivalent to a single geologic, lithographic, or stratigraphic unit; two contiguous layers of sand and limestone, for instance, may form a *single* aquifer. What is important in the definition is that

(1)  the part of the formation that constitutes the aquifer is saturated with water. An unsaturated permeable layer does not constitute an aquifer.

(2)  the variation of the permeability inside the formation, vertically or laterally, is restricted, so that two zones of the formation may not be separated by a zone of low permeability, through which the flow would be very small. For instance, a sand and a limestone layer, separated by a clay or marl layer, would constitute two aquifers. These two aquifers would communicate by *leakage* (see Section 5.3.g) through the layer of low permeability. We shall come back to that in Section 6.1.3.

## 6.1. Aquifer Types

### 6.1.1. *Unconfined Aquifers*

The term "unconfined aquifer" as opposed to "confined aquifer" will be defined in Section 6.1.2, but we shall begin by looking at a few examples of unconfined aquifers.

(a) *Valley aquifer.* In temperate climates, if the soil is assumed to be uniformly porous and permeable, we know (Chapter 1) that rainwater is infiltrated and saturates the rock up to a certain level, called the free surface, or water table. This saturated zone is called an aquifer from its free surface down to its lower limit (e.g., impermeable bedrock, or layer of low permeability separating it from the next aquifer). The overlying unsaturated zone, above the water table, forms in fact a continuum with the aquifer (the pressure is continuous through the water table, see Section 2.2.d) but is usually not considered as part of the aquifer, strictly speaking.

In the aquifer, the water flows toward the outlets, which are the low points in the topography (springs, streams in the surface flow network). Recharge occurs over the whole surface of the aquifer.

The chalk aquifers in the North of France or South of England are examples, as shown in Fig. 6.1. This cross-section shows the flow lines* and the lines of equal hydraulic head, which are called equipotential lines (or equipotential surfaces in three dimensions). If the permeability is isotropic, the flow lines are at right angles to the equipotential lines, according to Darcy's law. The slope of the free surface then indicates the flow direction of the aquifer, but the water flows through the whole thickness of the aquifer. Only the velocities are greater on the surface than at depth, since the distances are shorter, while the hydraulic head differences remain the same.

Only the deepest valleys drain the aquifer; the others are called dry valleys. The outlet is not a single point in space; it is a whole face of the aquifer from which water emerges and wells up. It is called the seepage face.

In chalk there is, strictly speaking, no actual bedrock, because the chalk is very thick (several hundred meters) in certain areas, and only the upper portion (10–30 m, for example) is fractured, weathered, and permeable,

---

* The flow lines give the direction of the velocity in the aquifer at a given time $t$. If the flow is steady, i.e., does not vary with time, the flow lines are constant. A particle of water (or of a tracer) would then follow a trajectory in the medium identical to the flow line that passes through the initial position of that particle. Such a trajectory is called a streamline or a flow path.

However, in transient flow conditions, i.e., when the flow varies with time, the flow lines also vary. At each time, they only show the direction of the velocity at each point. The trajectory of a particle in such a system is still called a streamline., but is no longer identical to any of the flow lines.

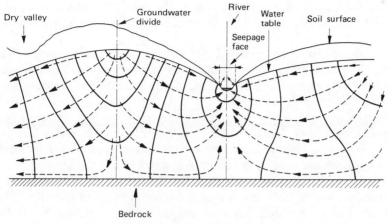

**Fig. 6.1.** Unconfined valley aquifer.

whereas the undisturbed chalk at depth has a very low permeability. The top of this undisturbed chalk is then taken as the lower limit of the aquifer.

Let us suppose that a piezometer, open only at the bottom, is installed in such an aquifer, to measure the local hydraulic head (and not the average head over the whole thickness of the aquifer). If the piezometer is drilled vertically into the zone of the seepage face, the deeper the piezometer, the higher the head: due to the upward direction of the flow, the head increases with depth, as shown by the equipotential lines. The situation is reversed at the summits of the free surface, where the flow is diverging: the head decreases with depth. Between these two limits, the head is more or less constant on a given vertical line. However, even at these limits, the variations in head in the vertical direction are very small, almost negligible.

The cross section of Fig. 6.1 is, in fact, greatly distorted: the vertical scale is perhaps 10 to 100 times greater than the horizontal one. If this cross-section is drawn with the same scale in both directions, it becomes Fig. 6.2.

The equipotential lines are, in fact, almost vertical. The assumption is often made that, in practice, the velocities in aquifers are virtually parallel to the free

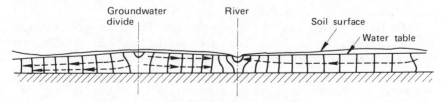

**Fig. 6.2.** Unconfined valley aquifer without scale distortion.

surface (i.e., close to horizontal), except in the vicinity of the outlets or of the groundwater divide lines. This is what we call Dupuit's hypothesis in Section 5.1.

The piezometric contour map (Fig. 6.3) shows such a valley aquifer in two dimensions. It is called a valley aquifer precisely because it is only drained by the valleys. On a larger scale, there is therefore a succession of small units, each of which is drained by a stream.

Piezometric maps are drawn in two dimensions and in principle show the lines where the equipotential surfaces intersect the free surface. However, since the equipotential surfaces are in practice almost vertical, the piezometric map gives approximately the hydraulic head at any depth in the aquifer.

The divide line between two valleys (dotted line on Fig. 6.3) separates the aquifer into several units. Each unit is drained by a given river; i.e., all the groundwater in that unit flows toward that river. This line is called the groundwater divide line and is drawn by selecting a set of base points on the river network (e.g., the tributaries of the rivers) and following from each of them in the upstream direction the groundwater flow line that ends up at this point. If the permeability is isotropic in the plane, which we shall assume most of the time, these flow lines are simply orthogonal to the equipotential lines. They eventually reach the summit of the aquifer, i.e., the point with the highest piezometric head. Each unit identified in this way is called an underground watershed; it is often quite close to the topographical watershed for the surface water. Like topographical watersheds, the position and number of the underground watersheds can be modified by selecting a different set of base points on the river network.

In the course of the year, the level of the water table of the aquifer varies by a few meters because, as we have seen (Chapter 1), it is fed by rainfall only in the winter: it decreases in summer and rises again after the autumn rains. If the water table is far from the soil surface (e.g., 10–30 m), it takes quite a long time for the infiltration to cross the unsaturated zone and the water table is at its lowest in October and November, and at its highest in April and May, for example. This kind of aquifer is also called phreatic aquifer (from the Greek phreatos, well), which simply means that it is the first aquifer encountered when a well is dug and therefore the most easily exploited. This type of valley aquifer is quite common and can be found in many types of rocks, such as sands, sandstone, limestone, tuffs, etc.

In the United States, the High Plains in the Great Plains are a good example of such aquifers (Fetter, 1980). The rocks can be the Ogallala formation (pliocene deposits eroded from the Rocky Mountains) or the Sand Hills (aeolian sands). Recharge occurs through the surface, and drainage occurs by the rivers, which sometimes cut the formation down to the bedrock, thus isolating different units. These aquifers are heavily developed for irrigation.

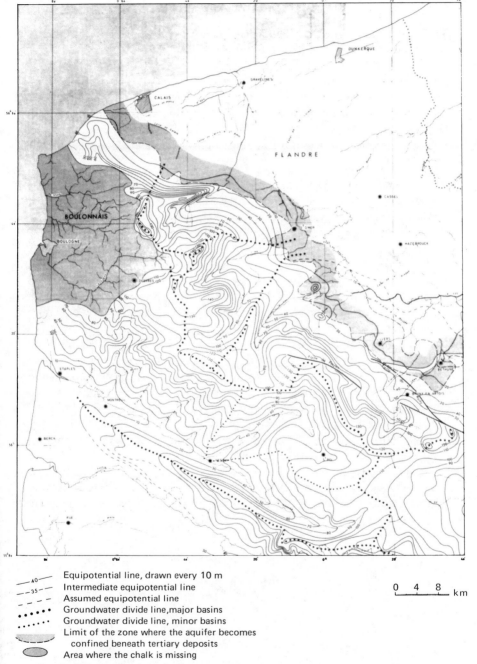

Equipotential line, drawn every 10 m
Intermediate equipotential line
Assumed equipotential line
Groundwater divide line, major basins
Groundwater divide line, minor basins
Limit of the zone where the aquifer becomes
   confined beneath tertiary deposits
Area where the chalk is missing

**Fig. 6.3.** Piezometric contour map of the chalk aquifer in Northern France. [From Cottez and Dassonville (1965).]

The Atlantic and Gulf Coastal plains also contain such aquifers, in continental or marine sediments or in glacial deposits towards the North.

Toth (1963) has studied the influence of the thickness of the aquifer and of the position of the streams on the shape of the flowlines and equipotential lines in a cross section. Freeze and Witherspoon (1967) have also studied numerically the effect of topography and permeability variations on the flow and equipotential lines. Freeze (1971) studied, in three-dimensions, the flow in both the saturated and unsaturated zones of a small underground watershed.

(b)   *Valley aquifers in arid zones.*   In arid zones (Fig. 6.4) rainfall is much lower than potential evapotranspiration, and surface recharge is almost zero. However, in the valleys, rivers may carry water from the mountains or flash floods may bring large quantities of water for a short time. This water usually infiltrates through the river bed into the aquifer and constitutes the only recharge mechanism. Therefore, the water table is higher beneath the valleys than elsewhere, contrary to what happens in humid zones. This situation occurs whenever the rainfall drops below 500 mm/yr in hot climates (e.g., in Spain, North Africa, Arabian Peninsula, etc.). In the United States, the sediment-filled basins Southwest of the Rocky Mountains are good examples of such aquifers.

It is important that river beds remain permeable so that water can infiltrate; if silt is deposited, the river beds could eventually become clogged. In natural systems, this clogging is prevented by erosion during floods. When dams are built, it is therefore important to create artificial floods in the stream, from time to time, to erode the silt.

Seen on a map, the rivers in desert-type climates, called wadis in Arabic, may never reach the sea. The flood water may be dispersed in the low plains, infiltrate, and later evaporate. This evaporation leaves the dissolved salts, giving some water in the low plains with very high salt content. Alternatively, the flood water may reach a depression, called a chott or a sebkha in Arabic (i.e., salt flat), creating a temporary lake, where the water eventually

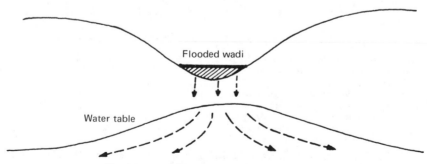

**Fig. 6.4.** Unconfined valley aquifer in arid region.

evaporates, leaving a salty crust on the surface (e.g., the salt lakes in Utah). These lakes are sometimes below sea level.

In both cases, the aquifers generally flow towards the same low plains or lakes. The water also evaporates there and becomes brackish.

Some of the water infiltrated into the river bed during floods may be consumed by the vegetation growing along the streams. The proportion of water consumed in this way may become very important in desert-type zones. Thus, surprisingly enough, recharge to aquifers through streams can be larger in extreme types of desert climates, where there is not enough water for vegetation to grow along the streams! One must remember that in desert climates, rainfall is very erratic, and one big storm (e.g., 300 mm) every 30 years may constitute the only recharge episode. The development of these aquifers must make allowances for this variability in the recharge.

In tropical zones, the two types of recharge may alternate between surface recharge during the rainy season and stream recharge in the dry season.

(c)    *Alluvial aquifer.*    This is an unconfined aquifer situated in the alluvial deposits found along the course of a stream. The water in the aquifer is generally in equilibrium with that of the stream, which alternately drains and recharges it.

This is, for example, the case of the Rhine river plain (Fig. 6.5) between France and Germany, which is a rift filled with recent alluvial deposits. The alluvial deposits are around 100 m thick in some places and consist of coarse sand, gravel, and pebbles with high permeability. These materials are saturated with water almost all the way to the surface, and they form one of the largest aquifers in France.

Virtually every stream has left fluvial deposits along its bed that link it to an alluvial aquifer. In the United States, such alluvial aquifers are found, for instance, on the Colorado plateau and in the Tennessee valley. The aquifer may vary in size at different points, as in Fig. 6.6. At the entrance of an alluvial plain, the water level in the stream is higher than that of the aquifer. The stream feeds the aquifer; the equipotential lines are close together (fast, diverging flow). In the middle of the plain, the flow is slower, and the river and the aquifer are at equilibrium. Downstream the situation is reversed, as the

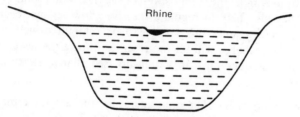

**Fig. 6.5.** Cross section of the Rhine valley.

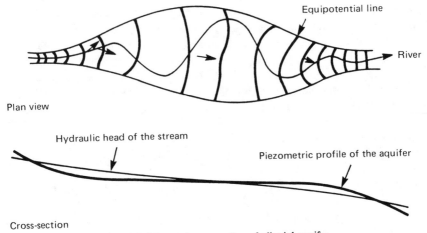

**Fig. 6.6.** Map and cross section of alluvial aquifer.

narrowing of the plain causes the water of the aquifer to drain toward the stream. This region is often marshy, as the aquifer surface is then very close to that of the soil surface and above that of the river. The flow in the aquifer is sometimes referred to as the "underflow" of the river.

Finally, it may happen that the stream bed is sealed by small particles, which break this connection. This is often the case of rivers regulated by dams, as already described. If the river is polluted, this silty layer may become anaerobic and bacteria may cause ammonia to be formed. The quality of the aquifer water may then deteriorate, even if only small amounts of water percolate through the river bed. However, in this anaerobic zone, some denitrification can simultaneously occur, which may sometimes be beneficial if the river water is heavily loaded with nitrates. In general, alluvial aquifers are very sensitive to the pollution of their rivers. Some pollutants carried by the river may be filtered or adsorbed (see Chapter 10), but many others will move with the water and reach the producing wells. Because of their high permeability and good recharge by the streams, alluvial aquifers are often very heavily developed. They generally produce better water than the streams themselves by averaging the composition of the river water (plus filtration and sorption). They also help to regulate the river flow regime: because of the reserve stored in the aquifer, it is possible to exploit it intensively in summer at low flows, while recharge will fill the reserves during the next winter. Such methods are in frequent use in Colorado (see Illangasekare and Morel-Seytoux, 1982).

(d) *Perched aquifers.* These aquifers lie on an impermeable lower formation, and they are not connected to a stream which feeds or drains them.

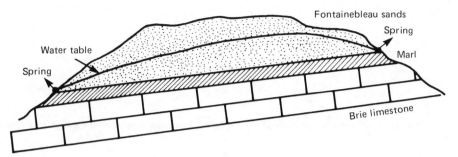

**Fig. 6.7.** Cross section of a perched aquifer.

An example is the unconfined aquifer in the Oligocene Fontainebleau sands (south of Paris, France), as in Fig. 6.7. On both sides of the formation, there are lines of springs. Seen from above this gives, for example, the map of Fig. 6.8. The largest springs are found in the valleys, at the lowest points of the underlying impervious formation.

If, under the impermeable (or less permeable) layer, in this case the marls, there is another unconfined aquifer (e.g., in the limestone), the upper one is said to have a "perched water table." The underlying aquifer is in fact recharged vertically by leakage through the marls, which have a low permeability (see Sections 5.3.g and 8.3.1).

Perched water tables can also be found in alluvial deposits, when a clay lens creates a local layer of low permeability inside the unsaturated zone overlaying an unconfined aquifer. The extent of the perched water table will be limited to that of the clay lens, as shown in Fig. 6.9. It may drain by leakage through the clay or laterally. These perched water tables may be permanent or only form during recharge in winter. When drilling a well, it is important not to confuse a local perched water table with the regional aquifer free surface.

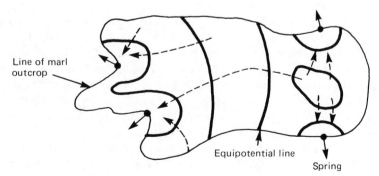

**Fig. 6.8.** Piezometric map of a perched aquifer.

Soil surface

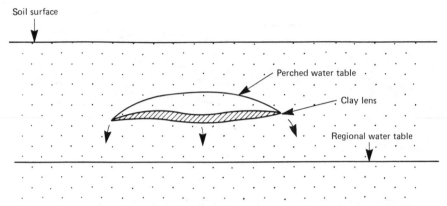

**Fig. 6.9.** Perched water table above a clay lens in alluvial deposits.

When the bottom of the clay lens has been drilled through, the perched water will generally drain through the drill hole toward the lower layer, and no water level will be found in the well until the true water table is reached.

(e) *Glacial deposits.* During the Pleistocene, Northern Europe, the Northern United States, and Canada were covered with glaciers, which deposited large amounts of sediment on the surface.

The lacustrine sediments were deposited in meltwater lakes; they are formed by silt and clay, and have a very low conductivity.

Glacial till, which is by far the most widespread sediment, was laid down directly by the glaciers when they melted. This means that it contains particles with a very wide range of size, from clay to large boulders. In general, glacial tills have a high clay content and a low permeability, although sets of thin vertical fractures may increase the regional hydraulic conductivity by a factor of up to 1000 (Freeze and Cherry, 1979) compared to the values measured on cores (in the range $10^{-10}-10^{-12}$ m/s). However, sandy tills may form local aquifers here and there.

Glaciofluvial deposits were laid down by subglacial streams, and by rivers during interglacial periods. They consist of sand and gravel and can be highly permeable. They are often buried inside a thick till layer and difficult to locate. Their shape can be that of a narrow valley, straight or meandering, or that of an extended thin strip. The subglacial stream sediments deposited by the last glacier sometimes lie on top of the general till surface; they form meandering ridges in the topography, which are called eskers (from the Irish "eiscir"). Because of their topographic position, they are largely unsaturated, being drained on the sides, but they may contain some water in the middle.

Glacial outwash is formed by sediments brought by the subglacial streams and the moraines in front of the glaciers. It forms interbedded layers of sand, gravel, occasional clay, and silts, and can contain good aquifers, generally valley aquifers. They are often exploited as gravel pits. Cape Cod and Long Island, in the United States, are good examples of glacial outwash.

(f) *Permafrost.* At northern latitudes (e.g., in Alaska, and the northern territories in Canada, Northern Sweden, Siberia), the ground can be frozen down to several hundred meters. This frozen ground is called the permafrost. Depending on the salinity of the water and the nature of the soil, the mean annual temperature may have to be significantly less than 0°C to cause the formation of permafrost. The permeability of the permafrost is almost negligible, even in sands or coarse material. In the summer, the upper layer of the permafrost (e.g., 1 m) may thaw (melt), but as the rest of the layer is impervious, the drainage is poor, and the soil is marshy (tundra). In a hilly topography, this wet soil may start to flow, and create mud slides towards the valleys. When the great pipeline from Alaska to Washington was built through Canada, great care was taken that the pillars, on which the pipe is laid and which are anchored in the permafrost, would not melt the ground, thus jeopardizing its stability. To achieve this, heat exchangers were installed on *each* pillar so that the heat brought by the pipe can radiate toward the atmosphere.

Aquifers may sometimes be found beneath the permafrost if it is not too thick, e.g., alluvial valleys, or alluvial fans, beneath lakes where the permafrost is thinner or missing. Because of the permafrost, the recharge to these aquifers may be poor. The aquifer layer can also become confined (Section 6.1.2) beneath the permafrost. Wells can then be artesian. If this groundwater discharges naturally at the surface, large cones of ice called pingos are formed in the winter.

(g) *Karstic systems, limestone aquifers.* We have seen in Section 4.1.e that, in fractured limestone, the dissolution of carbonates by carbonic acid present in the atmosphere and in the top soil creates enlarged fractures, conduits, caverns, or caves. This is called the karstic regime. Very often, the surface water network communicates with the groundwater through numerous systems of sinkholes, losses, and resurgences (i.e., outlets where the water reappears). Under a limestone plateau, the karstic system is defined by its base level, i.e., the elevation of the downstream outlet(s). The groundwater flows through the system (conduits, fractures) toward these outlets, which are generally limited in number (springs). The elevation of the water (the head) in this system is, of course, always higher than that of the outlets, but the gradient is usually very small. In rainy seasons, karst can have floods very similar to surface-water networks.

In the "blocks" between the conduits (or drains) of the network, recharge water from the surface may percolate slowly through a finer network of joints or fractures. This brings a delayed flux to the main system, where flow is very rapid. This secondary system of fractures builds up the reserves of the karst and makes it possible for karstic springs to continue to deliver a decaying flow in dry seasons. This base flow may, however, sometimes dry up. The water in the blocks, above the water level in the drains, will saturate the smaller fissures, while the larger ones will dry out. This is due to surface tension, as greater capillary pressure is needed to drain a small fissure than a larger one. The low-permeability limestone matrix itself, between the fractures, can be saturated or unsaturated.

Beneath the "base level" of the karst, all the fractures are filled with water and form a continuous aquifer. The question is whether or not this zone has open fractures. Generally, dissolution does not occur much below the base level, because the water has already become saturated with carbonate; therefore the answer is no. However, the base level may have changed during geological time (e.g., change in sea level, change in river level due to sediment deposits, etc.). Thus, beneath the present base level, there may exist a "paleokarst" where a real aquifer can be found.

Choosing a well site in a karstic system is very tricky: if a conduit is not found, the water available (in the blocks) is negligible. The best location for the well is on a fracture or, even better, at the intersection of two fractures. One tries to detect the fractures with the help of aerial photographs or geophysical measurements (electric resistivity, etc.). Before abandoning a dry well, one may try to inject tons of acid (HCl) to open existing fractures, in the hope that they will eventually connect the well with a drain. Alternatively, dynamite blasting in the well may create such a fracture. One may also try to deepen the well beneath the base level in the hope that an open network of fractures will be found.

Karstic systems are very common all around the Mediterranean. Some of the springs are even found off-shore, beneath the sea level. They were formed during the ice age, when the sea level was lower, as was the base level. Attempts are being made to exploit these submarine springs before they mix with sea water [see for example, Potié (1973)]. In the United States, karstic systems are found, for instance, in Kentucky, Florida, and the Dakotas.

Not all limestone layers are truly karstic. Dolomitic rocks, for instance, tend to be naturally permeable without being fractured, and they are much less soluble. Dissolution features occur along the fractures, and locally increase the hydraulic conductivity, but they usually form continuous aquifers and not networks of conduits. The same thing is true for chalk, or for marly limestone. In valley aquifers in such terrains, the hydraulic conductivity is generally higher near the valleys than beneath the plateaus, because the drainage in the rivers has locally increased the conductivity by dissolution.

Karstic-type features can also be found in evaporites, when they are in contact with water. However, the dissolution is much more rapid, and general subsidence or local collapses of caverns into sinkholes reaching the surface can be observed after a few years (see Johnson, 1981). Salt layers have themselves a very low hydraulic conductivity ($10^{-12}$–$10^{-16}$ m/s) unless they contain thin, interbedded sediment layers (clay, silts, etc.).

(h) *Volcanic rocks.* Volcanic tuffs and ashes are generally highly porous, but their permeability is quite low. In Nevada, for instance, very thick layers of welded and nonwelded tuffs are found up to 600 m above the water table. They form an unsaturated zone where water can slowly infiltrate (but the recharge is very small in Nevada). The weathering of these tuffs produces clay, which decreases their permeability even more.

Lava flows (e.g., basalts) tend, in general, to be more permeable. We have already described, in Section 4.1.e, the basalt columns, which are highly conductive. Gases escaping from lava create bubbles and pores. The cooling also creates joints. Within successive lava flows sediment deposits can form interbedded, highly permeable layers. Finally, during cooling, the upper surface solidifies first and can form a "bridge" under which the lava continues to flow. Openings are thus created in the direction of the lava flow. For all these reasons, lava can be highly to moderately permeable.

In the volcanic Canary Islands (trachytes and basalts), the moderately permeable rock is exploited by man-made blind tunnels several kilometers long, which dip slightly towards the entrance so that the collected water flows out by gravity. Flow rates per tunnel can reach several liters per second. The same system is also used in the Hawaiian Islands (Fetter, 1980) and is called Maui tunnels, but the reason for them is different: the basalt is much more permeable, but conventional wells would produce salt water, because there is only a thin lens of fresh water on top of salt water (see Section 9.4). On the Columbia plateau, in the states of Washington, Oregon, and Idaho, the basalt layers (which are up to 3000 m thick) can be very permeable and form good aquifers. At the same time, very compact basalt layers are found at depth, where the hydraulic conductivity, measured at a local scale, is in the range $10^{-8}$–$10^{-11}$ m/s (Freeze, 1979). A shaft is being sunk in the Hanford, Washington, area to study such layers in more detail, with a view to using them as possible radioactive waste repositories. However, it is highly debatable whether such thick layers have indeed, at a large scale, such a low hydraulic conductivity.

(i) *Crystalline rocks.* Granitic and metamorphic rocks generally have a very low permeability if they are not fractured or if their fractures are sealed. Because fractures have a tendency to close with depth (see Section 4.1.e), wells are usually not drilled below 50 or 100 m, unless a crushed zone is known to exist at depth. Here again, aerial photographs help locate the position of fracture intersections, where the wells can be put down. Geophysical resistivity measurements are also useful. Flow rates per well are usually small

(1–10 m³/h, occasionally up to 30 m³/h). Most of the Precambrian shield of Central and West Africa contains such mediocre aquifers, as do the northeastern United States and Canada.

Some localized aquifers may also be found in areas where weathering has occurred (arena sands, laterite, etc.), not to mention alluvial deposits.

(j) *Coastal aquifers.* These aquifers bring terrestrial fresh water into contact with marine salt water; we shall study the related mechanisms in Section 9.4.

### 6.1.2. Confined Aquifers

An aquifer is said to be confined if it is overlaid by a formation with low (or zero) permeability and if the hydraulic head of the water it contains is higher than the elevation of the upper limit of the aquifer (Fig. 6.10). When a well or a piezometer is drilled into such an aquifer, the water wells up suddenly in the borehole as soon as the impermeable upper limit of the aquifer is broken through. The water contained in the aquifer is in fact at a pressure higher than that of the atmosphere: hence the term confined aquifer.

If this pressure is sufficient for the water to reach the ground surface and well up (i.e., the piezometric head is higher than the elevation of the ground), the confined aquifer is said to be "artesian" and the well "artesian" or "flowing" (from the province of Artois, France, where the phenomenon was first observed). An example of artesian conditions is shown in Fig. 6.11.

These artesian conditions may, however, disappear with time, if the aquifer is exploited, because the hydraulic head in the aquifer decreases.

A water table or unconfined aquifer, as opposed to a confined aquifer, is one where the piezometric surface coincides with the free surface of the aquifer, which is overlaid by an unsaturated zone, as in Fig. 6.1 or 6.2.

The conceptual surface joining the water levels in all the piezometers is called the piezometric level (or surface) or piezometric head. It has no physical

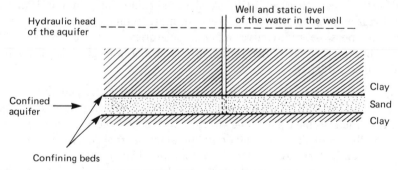

**Fig. 6.10.** Cross section of a confined aquifer.

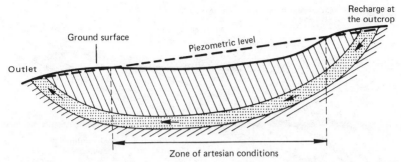

**Fig. 6.11.** Example of a confined aquifer: cross section of the Gironde eocene sands (France).

significance. On the contrary, for unconfined aquifers, this piezometric surface coincides with the water table.

(a) *Multilayered systems.* In the example of the Fontainebleau sands given in Section 6.1.d, the aquifer in the underlying limestone is confined over most of the area. Figure 6.12 illustrates this.

In large sedimentary basins (e.g., the Paris basin in France, the Gulf Coastal Plain or the Illinois–Wisconsin basin in the United States, or the continental formations in North Africa), successive layers of sands, sandstones, clays, marls, limestone, dolomites, evaporites, etc. can be found. Except for the first layer, all others form confined aquifers and confining beds. They are referred to as multilayered systems. They are generally very productive, and wells 2000 m deep or more can be drilled. In general, the deeper the aquifer, the higher the head, because deeper aquifers generally outcrop at a higher elevation on the periphery of the basin, and therefore their initial head is higher. At great depth, the water is hot due to the geothermal gradient and can be used as a geothermal resource. In the Paris basin, the Dogger aquifer at 1800 m

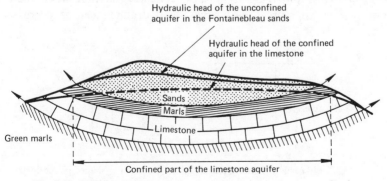

**Fig. 6.12.** Cross section of a two-layer system, confined and unconfined.

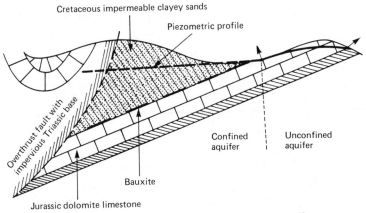

**Fig. 6.13.** Cross section of a confined–unconfined aquifer.

produces water at 70°C; in the Sahara, the Albian aquifer at 2000 m produces water at 60°C, which must be cooled in atmospheric towers, before it can be used for irrigation. Deep aquifers, however, often contain brackish waters.

(b) *Unconfined aquifers becoming confined.* Figure 6.13 is a schematic cross-section of the aquifer in the Jurassic dolomite near Brignoles (Var, France), which contains a top layer of bauxite lenses. The extraction of this bauxite poses serious problems of mine drainage. The aquifer is unconfined at the dolomite outcrops but becomes confined as soon as the dolomite is covered by the impermeable cretaceous clays.

A confined aquifer can be compared to a U-shaped permeameter as in Fig. 6.14. The head is always above the upper limit of the permeable medium.

(c) *Difference between confined and unconfined aquifers when the piezo-metric surface is lowered.* It must be remembered that when there is a drawdown in an aquifer (i.e., when its hydraulic head is decreased by

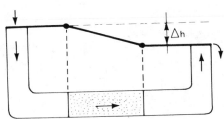

**Fig. 6.14.** Analogy between a confined aquifer and a U-shaped permeameter.

withdrawals) (1) the saturated thickness decreases in an unconfined aquifer, causing a reduction in the transmissivity (permeability × saturated thickness) and of the area open to flow; and (2) none of this happens in the case of a confined aquifer. The area open to flow stays the same as well as the transmissivity, as long as the drawdown does not push the piezometric surface down below the upper limit of the aquifer (in which case the aquifer becomes unconfined).

Any of the different formations that we have considered in Section 6.1.1 can constitute confined aquifers if there exists a confining layer to cover it and if the hydraulic head in the aquifer becomes higher than the lower limit of this confining bed.

### 6.1.3. Media with Low Permeability

Strictly speaking, these media do not constitute aquifers as they cannot be exploited. However, as a general rule, they contain water and form either an unconfined system, if the layer in question outcrops, or a confined one, if the layer lies at depth beneath a less permeable formation. Media with low or very low permeability should never be taken to "have no water"; instead, one must remember that the medium is probably saturated with water, which flows out very slowly or scarcely at all owing to the low permeability of the medium.

When a mine is opened and air starts to circulate, the medium may well dry out and any sign of flow may disappear. This is why salt mines, for example, although pronounced completely dry, may in certain cases be considered as water-saturated media, since the flow of water passing through is so small that it evaporates when it enters the galleries.

This type of medium plays a significant role in numerous problems, where water content is of importance:

(1) Civil engineering: consolidation, compaction, seepage force, and stability.

(2) Hydrogeology: recharge of deep aquifers through aquitards by leakage.

Generally, we distinguish between (1) aquitards, which are less permeable beds from which water cannot be produced economically through wells, but where the flow is significant enough to feed adjacent aquifers through vertical leakage, and (2) aquicludes, which have *very low* permeability and cannot give rise to any appreciable leakage, at least on a small scale (e.g., during a pumping test). Leakage through them may, however, not be completely negligible over very large areas.

## 6.2. Aquifer Reserves

(a)  *Unconfined aquifers.*   If there is a drawdown $\Delta h$ of the free surface (or the piezometric surface) of an unconfined aquifer (Fig. 6.15), the liberated volume of water is obviously the product of the volume comprised between the two successive positions of the free surface and the specific yield $\omega_d$ of the aquifer.

However, this volume is not immediately available as the moisture profile of the unsaturated zone must have time to decline by $\Delta h$, as shown in Fig. 6.16. If the final profile is a parallel shift of the initial profile, the liberated volume is indeed $\omega_d \, \Delta h$ per unit surface area.

The time needed for the movement of the profile depends on the grain size of the porous medium: see, for example, the table from King's experiment in Section 2.3.c.

The reserve of an unconfined aquifer is therefore given by the differences between the present piezometric surface and the piezometric surface to which it is acceptable to lower the water level in the aquifer; this difference is then multiplied by the area and the porosity. An example is shown in Fig. 6.17. One could, however, decide on other piezometric surfaces of maximum drawdown (e.g., the suitable depth of wells).

(b)  *Confined aquifers.*   Imagine an elementary volume of a confined aquifer, the hydraulic head of which is lowered by $\Delta h$ as shown in Fig. 6.18.

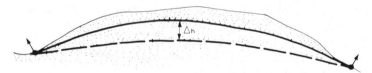

**Fig. 6.15.**  Drawdown in an unconfined aquifer.

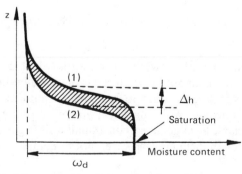

**Fig. 6.16.**  Shift of the mositure profile by a drawdown $\Delta h$ in an unconfined aquifer.

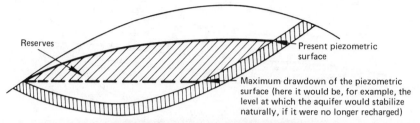

**Fig. 6.17.** Reserve in an unconfined aquifer.

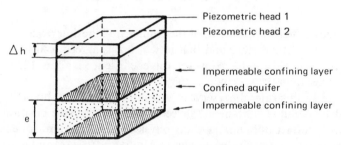

**Fig. 6.18.** Drawdown $\Delta h$ in a confined aquifer.

The variation in hydraulic head $\Delta h$ produces no dewatering of the confined aquifer. However, we have seen in Chapter 5, when we established the diffusion equation for a confined aquifer, that this decrease in hydraulic head causes a "production" of water under the influence of two phenomena (see Section 5.3):

(1)   Decompression of the water: term $\omega \beta_l$ (compressibility coefficient of water; $\omega$ = total porosity).

(2)   Compaction of the porous medium: term $\alpha - \omega \beta_s$ (compressibility coefficient of the porous matrix, minus $\omega$ times the compressibility coefficient of the solid grains).

Both these effects are combined in the definition of the storage coefficient,

$$S = \rho \omega g e (\beta_l - \beta_s + \alpha/\omega)$$

where $e$ is the thickness of the aquifer, $\rho$ is the mass per unit volume of water, $g$ is the acceleration due to gravity, and $S$ is dimensionless.

By definition, the volume of water produced by the variation in the hydraulic head $\Delta h$ per unit surface area (in the horizontal plane) of a confined aquifer is $V = S \Delta h$. In other words, in the case of a confined aquifer, the storage coefficient $S$ plays the same role as the specific yield $\omega_d$ in the case of an unconfined aquifer. The reserve of a confined aquifer is then the product of

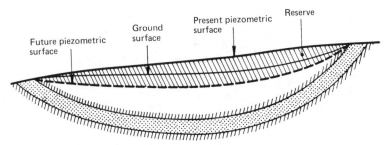

**Fig. 6.19.** Reserve of a confined aquifer.

the storage coefficient $S$, the area of the aquifer, and the difference between the present piezometric surface and that to which it is agreed to draw down the head in the confined aquifer. However, this coefficient $S$ is about 1000 to 10,000 times smaller than the specific yield $\omega_d$.

Figure 6.19 illustrates the reserve for a confined aquifer. It should be pointed out that the volume contained between the two successive positions of the piezometric surface does not have any physical meaning. Here, for example, it is located in part in the air (the aquifer was initially artesian) and in part in the first few meters of the soil.

Furthermore, we must remember that if there is a drawdown of the piezometric head in a confined aquifer below the upper limit of the aquifer, it becomes unconfined; the additional reserve, which then becomes available, is calculated in the same way as that of an ordinary unconfined aquifer. An example is given in Fig. 6.20. If the piezometric surface in this aquifer is lowered from (1) to (2), the volume withdrawn is the area $A$ multiplied by the specifie yield plus the area $B$ multiplied by the storage coefficient.

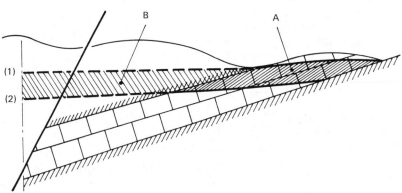

**Fig. 6.20.** Reserve in a confined–unconfined aquifer.

## 6.3. Usual Boundary Conditions and Initial Conditions

In Chapter 5 we established the diffusion equation, which is a partial differential equation of elliptic type in steady flow ($\partial h/\partial t = 0$) or of parabolic type in transient flow ($\partial h/\partial t \neq 0$). We have now seen the different types of aquifers encountered in the field, which are the domains in which we shall try to integrate the diffusion equation.

However, in order to do this, we must first define the boundary conditions of these domains of integration. In mathematics, we have three types of boundary conditions:

(1) Dirichlet's conditions, which concern the dependent variable: $h$ is prescribed on the boundary.

(2) Neumann's conditions, which concern the first derivative of the dependent variable: $\partial h/\partial n$ is prescribed.

(3) Fourier's conditions, which concern $h$ and $\partial h/\partial n$ such that $h + \alpha(\partial h/\partial n)$ is prescribed.

We shall add a fourth type: the conditions on a free surface or on a seepage face, which are double boundary conditions. Then we shall examine the problem of initial conditions.

(a) *Prescribed head boundaries.*   Dirichlet's conditions are imposed on a boundary if the hydraulic head on the boundary is independent of the flow conditions in the aquifer. This is generally the case where there is contact between the aquifer and a free expanse of water (sea, lake, river, etc.) Figure 6.21 illustrates this. Along the contact area (A) of the aquifer–river, the potential (hydraulic head) is constant and imposed by the water level in the river. The river may recharge or drain the aquifer.

Hence, on a map, a river may be a prescribed head boundary of an aquifer. Of course, the hydraulic head in the river varies along its course and

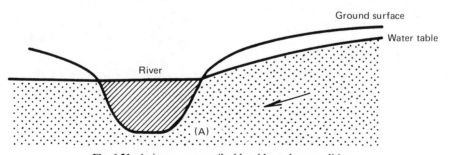

**Fig. 6.21.** A river as a prescribed head boundary condition.

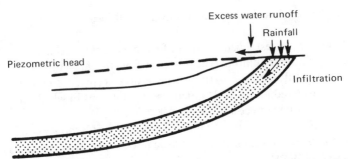

**Fig. 6.22.** An outcrop of an aquifer forming a prescribed head boundary condition.

sometimes with time as well, but these variations are controlled by the surface conditions, not by the flow in the aquifer.

An outlet of an aquifer (line of springs) may also be considered as a prescribed head boundary: i.e., that of the water level in the spring as long as the aquifer flows outwards.

The outcrops of an aquifer layer (Fig. 6.22) can also, in certain cases, play the part of a constant head boundary, as long as the infiltration rate of the rainfall on the outcrops is higher than the flux of water flowing toward the center of the aquifer. In other words, the aquifer layer on the outcrops is assumed to be always "waterlogged," as the excess of infiltrated water is drained by the surface stream network on the outcrop.

(b)  *Prescribed flux boundaries.*  This is a Neumann condition. If we impose the value of the normal hydraulic head gradient $\partial h/\partial n$, on the boundary, this is, according to Darcy's law, equal to imposing the value of the flux $-K\,\partial h/\partial n$ or $-T\,\partial h/\partial n$ on this boundary:

We distinguish between

(1)  No-flow boundaries: $\partial h/\partial n = 0$. For example, the contact between an aquifer formation and an impermeable layer, a fault*, or a confining bed as shown in Fig. 6.23.

---

* A fault is not always a no-flow boundary. Some faults may let water flow through the surrounding crushed zone or put the aquifer in contact with another permeable layer. Such faults do not constitute boundaries.

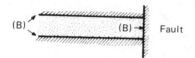

**Fig. 6.23.** A fault as a no-flow boundary condition.

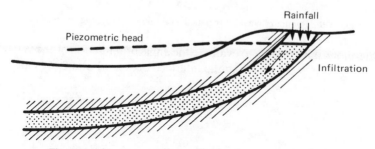

**Fig. 6.24.** An outcrop as a prescribed flux boundary condition.

(2)   Boundaries with a prescribed nonzero flux, such as (a) an outcrop in a zone where the rainfall infiltration rate is lower than the potential of the aquifer to "soak" it up (Fig. 6.24) [the prescribed flux is equal to the infiltration rate: it is the infiltration rate of the rainfall that determines the incoming flow], or (b) a withdrawal with a prescribed production rate in an exploitation (wells, ditches, etc.), which also constitutes a boundary with a prescribed flux (Fig. 6.25).

$$\int_{(F)} K \frac{\partial h}{\partial n} d\sigma = Q$$

The contact surface between two adjacent media cannot, in principle, be regarded as a boundary for either medium. Indeed, if the hydraulic conductivities $K_1$ and $K_2$ are isotropic, we can write two conditions at the interface:

$$h_1 = h_2 \qquad \text{(equality of hydraulic head)}$$

$$K_1 \frac{\partial h_1}{\partial n} = K_2 \frac{\partial h_2}{\partial n} \qquad \text{(equality of flux)}$$

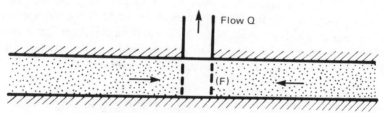

**Fig. 6.25.** Prescribed flux in a well or ditch.

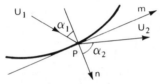

Fig. 6.26. Boundary between two media of differing hydraulic conductivity.

Furthermore, it is possible to evaluate the angle of refraction of the velocity $U$ (Fig. 6.26). As $h_1 = h_2$ all along the interface between 1 and 2, we can write:

$$\frac{\partial h_1}{\partial m} = \frac{\partial h_2}{\partial m}$$

and as

$$U_1 \sin \alpha_1 = -K_1 \frac{\partial h_1}{\partial m} \qquad\qquad U_2 \sin \alpha_2 = -K_2 \frac{\partial h_2}{\partial m}$$

$$U_1 \cos \alpha_1 = -K_1 \frac{\partial h_1}{\partial n} \qquad\qquad U_2 \cos \alpha_2 = -K_2 \frac{\partial h_2}{\partial n}$$

we get

$$\frac{U_1 \sin \alpha_1}{K_1} = \frac{U_2 \sin \alpha_2}{K_2} \qquad \text{and} \qquad U_1 \cos \alpha_1 = U_2 \cos \alpha_2$$

Hence

$$\frac{tg\alpha_1}{K_1} = \frac{tg\alpha_2}{K_2}$$

In some cases, however, it is possible to consider the interface as a prescribed flux or head boundary when the hydraulic conductivity contrast is large. As an example, consider a highly permeable alluvial aquifer deposited in a much less permeable bedrock. The bedrock receives recharge on the plateaux on each side of the valley; this recharge flows through the bedrock toward the central alluvial aquifer with high gradients and low velocities. The flux at the bedrock–alluvium interface can then be considered as prescribed for the alluvial aquifer since this flux depends very much on the flow conditions in the bedrock and very little on the flow conditions in the alluvial aquifer. Inversely, the interface could be considered as a prescribed head boundary for the bedrock because the head in the alluvium will depend very little on the flow in the bedrock.

(c) *Fourier's conditions.* Take a stream draining (or feeding) a water table aquifer, but with a low permeability silt layer, deposited on the bottom of the

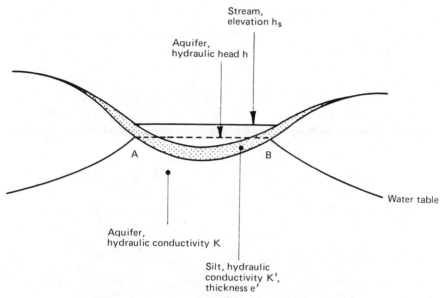

**Fig. 6.27.** Fourier's condition between an aquifer and a stream.

stream (Fig. 6.27). The difference in hydraulic head $\Delta h = h_{\text{stream}} - h_{\text{aquifer}}$ across the silt layer (denoted by $h_s - h$) creates the necessary gradient for a certain flow $q$ per unit surface area of contact between aquifer and stream, in accordance with Darcy's law.

$$q = K' \frac{\Delta h}{e'} = K' \frac{h_s - h}{e'}$$

However, when evaluated in the aquifer according to Darcy's law, this flux is given by

$$q = -K \frac{\partial h}{\partial n}$$

where **n** is the normal line to the contact surface oriented towards the stream. By equating the two expressions, we get

$$-K \frac{\partial h}{\partial n} + \frac{K'}{e'} h = \frac{K'}{e'} h_s$$

which is a Fourier condition definition. However, this condition is used much less frequently than the two previous ones.

(d) *Free surface.* Two conditions define a free surface (see Fig. 6.28):

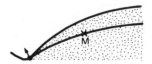

**Fig. 6.28.** Free-surface condition. The hydraulic head in M is equal to the elevation of the water table in M.

(1)    The pressure $p$ is equal to the atmospheric pressure at any point $M$ of the free surface (see Section 2.2.d). Expressed in hydraulic head, 0 is by convention taken as the atmospheric pressure, and we write

$$h = z$$

(2)    The free surface is a no-flow boundary, if the aquifer is not recharged through its surface:

$$\frac{\partial h}{\partial n} = 0$$

Thus a second condition is imposed on the same surface.

However, sometimes the aquifer is recharged through its free surface and the flux that transits across it is prescribed (e.g., mean annual recharge), so that

$$\frac{\partial h}{\partial n} = a$$

where $\mathbf{n}$ is the normal line oriented outward. The situation is the same if evaporation takes away water from the aquifer ($a$ is then negative).

The whole problem with the free surface is that we do not *a priori* know its position. We have to find by successive approximations a surface in space that simultaneously satisfies

$$h = z \quad \text{and} \quad \partial h/\partial n = a$$

Consequently, this problem is quite an intricate one. Usually an estimated position of the free surface is chosen initially, which then determines the boundary of the domain of integration. On this boundary the hydraulic head ($h = z$) is prescribed, and after integrating the equation we verify that the calculated flux $K \, \partial h/\partial n$ is correct. If it is not correct, the position of the free surface is moved in the desired direction and the calculation is repeated.

Another way of solving the problem is to consider the free surface not as a boundary of the flow but as part of a continuum comprising the saturated aquifer and the overlying unsaturated zone up to the soil surface. The diffusion equation of the unsaturated medium must now be solved (see Section 9.2). The free surface then becomes the area where the points of zero pressure are located.

In transient states and for a homogeneous medium, the diffusion equation in an aquifer bounded by a free surface becomes

$$K_x\frac{\partial^2 h}{\partial x^2} + K_y\frac{\partial^2 h}{\partial y^2} + K_z\frac{\partial^2 h}{\partial z^2} = 0 \qquad \text{in the aquifer}$$

$$K_x\left(\frac{\partial h}{\partial x}\right)^2 + K_y\left(\frac{\partial h}{\partial y}\right)^2 + K_z\left(\frac{\partial h}{\partial z}\right)^2 = \omega\frac{\partial h}{\partial t} + (K_z - q)\frac{\partial h}{\partial z} - q$$

on the free surface, where $q$ is the flux (volume per unit horizontal surface area per unit time) exchanged between the aquifer and the outside (evaporation, infiltration) across the free surface, and $q$ is positive if it is withdrawn.

(e) *Seepage face.* When the water in an aquifer seeps outward along an outlet surface (Fig. 6.29), the contact surface (S) is called a seepage face. The boundary conditions are (1) $h = z$, since the pressure is by definition equal to the atmospheric pressure, and (2) $\partial h/\partial n < 0$ where **n** is directed outward. Indeed, the flow in the aquifer goes outward.

The seepage face poses the same kind of problems as the free surface: although the elevation $z$ along the seepage face is known, it is necessary to determine, by successive approximations, the points A and B where the seepage face begins and ends, respectively, and where the free surface starts.

Usually the position of the surface is imposed, and subsequently one checks that the flow is indeed outward.

Free-surface and seepage-face conditions can also be treated graphically or analytically in two dimensions by the hodograph method, if the flow is in a steady state. This consists of representing the flow in the hodograph plane, the axes of which are the components of the filtration velocity $U_x$ and $U_y$ [see Polubarinova-Kochina (1962), Bear (1972), and Strack (1985).]

(f) *No boundary conditions.* Finally, in certain cases, when the domain of integration is assumed to be infinite, it is possible to disregard boundary

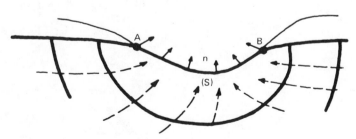

**Fig. 6.29.** Seepage face condition.

conditions. The boundary conditions are then prescribed at infinity without any need to define their character.

This kind of situation is very often created when one looks for analytical solutions to the diffusion equation, whereas numerical or analog methods are better suited to cases where the boundary conditions are known at finite distances. Examples will be given in other chapters.

(g) *Initial conditions.* It is worth remembering that for transient-state problems ($\partial h/\partial t \neq 0$, parabolic equation), it is also necessary to define the initial conditions of the problem, i.e., the value of the hydraulic head $h$ at all points of the domain for $t = 0$.

# Chapter 7

# Steady-State Solutions
# of the Diffusion Equation

## 7.1. General Properties of the Diffusion Equation

(a)  *There is only one solution.*   Let $D$ be a given domain of integration of the diffusion equation equipped with its boundary and initial conditions. It can be shown that, if $h$ satisfies these boundary and initial conditions as well as the diffusion equation

$$\text{div}(K\,\mathbf{grad}\,h) = S_s\frac{\partial h}{\partial t} + q$$

then $h$ is the unique solution to the problem. This is true in both a steady and a transient state.

In this and the following chapter, we shall give a few analytical solutions to the diffusion equation. As this equation is identical to the heat equation, many other solutions can be found in books on heat conduction. (One of the most widely used reference books in hydrogeology is therefore *Conduction of Heat in Solids*, by Carslaw and Jaeger (1959, and its later editions). The current analytical methods for integrating this equation in order to forge new solutions are based on the use of Fourier and Laplace transforms and of conformal mapping.

(b)  *Principle of superposition.*   A fundamental remark must be made before we turn to the solution of the diffusion equation: this equation and its boundary conditions are linear in $h$.

Therefore, in a domain $D$ with boundaries that are stable in space or at infinity, the equation

$$\text{div}(K \, \textbf{grad} \, h) = S_s \frac{\partial h}{\partial t} + q$$

is linear in $h$ and $q$. Consequently, if $(h_1, q_1)$ and $(h_2, q_2)$ are two special solutions of the diffusion equation that satisfy the given boundary conditions, then $(\forall \alpha, \beta) \; \alpha h_1 + \beta h_2$ is a solution of the same equation with the flows $\alpha q_1 + \beta q_2$ and their resulting boundary conditions $(\alpha h_1 + \beta h_2)$ on the prescribed head boundary condition, $(\alpha \, \partial h_1 / \partial n + \beta \, \partial h_2 / \partial n)$ on the prescribed flux boundary condition.

*Example.* Assume that, in a domain $D$, an aquifer with steady-state flow satisfies

$$\text{div}(T \, \textbf{grad} \, h_0) = q_0$$

where $q_0$ is the distribution of the source term in space. If this flow is disturbed, for example by the installation of a well with a production rate $q$ starting at time $t = 0$ at a given point M, the distribution of the hydraulic heads $h$ in the aquifer is a solution of the equation*

$$\text{div}(T \, \textbf{grad} \, h) = S \frac{\partial h}{\partial n} + q_0 + q$$

where $h$ satisfies the same boundary conditions as $h_0$ and has as initial condition $h = h_0$ for $t = 0$.

Let us then define the drawdown in the aquifer by

$$s = h_0 - h$$

Substituting $h = h_0 - s$ in the preceding equation, we get

$$\text{div}[T \, \textbf{grad}(h_0 - s)] = S \frac{\partial}{\partial t}(h_0 - s) + q_0 + q$$

Because of the linearity, we write

$$\text{div}(T \, \textbf{grad} \, h_0) - \text{div}(T \, \textbf{grad} \, s) = S \frac{\partial h_0}{\partial t} - S \frac{\partial s}{\partial t} + q_0 + q$$

or yet, taking into account the first relation that $h_0$ satisfies, and keeping in mind that $\partial h_0 / \partial t = 0$ (steady state),

$$\text{div}(T \, \textbf{grad} \, s) = S \frac{\partial s}{\partial t} - q$$

---

* Here, $q_0$ and $q$ represent the spatial distribution of the algebraic source term in the aquifer; as the new source term $q$ is in fact a sink representing a single well located in M and zero elsewhere, it should be written $q\delta(M)$ where $\delta(M)$ is the Dirac $\delta$ function (zero everywhere but in M, where $\delta = \infty$, and with an integral over space equal to 1). We will keep the simplified notation $q$ here.

The drawdown $s$ satisfies the diffusion equation with the following conditions:

(1)  Initial: $s = 0$ for $t = 0$.

(2)  Prescribed head boundaries: $h =$ constant, so $s = 0$, since $h = h_0 - s =$ constant.

(3)  Prescribed flux boundaries: $\partial h / \partial n =$ constant, so $\partial s / \partial n = 0$; since

$$\frac{\partial h}{\partial n} = \frac{\partial h_0}{\partial n} - \frac{\partial s}{\partial n} = \text{constant}$$

In other words, the drawdown $s$ due to the pumping $q$ satisfies an equation in the domain $D$, where the boundary and initial conditions are very much simplified compared to those of the original problem. Moreover, a drawdown $\alpha s$ would correspond to the production of $\alpha q$: by calculating only one solution $s$ of the drawdown for a given flow $q$, it is possible to give an infinite number of solutions $h = h_0 - \alpha s$ to the problem of pumping at any arbitrary rate $\alpha q$. Accordingly, it is possible to add together the influence (i.e., the drawdown) of production in several different wells.

We shall use this property of linearity very often in order to superimpose known solutions (the method of images, for example), or even to fashion a new solution by *integration* of a given solution.

However, one must remember that, strictly speaking, the diffusion equation in two or three dimensions is only linear for a confined aquifer; in an unconfined aquifer, the transmissivity $T$ may vary with the hydraulic head $h$, causing the equation to become nonlinear and making it impossible to rigorously apply the method of superposition.

Furthermore, in a vertical cross-section, this method cannot be applied to an unconfined aquifer, because the position of the free surface varies and the domain of integration is no longer stable.

The problems of the unconfined aquifer are therefore more intricate. We shall see later on that the best way of treating them is sometimes to take the overlying unsaturated zone into account. This does not mean, however, that the problem of nonlinearity is solved.

Strack (1985) was however able to show that the method of superposition can still be applied to unconfined flow conditions in steady state in two dimensions by using as variable the "discharge potential" $\phi = \frac{1}{2}Kh^2 +$ const, where $K$ is the hydraulic conductivity of the aquifer, and $h$ the head measured above the elevation of the impervious base of the aquifer (assumed to be horizontal, as in Section 5.1, where we obtained a diffusion equation in $h^2$). See also Section 7.5.

(c)  *Anisotropy.* We shall mainly study analytical solutions in homogeneous isotropic media. The problems of anisotropic media may be expressed as equivalent isotropic ones by stretching the coordinates.

If $K_x$, $K_y$, and $K_z$ are the three diagonal components of the hydraulic conductivity tensor in the principal directions of anisotropy, then Darcy's law and the diffusion equation (assuming that these conductivities are uniform in space) give

$$U_x = -K_x \frac{\partial h}{\partial x} \qquad U_y = -K_y \frac{\partial h}{\partial y} \qquad U_z = -K_z \frac{\partial h}{\partial z}$$

$$K_x \frac{\partial^2 h}{\partial x^2} + K_y \frac{\partial^2 h}{\partial y^2} + K_z \frac{\partial^2 h}{\partial z^2} = S_s \frac{\partial h}{\partial t}$$

where $S_s$ is the specific storage coefficient of the aquifer.

A change of coordinates yields

$$x' = \sqrt{\frac{K}{K_x}} x \qquad y' = \sqrt{\frac{K}{K_y}} y \qquad z' = \sqrt{\frac{K}{K_z}} z$$

where $K$ is an arbitrary coefficient with the dimensions of a hydraulic conductivity:

$$\frac{\partial h}{\partial x'} = \frac{\partial h}{\partial x} \frac{dx}{dx'} = \sqrt{\frac{K_x}{K}} \frac{\partial h}{\partial x}$$

and

$$\frac{\partial^2 h}{\partial x'^2} = \frac{\partial}{\partial x}\left(\frac{\partial h}{\partial x'}\right) \frac{dx}{dx'} = \frac{K_x}{K} \frac{\partial^2 h}{\partial x^2}$$

Therefore, the diffusion equation becomes (in the new coordinate system):

$$\frac{\partial^2 h}{\partial x'^2} + \frac{\partial^2 h}{\partial y'^2} + \frac{\partial^2 h}{\partial z'^2} = \frac{S_s}{K} \frac{\partial h}{\partial t}$$

which is an ordinary Laplace equation in the new axis system. It must be noted that, with anisotropy, the equipotential lines and the flow lines are no longer at right angles in the system of real coordinates $x$–$y$–$z$, while they are at right angles in the system $x'$–$y'$–$z'$. The velocity components in the new system are

$$U'_x = -K \frac{\partial h}{\partial x'} \qquad U'_y = -K \frac{\partial h}{\partial y'} \qquad U'_z = -K \frac{\partial h}{\partial z'}$$

Hence, we deduce that

$$U_x = \sqrt{\frac{K_x}{K}} U'_x \qquad U_y = \sqrt{\frac{K_y}{K}} U'_y \qquad U_z = \sqrt{\frac{K_z}{K}} U'_z$$

If we calculate the flux $Q'$ of the vector $\mathbf{U}'$ across an arbitrary surface $\Sigma'$,

$$Q' = \int_{\Sigma'} \mathbf{U}' \cdot \mathbf{n}\, d\sigma' = \int_{\Sigma'} (U'_x J'_1 + U'_y J'_2 + U'_z J'_3)\, du\, dv$$

where $J'_i$ is the direction cosine of the normal line to $\Sigma'$ and $u$, $v$ are arbitrary parametric coordinates of the surface $\Sigma'$.

It is then clear that if we try to calculate the flux of the vector **U** in the homologous surface $\Sigma$, defined by the same parametric coordinates $u, v$, we get the following relations between the Jacobian functions (direction cosines) $J_1$ and $J'_1$ of the two sufaces $\Sigma$ and $\Sigma'$:

$$J'_1 = \frac{D(y', z')}{D(u, v)} = \sqrt{\frac{K^2}{K_y K_z}} J_1 \quad \text{with} \quad J_1 = \frac{D(y, z)}{D(u, v)}$$

If $U_x$ is substituted for $U'_x$ in the above integral, we see that

$$Q' = \sqrt{\frac{K^3}{K_x K_y K_z}} Q \quad \text{or} \quad Q = \sqrt{\frac{K_x K_y K_z}{K^3}} Q'$$

which gives the relation between the flows in the anisotropic system and the equivalent isotropic system. In order to make these flows identical, we only have to take $K = \sqrt[3]{K_x K_y K_z}$. The same problem would arise for $S_s$ if a different value were chosen for $K$.

Using transmissivities in two dimensions, an identical development would give $T = \sqrt{T_x T_y}$ and we would define the change of coordinates to be $x' = \sqrt{T/T_x}\, x$ and $y' = \sqrt{T/T_y}\, y$.

## 7.2. Parallel Flow: First Solution in a Steady State

An aquifer with parallel (or uniform) flow is an aquifer where the velocity is a constant (in intensity and direction) at all points. The hydraulic head satisfies

$$h = ax + by + cz + d$$

which is a solution of

$$\nabla^2 h = 0$$

and which indeed gives a constant velocity $U_x = -Ka$, $U_y = -Kb$, $U_z = -Kc$. The constants are identified using the boundary conditions. A polynominal expression of the second degree is a solution of the problem $\nabla^2 h = q$ (constant infiltration). Of course, the velocity is no longer uniform.

## 7.3. Two-Dimensional Solutions in Radial Flow

(a) *Dupuit's elementary solution.* In polar coordinates $(r, \theta)$ in two dimensions, the Laplace operator is written

$$\nabla^2 h = \frac{1}{r} \frac{\partial}{\partial r}\left( r \frac{\partial h}{\partial r} \right) + \frac{1}{r^2} \frac{\partial^2 h}{\partial \theta^2}$$

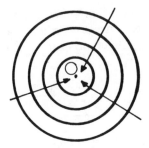

Fig. 7.1. Radial flow toward the origin.

An elementary solution to the problem is the one that depends only on $r$: $\partial^2 h/\partial\theta^2 = 0$. Hence,

$$\frac{1}{r}\frac{\partial}{\partial r}\left(r\frac{\partial h}{\partial r}\right) = Q$$

This is easy to integrate, giving:

$$h = a\ln r + b$$

where $r\,\partial h/\partial r = a$, $a$ and $b$ are constant, and ln is the natural logarithm.

If we look at this solution in two dimensions, we find that it is a flow converging radially on the origin (Fig. 7.1). The equipotentials ($h$ constant) are circles. If we calculate the flow crossing a given equipotential line at the distance $r$ from the origin, then, according to Darcy's law,

$$\text{Flow} = \int_0^{2\pi} T\frac{\partial h}{\partial r}\,d\theta = 2\pi Ta = \text{const} = Q$$

This constant flow then represents the flow rate $Q$ withdrawn from the aquifer at the point of origin, for example in a borehole with a given radius $r_0$, as shown in Fig. 7.2. The elementary solution just given is therefore that of a well in a confined aquifer. The constant $a$ is given by the flow rate produced by

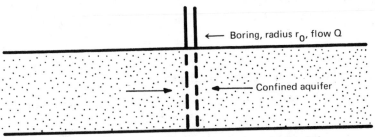

Fig. 7.2. Cross section of a confined aquifer through a borehole.

the well according to the integral of the flow:

$$a = \frac{Q}{2\pi T}$$

The constant $b$ is given by the boundary conditions. The simplest boundary condition is obtained by imposing the value $h$ at a distance $R$ from the point of origin:

$$h = H, \qquad r = R$$

Then

$$b = H - \frac{Q}{2\pi T}\ln R$$

Finally, the hydraulic head $h$ in the vicinity of the borehole is given by

$$h(r) - H = \frac{Q}{2\pi T}\ln\frac{r}{R}$$

which is Dupuit's or Thiem's formula.

This formula corresponds exactly to the problem of "the well on an island:" the boundary condition $h = H, r = R$ is only satisfied for a confined aquifer on a circular island, as in Fig. 7.3.

However, in reality, the water level in a borehole in any aquifer often stabilizes after some time (arriving at steady state) for a number of reasons, which we shall examine later (recharge boundary, leakage). The profile of the hydraulic head, depending on the distance from the boring, is then a logarithmic function (Fig. 7.4), which allows us to define a "fictitious radius of action" $R$ corresponding to the distance from the borehole where the drawdown $(h_{initial} - h)$ would be zero. This is mostly quite far from the physical reality, but it is often used in practice. It will be discussed again in Section 8.1.3.

(b)  *Well in an unconfined aquifer.*  We have seen in Section 5.1 that, if the bedrock is horizontal and the velocity is assumed to be always parallel to it, the

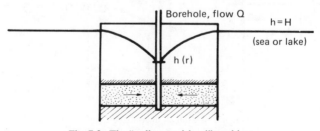

**Fig. 7.3.** The "well on an island" problem.

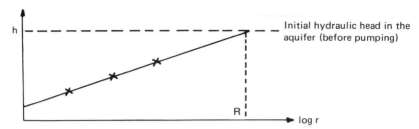

**Fig. 7.4** Hydraulic head in an aquifer as a function of the distance to a well. Observations are marked with × 's.

diffusion equation for an unconfined aquifer in steady state may be written

$$\nabla^2 h^2 = 0$$

By repeating the above reasoning, we can immediately deduce that for the radial problem, the square of the hydraulic head is a logarithmic function of the radius. More precisely,

$$h_0^2 - H^2 = \frac{Q}{\pi K} \ln \frac{r_0}{R}$$

where $R$ is the radius of action already defined, $r_0$ the radius of the well, $K$ the hydraulic conductivity, $H$ the hydraulic head at the boundary, and $h_0$ the hydraulic head in the aquifer around the well (Fig. 7.5.). This is known as the Dupuit–Forchheimer formula.

Observe that, in reality, the surface of the water in the well does not exactly correspond to the free surface in $h_0$. There is a certain length of seepage face in the borehole and head losses due to the well screen, which have not been taken into account here. The piezometric profile in the aquifer is given by

$$h^2 = H^2 + \frac{Q}{\pi K} \ln \frac{r}{R}$$

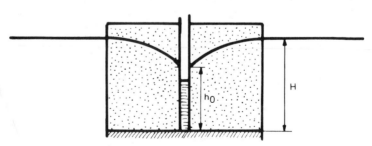

**Fig. 7.5.** Well in an unconfined aquifer.

**Fig. 7.6.** A two-well system.

In fact, the Dupuit assumption that the velocities are always horizontal becomes less and less acceptable closer to the well. The true piezometric profile lies above the Dupuit approximation.

(c)  *Method of images.*  Take two different wells with centers O and O', each producing at a steady rate of $Q$ and $Q'$, respectively. We want to find the hydraulic head at all points M of the domain (Fig. 7.6). According to the principle of superposition, this could be done by adding up the elementary logarithmic solutions of all these wells. At M, we can write

$$h_M = \frac{Q}{2\pi T} \ln r + \frac{Q'}{2\pi T} \ln r' + \text{const}$$

We identify the constant from the boundary conditions, if this is possible.

*First special case: Prescribed head boundary.*   Assume that in the well O' a flow of $-Q$ is produced; i.e., in fact, the flow $Q$ is injected. The solution becomes

$$h_M = \frac{Q}{2\pi T} \ln \frac{r}{r'} + \text{const}$$

If we study the points M where $r = r'$ (i.e., on the mediator of OO'), as in Fig. 7.7, we see that $h_M$ is constant. In other words, a constant hydraulic head is prescribed on the mediator of OO'.

This means that we have found an exact solution to the following problem: a *single* well O, situated at a distance $d$ from an infinite straightline boundary with a prescribed head $h = H$ (see Fig. 7.8):

$$h = \frac{Q}{2\pi T} \ln \frac{r}{r'} + H$$

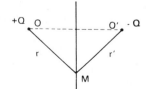

**Fig. 7.7.** Mediator of the OO' segment.

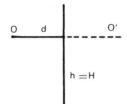

Fig. 7.8. Prescribed head boundary, real well and image well.

where $r$ is the distance to the real borehole O and $r'$ is the distance to the fictitious point O' symmetrical to O with respect to the boundary.

This solution is indeed the unique solution of the present problem, because it satisfies the boundary conditions and the diffusion equation.

The only reservation we make is that, at the well O, the radius of the borehole $r_0$ must be negligible compared to the distance $2d$ between O and O', so that the hydraulic head $h_0$ in the well O is really a constant around its circumference. If this is not the case, O and O' are no longer the centers of the boreholes, but the poles of the pencils of circles, i.e., the positions of the points, where the ratio $r/r'$ is constant.

Usually, the fictitious point O' is called the image well of the well O, an image that has an opposite sign, because the flow of the fictitious image well is the opposite of that of the real well.

We must, however, remember that the above solution also describes the case of two wells with the same flow rate but of opposite sign in an infinite medium.

*Second special case: No flow boundary.* In the initial expression with two borings O and O', we now let $Q' = Q$.

$$h_M = \frac{Q}{2\pi T} \ln rr' + \text{const}$$

It is immediately obvious from symmetry that on the mediator of OO',

$$\frac{\partial h}{\partial n} = 0$$

This may easily be demonstrated by switching to Cartesian coordinates $r^2 = x^2 + y^2$ and calculating $\partial h/\partial x$.

Thus we have found the analytical solution of the problem of a single well O situated at a distance $d$ of an infinite straight-line boundary with a no-flow boundary condition $\partial h/\partial n = 0$ (Fig. 7.9):

$$h = \frac{Q}{2\pi T} \ln rr' + \text{constant}$$

where $r'$ is the distance to the "fictitious image well" symmetrical to the well O

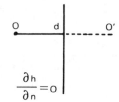

**Fig. 7.9.** No-flow boundary, real well and image well.

$$\frac{\partial h}{\partial n} = 0$$

with respect to the boundary, but this time, the image well has the same sign (flow $+ Q$) as the well O.

The same remark applies to the relation between the radius $r_0$ of the boring compared to $d$ and the description of the solution for two real boreholes in an infinite medium.

*Several boundaries.*    By this method of images (a well giving rise to an image with respect to a boundary) it is therefore possible to describe a problem with several boundaries.

*First example.*    Alluvial half-aquifer: two parallel boundaries, one with a prescribed head (the river) and the other with no flow (the hillside), as shown in Fig. 7.10. However, each fictitious image well gives rise to another fictitious image (of the same or opposite sign) with respect to the other boundary, thus producing an infinite double series of images farther and farther away. In practice, only a few terms are used.

*Second example: Confluence of two rivers (prescribed head boundaries).*    If the angle of the two boundaries is exactly $2\pi/n$ ($n$ integer), $n$ fictitious images arranged in a circle are generated as in Fig. 7.11.

There are numerous examples of the use of the method of images.

(d)    *Well line.*    It is sometimes useful to imagine an infinite series of wells separated from each other by a distance $a$ and producing at the same rate $Q$ in an infinite aquifer. The solution is obviously found by adding an infinite number of elementary solutions. However, the symmetry of the flow may also

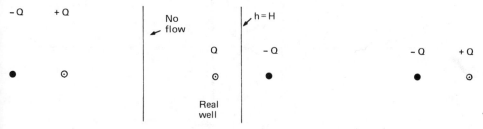

**Fig. 7.10.** Infinite series of image wells for a two-boundary system.

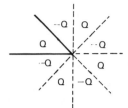

**Fig. 7.11.** Intersecting prescribed head boundaries.

be used to advantage, as the mediators of the segments joining two nearby wells are flow lines marking the limits of the flow towards each of the wells (Fig. 7.12). The flow is then a succession of identical "modules" defined for instance in

$$x \in \left\{ -\frac{a}{2}, +\frac{a}{2} \right\}$$

Schneebeli (1966) has shown that, in such a module, the elementary solution is expressed by

$$h = \frac{Q}{4\pi T} \ln \frac{\cosh(2\pi y/a) - \cos(2\pi x/a)}{2}$$

When $y$ becomes large compared to $a$ (in practice, $y > a$), the cosh term becomes predominant compared to the cos term and we can write

$$h \simeq \frac{Q}{2aT} \left( y - a\frac{\ln 2}{\pi} \right)$$

This is equal to a uniform flow parallel to the $y$ axis with the constant gradient

$$\frac{\partial h}{\partial y} = \frac{Q}{2aT}$$

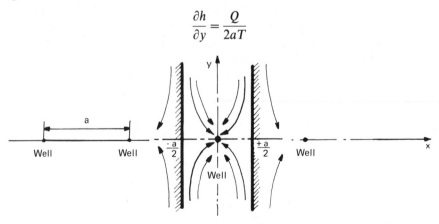

**Fig. 7.12.** Line of wells.

As soon as one moves away from the well line ($y > a$), one can therefore represent it as a continuous drainage ditch, located along the $x$ axis at $y = 0$, and withdrawing the same flow from the aquifer as the line of wells with a constant prescribed head given by

$$h(y = 0) = -\frac{Q}{2\pi T}\ln 2$$

taking as a reference $h = 0$ for $y = 0, x = \pm a/2$ in the complete exact solution above.

This procedure is often useful in drawdown projects, when the aim is to pass from a well line to a ditch or vice versa. It is easy to generalize to the case where the well line is parallel to a boundary by means of the method of images.

(e) *Characteristic curve of a well.* In steady state the flow rate of a given well may be expressed as a function of the drawdown (initial hydraulic head minus that of the stabilized state) in the borehole:

$$Q = 2\pi\frac{(H - h)}{\ln(R/r_0)}$$

in confined aquifers (Dupuit's formula) where $r_0$ is the well radius, and in unconfined aquifers,

$$Q = \pi K\frac{(H^2 - h^2)}{\ln(R/r_0)}$$

Hence, we deduce that the curve describing the evolution of the flow $Q$ versus the stabilized drawdown $s = H - h$ should be a line for a confined aquifer and a parabola for an unconfined one.

In reality, the "characteristic curve" of a well, which gives the drawdown $s$ versus $Q$, always has a parabolic shape, as in Fig. 7.13.

There are always quadratic head losses (nonnegligible term $v^2/2g$) in the first 10 or 20 cm surrounding a well, in the filtering gravel pack, and in the central well screen as shown in Fig. 7.14. The characteristic curve of the well describing this quadratic loss of hydraulic head is particularly useful for determining the power of a pump in order to obtain a given production rate.

**Fig. 7.13.** Characteristic curve of a well.

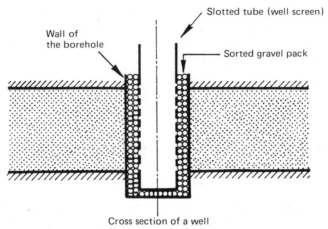

Cross section of a well

**Fig. 7.14.** Gravel pack and screen of a well.

In practice, it is admitted that the quadratic form of the head losses in the formation, in the well screen, and even in the well casing allows us to formulate a law of variation in the *stabilized* drawdown with the flow rate, which has the following form:*

$$s = AQ + BQ^2$$

Therefore, tests are made on the well at several flow rate steps, each of them sufficiently long for the water level to be fairly stabilized (after a while, $s$ does not vary much with time; each step lasts a few hours). Then $s/Q$ is plotted versus $Q$. This should be a line of slope $B$ and vertical intercept $A$. Walton (1970) characterizes the state of the well by the value of $B$:

$$B < 675 \quad \text{m/(m}^3\text{/s)}^2 \quad \text{good well, highly developed[†]}$$

$$675 < B < 1350 \quad \text{m/(m}^3\text{/s)}^2 \quad \text{mediocre well}$$

$$B > 1350 \quad \text{m/(m}^3\text{/s)}^2 \quad \text{clogged or deteriorated well}$$

$$B > 5400 \quad \text{m/(m}^3\text{/s)}^2 \quad \text{well that cannot be rehabilitated}$$

* See Note Added in Proof at the end of this chapter.

[†] If a well in an alluvial medium is to be developed, the fine particles in the formation around the borehole are set in motion through alternating pumping and injection so that they may be extracted by pumping. In this way, the permeability of the sediment close to the well is increased and the quadratic losses in hydraulic head decrease. In a limestone formation, the quality of the well is improved by injection of acid (HCl), which dissolves the rock and opens the fractures. In a fractured medium, blasting may also be used to increase fracturing locally. In a formation containing clay particles or drilling mud, polyphosphates are used to remove the clay.

## 7.4. Elementary Solution in Spherical Coordinates

In spherical coordinates in three dimensions $(\rho, \theta, \phi)$, the Laplace operator is written as

$$\nabla^2 h = \frac{1}{\rho^2} \frac{\partial}{\partial \rho} \left( \rho^2 \frac{\partial h}{\partial \rho} \right) + \frac{1}{\rho^2 \sin \theta} \frac{\partial}{\partial \theta} \left( \sin \theta \frac{\partial h}{\partial \theta} \right) + \frac{1}{\rho^2 \sin^2 \theta} \frac{\partial^2 h}{\partial \phi^2}$$

A solution that depends only on the distance $\rho$ at the origin satisfies

$$\nabla^2 h = \frac{\partial}{\partial \rho} \left( \rho^2 \frac{\partial h}{\partial \rho} \right) = 0$$

that is,

$$h = -\frac{a}{\rho} + b$$

It can also be shown that this solution is a flow converging on the origin, which corresponds to a constant withdrawal $Q$ in all spheres of radius $R$ centered on the origin. The flow $Q$ is

$$Q = 4\pi a$$

As an example of how to calculate a new solution by *integration* of an elementary solution, we look for the solution $h$ that corresponds to a withdrawal at a constant flow rate on a segment of the line $z = \pm C$ with a constant withdrawal density $dQ = \lambda \, d\xi$ on this segment.

The elementary solution for a withdrawal at a point $\xi$ of the segment $(+C, -C)$ of the $z$ axis is

$$h = \frac{dQ}{4\pi \sqrt{x^2 + y^2 + (z - \xi)^2}}$$

whence by integration,

$$H = \int_{-C}^{+C} \frac{\lambda \, d\xi}{4\pi \sqrt{x^2 + y^2 + (z - \xi)^2}}$$

$$= \frac{\lambda}{4\pi} \ln \frac{z + C + \sqrt{x^2 + y^2 + (z + C)^2}}{z - C - \sqrt{x^2 + y^2 + (z - C)^2}}$$

## 7.5. Complex Potential in Two Dimensions

If the hydraulic conductivity $K$ (or the transmissivity $T$) is constant, uniform, and isotropic, the velocity potential $\phi = Kh$ (or $Th$) is defined.

Darcy's law and the steady-state diffusion equation become, as functions of $\phi$,

$$\mathbf{U} = -\mathbf{grad}\, \phi$$

$$\nabla^2 \phi = 0$$

It is then possible to define a conjugate function $\psi$ called the "stream function" by

$$\frac{\partial \psi}{\partial x} = -\frac{\partial \phi}{\partial y}, \qquad \frac{\partial \psi}{\partial y} = \frac{\partial \phi}{\partial x}$$

This definition is possible because $\nabla^2 \phi = 0$, which means that $\nabla^2 \psi = 0$. The above conditions are Cauchy's conditions on the two functions $\phi$ and $\psi$, which define an analytical function $\Gamma$,

$$\Gamma = \phi + i\psi$$

which is an analytical function of the complex variable $z = x + iy$ (and not of $x$ and $y$ separately; cf. Cauchy's conditions). The function $\Gamma$ is called the complex potential of the flow.

Why is $\psi$ called the stream function? This can easily be explained. Let P and P' be two neighbouring points of the complex plane, as in Fig. 7.15.

Now calculate the flow crossing the segment PP' using

$$dQ = \mathbf{U} \cdot \mathbf{n}\, ds$$

The components of $\mathbf{U}$ and $\mathbf{n}\, ds$ are

$$\mathbf{U} \quad \begin{cases} -K\dfrac{\partial h}{\partial x} = -\dfrac{\partial \phi}{\partial x} = -\dfrac{\partial \psi}{\partial y} \\[2mm] -K\dfrac{\partial h}{\partial y} = -\dfrac{\partial \phi}{\partial y} = \dfrac{\partial \psi}{\partial x} \end{cases}$$

and

$$\mathbf{n}\, ds \quad \begin{cases} -dy \\ dx \end{cases}$$

Hence

$$dQ = \frac{\partial \psi}{\partial y}\, dy + \frac{\partial \psi}{\partial x}\, dx = d\psi$$

Thus, between two points A and B as in Fig. 7.16, the flow crossing *any* curve that joins A to B is

$$\text{flow} = \psi(\text{B}) - \psi(\text{A})$$

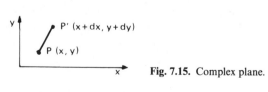

Fig. 7.15. Complex plane.

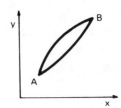

**Fig. 7.16.** Complex plane.

Hence, the lines of constant $\phi$ are equipotentials of the flow, and the lines of constant $\psi$ are the stream lines of the flow, whence the name stream function for $\psi$.

*Example.*  The elementary radial solution in two dimensions is very easily expressed in terms of the complex potential:

$$z = re^{i\theta}$$

$$\Gamma(z) = \frac{Q}{2\pi} \ln z = \frac{Q}{2\pi} \ln r + i\frac{Q\theta}{2\pi} = \phi + i\psi$$

Therefore, the equipotentials are $\phi = (Q/2\pi)\ln r$. We recognize the same expression given already if we remember that the velocity potential is $\phi = Kh$ (or $Th$).

The complex potential is valuable in that it permits the use of a number of analytical methods of transformation. In particular, conformal mapping (inversion, for example), which retains the angles, may be applied to this type of problem and makes it possible to find simple analytical solutions to problems that appear not to have any. See Polubarinova-Kochina (1962) Bear (1972), or Strack (1985) for this type of approach in mathematical hydrogeology.

The underlying principle of the process is as follows. The complex $x$–$y$ plane is transformed into a plane $u$–$v$, where the given flow problem has a known potential $\Gamma$. By inverse mapping, we obtain the complex potential $\Gamma'$ in the initial plane $(x$–$y)$. The elementary solution to the problem of Section 7.3.d, for example, has been calculated by Schneebeli (1966) using the following mapping:

$$\Gamma' = \sin\frac{\pi\Gamma}{a}$$

which transforms the infinite plane into a "module"

$$x \in \left\{ -\frac{a}{2}, +\frac{a}{2} \right\}$$

with the desired boundary. It is then sufficient to separate what is real from what is imaginary.

For unconfined flow conditions Strack (1985) uses the potential $\phi = \frac{1}{2}Kh^2 +$ const, assuming that the head is measured from the elevation of the horizontal impervious base of the unconfined aquifer. It is clear that

$$\nabla^2\phi = \frac{\partial}{\partial x}\left(\frac{1}{2}K\frac{\partial h^2}{\partial x}\right) + \frac{\partial}{\partial y}\left(\frac{1}{2}K\frac{\partial h^2}{\partial y}\right) = \frac{\partial}{\partial x}\left(Kh\frac{\partial h}{\partial x}\right) = \frac{\partial}{\partial y}\left(Kh\frac{\partial h}{\partial y}\right) = 0$$

is the diffusion equation for the unconfined aquifer, as was given in Section 5.1. The associated stream function $\psi$, defined by the same Cauchy conditions as above, also gives the flow in the aquifer. The method of the complex potential can therefore also be applied to unconfined flow conditions with this new definition of the potential. Once the potential has been determined, the head is calculated by

$$h = \sqrt{2\phi/K}$$

if the constant is taken to be zero.

Since the unconfined flow equation is linear in $\phi$, but not in $h$, the principle of superposition can be applied for $\phi$, but not for $h$; $h$ can only be calculated from the sum of the $\phi$'s.

*Example:*   The elementary solution for a single well is $\phi = (Q/2\pi)\ln r$; for a doublet of wells, one injecting and one pumping with the same flowrate, the potential is

$$\phi = \phi_1 + \phi_2 = \frac{Q}{2\pi}\ln r_1 - \frac{Q}{2\pi}\ln r_2 = \frac{Q}{2\pi}\ln\frac{r_2}{r_1}$$

The head distribution is then calculated from $h = \sqrt{2\phi/K}$.

## Note Added in Proof

In the petroleum industry, it is usual to assume that the medium surrounding the borehole has been modified by the drilling. A dimensionless "skin effect" $s_k$ is defined by $s_k = (k/k_s - 1)\ln(R_s/R)$, where $k$ is the intrinsic permeability of the formation, $k_s$ is that of the perturbed zone, $R$ is the radius of the well, and $R_s$ is the radius of the perturbed zone. The skin effect can be positive or negative, e.g., if the perturbed zone has been clogged by injection of mud, or, on the contrary, developed by the production or by other operations (e.g., acidification). The skin effect can be determined by interpretation of the pumping tests and recovery curves.

# Chapter 8

# Transient Solutions of the Diffusion Equation, Pumping Tests, and Measurements of Aquifer Properties

We shall try to find a few analytical solutions commonly applied to the diffusion equation in transient state. We derived in two dimensions (confined or unconfined aquifer subject to Dupuit's hypothesis),

$$\nabla^2 h = \frac{S}{T} \frac{\partial h}{\partial t}$$

or in three dimensions

$$\nabla^2 h = \frac{S_s}{K} \frac{\partial h}{\partial t}$$

Note that the general properties of the diffusion equation (uniqueness of solution, linearity, anisotropy) given in Section 7.1. hold true for the transient state as well.

## 8.1. Elementary Solutions in Radial Coordinates

In radial coordinates, we write the diffusion equation in two dimensions, and if we assume that the solution only depends on the distance $r$, we obtain

$$\frac{1}{r}\frac{\partial}{\partial r}\left(r\frac{\partial h}{\partial r}\right) = \frac{S}{T}\frac{\partial h}{\partial t}$$

Let us define

$$\alpha = \frac{T}{S} \quad \text{or} \quad \frac{K}{S_s}$$

which is called the aquifer diffusivity. One elementary solution is the Laplace solution*:

$$h = C\exp(-r^2/4\alpha t)t^{-n/2}$$

with $n = 1, 2$, or $3$,

$$r = x \qquad\qquad (n = 1)$$
$$r = \sqrt{x^2 + y^2} \qquad (n = 2)$$
$$r = \sqrt{x^2 + y^2 + z^2} \qquad (n = 3)$$

This corresponds to an impulse point injection of fluid at the origin, in an infinite aquifer, with initial condition $h = 0 \ \forall r$. In the following, other solutions are given.

### 8.1.1. *Theis's Solution*

Theis (1935) presented an integral solution (possible because of the linearity of the equations) of this elementary solution in two dimensions, which corresponds to a continuous point injection of fluid at the origin:

$$h(r,t) = \int_0^t C\frac{\exp(-r^2 S/4T\tau)}{\tau}\,d\tau$$

This is also the solution of the diffusion equation with boundary conditions prescribed at infinity and initial conditions $h = 0 \ \forall r$.

---

\* It is easily obtained by using the Laplace transform, which is a very efficient method for solving a number of transient problems.

With this solution we calculate the flow crossing a cylinder of radius $r$:

$$Q(r,t) = -2\pi rT\frac{\partial h}{\partial r} = -2\pi rT\frac{\partial}{\partial r}\left[ C\int_0^t \frac{\exp(-r^2 S/4T\tau)}{\tau}d\tau\right]$$

$$= \pi r^2 CS\int_0^t \frac{\exp(-r^2 S/4T\tau)}{\tau^2}d\tau$$

$$Q(r,t) = 4\pi TC\exp(-r^2 S/4Tt)$$

If $r = 0$, the flow rate injected at the origin is thus constant; if $r \to 0$ or $t \to \infty$, $Q \to 4\pi TC$.

The flow rate $Q(r_0,t)$ crossing the cylinder of radius $r_0$ representing a borehole is therefore constant if $r_0$ is negligible or $t$ is large: this solution, called the Theis solution, is consequently one that corresponds to injection (or pumping) at a constant rate in a well with a negligible radius, and $C = Q/4\pi T$.

$$h(r,t) = \frac{Q}{4\pi T}\int_0^t \frac{\exp(-r^2 S/4T\tau)}{\tau}d\tau$$

If we write

$$u = \frac{4Tt}{r^2 S}$$

then

$$h(r,t) = \frac{Q}{4\pi T}\int_{1/u}^\infty \frac{e^{-\tau}}{\tau}d\tau = \frac{Q}{4\pi T}\left[-E_i\left(-\frac{1}{u}\right)\right]$$

Here, $E_i$ is the exponential integral function, which is known and tabulated.

In practice, the so-called "Theis curve" is drawn as a function of the parameter $u$:

$$h(r,t) = \frac{Q}{4\pi T}W(u)$$

$W(u)$, the Theis function, is generally drawn on log–log paper. See Table 8.1 and Figs. 8.6 and 8.7.

Note that if $Q \geq 0$, $h$ grows with $u$ (or with $t$). This is then the case, when the flow $Q$ is injected, and $Q < 0$ corresponds to the case where the flow is withdrawn.

### 8.1.2. Jacob's Logarithmic Approximation

If $t$ is large, then so is $u$, and

$$-E_i(-1/u) \to \ln u - \gamma$$

where $\gamma$ is the Euler constant ($\gamma = 0.577$, $e^\gamma = 1.781$).

Table 8.1

The Theis Function, $W$ versus $u^a$

| u | 1.0 | 0.5 | 0.333 | 0.25 | 0.20 | 0.167 | 0.143 | 0.125 | 0.111 |
|---|---|---|---|---|---|---|---|---|---|
| × 1 | 0.219 | 0.049 | 0.013 | 0.0038 | 0.0011 | 0.00036 | 0.00012 | 0.000038 | 0.000012 |
| × $10^1$ | 1.82 | 1.22 | 0.91 | 0.70 | 0.56 | 0.45 | 0.37 | 0.31 | 0.26 |
| × $10^2$ | 4.04 | 3.35 | 2.96 | 2.68 | 2.47 | 2.30 | 2.15 | 2.03 | 1.92 |
| × $10^3$ | 6.33 | 5.64 | 5.23 | 4.95 | 4.73 | 4.54 | 4.39 | 4.26 | 4.14 |
| × $10^4$ | 8.63 | 7.94 | 7.53 | 7.25 | 7.02 | 6.84 | 6.69 | 6.55 | 6.44 |
| × $10^5$ | 10.94 | 10.24 | 9.84 | 9.55 | 9.33 | 9.14 | 8.99 | 8.86 | 8.74 |
| × $10^6$ | 13.24 | 12.55 | 12.14 | 11.85 | 11.63 | 11.45 | 11.29 | 11.16 | 11.04 |
| × $10^7$ | 15.54 | 14.85 | 14.44 | 14.15 | 13.93 | 13.75 | 13.60 | 13.46 | 13.34 |
| × $10^8$ | 17.84 | 17.15 | 16.74 | 16.46 | 16.23 | 16.05 | 15.90 | 15.76 | 15.65 |
| × $10^9$ | 20.15 | 19.45 | 19.05 | 18.76 | 18.54 | 18.35 | 18.20 | 18.07 | 17.95 |
| × $10^{10}$ | 22.45 | 21.76 | 21.35 | 21.06 | 20.84 | 20.66 | 20.50 | 20.37 | 20.25 |
| × $10^{11}$ | 24.75 | 24.06 | 23.65 | 23.36 | 23.14 | 22.96 | 22.81 | 22.67 | 22.55 |
| × $10^{12}$ | 27.05 | 26.36 | 25.96 | 25.67 | 25.44 | 25.26 | 25.11 | 24.97 | 24.86 |
| × $10^{13}$ | 29.36 | 28.66 | 28.26 | 27.97 | 27.75 | 27.56 | 27.41 | 27.28 | 27.16 |
| × $10^{14}$ | 31.66 | 30.97 | 30.56 | 30.27 | 30.05 | 29.87 | 29.71 | 29.58 | 29.46 |
| × $10^{15}$ | 33.96 | 33.27 | 32.86 | 32.58 | 32.35 | 32.17 | 32.02 | 31.88 | 31.76 |

$^a$ After Wenzel (1942).

In practice, as soon as $u = 4Tt/Sr^2 \geq 100$, the logarithmic approximation of the Theis formula can be used. This is also called Jacob's formula.

$$h(r,t) = \frac{Q}{4\pi T} \ln \frac{4Tt}{e^\gamma Sr^2} = \frac{Q}{4\pi T} \ln \frac{2.25\,Tt}{Sr^2}$$

For $u = 100, 50, 25, 10$, the errors in using Jacob's instead of Theis's formula are 0.3%, 0.5%, 1.4%, 5.2%, respectively.

On semilog paper, the response curve $h(t)$ at a given point is a straight line (as shown in Fig. 8.3).

A review of the basic assumptions of the Theis and Jacob formulas may be helpful. They are

(1)   Infinite, homogeneous, and isotropic medium.

(2)   Constant transmissivity (confined aquifer or, with approximation, unconfined aquifer with small drawdown; $S$ is then replaced by $\omega_d$, the specific yield).

(3)   Two-dimensional approximation, i.e., the hydraulic head does not vary in the third dimension: the velocity is parallel to the confining beds for a confined aquifer, or to the bedrock, assumed horizontal, for an unconfined aquifer (Dupuit's hypothesis).

(4)   Boring going through the entire thickness of the aquifer* (so that the problem remains two-dimensional), pumping at a constant rate with a negligible borehole radius.

* This is then called a fully penetrating well.

(5) Initial conditions $h(t, 0) = 0 \; \forall r$, i.e., an aquifer that is initially immobile. If this is not the case, according to the principle of superposition, the drawdown $s = h_0 - h$ satisfies the initial conditions if $h_0$ is a steady state.

*Note.* The variation of the hydraulic head around the well can be computed for large $t$ from

$$\frac{\partial h}{\partial t} = \frac{Q}{4\pi T} \frac{\exp(-Sr^2/4Tt)}{t} \simeq \frac{Q}{4\pi T} \frac{1}{t} \qquad \text{if } r \text{ is small}$$

Hence $\partial h / \partial t \to 0$ as $t \to \infty$: the hydraulic head variation becomes very slow in the vicinity of the well. Furthermore, as $\partial h / \partial t$ depends very little on $r$, the piezometric profile moves down while remaining parallel to itself in the vicinity of the well.

### 8.1.3. Relations between Transient and Steady States

In a steady state, Dupuit's formula, which gives the drawdown in a boring of radius $r_0$, is

$$s_{r_0} = \frac{Q}{2\pi T} \ln \frac{R}{r_0}$$

where $R$ is the radius of action of the well, i.e., the zone inside which the effect of the pumping is felt. Beyond $R$, the drawdown caused by the well is taken to be zero.

This notion is often accepted in practice. In most cases, this zone $R$ is fictitious. The drawdown is stabilized and a steady state is established through the influence of a boundary such as a river at some distance or of a leakage phenomenon (see Section 8.3) or simply of surface recharge for an unconfined aquifer.

However, in an aquifer that is not recharged by leakage, infiltration, or through a boundary, this radius of action around the well may be expressed as a function of the pumping time. We use Jacob's logarithmic approximation at the radius $r_0$ of the well itself to find

$$s_{r_0} = \frac{Q}{4\pi T} \ln \frac{2.25 \, Tt}{Sr_0^2} = \frac{Q}{2\pi T} \ln \frac{1.5\sqrt{Tt/S}}{r_0}$$

If this expression is compared to Dupuit's formula, it gives $R = 1.5\sqrt{Tt/S}$.

If the aquifer is infinite and not recharged, $R$ varies as $\sqrt{t}$. If $t$ is large, $R$ varies very slowly and it seems as if a steady state has been obtained.

Moreover, at a given time, a piezometric profile passing through the well actually has a logarithmic expression as given by Dupuit's formula (as long as

Jacob's approximation can be used, i.e., if head measurements are made not very far from the well).

### 8.1.4. *Application of the Principle of Superposition*

Just as in a steady state (Section 7.3.c), the principle of superposition may be used, either (1) to calculate the influence of several wells pumping in the same aquifer, (2) to describe artificially the influence of a straight line boundary (method of images), or (3) to study the recovery of the aquifer after the production has stopped.

(a) *Impervious straight-line boundary (no flow)*.   We shall use the drawdown as an example. The production is of the same sign in both the well and its image well (Fig. 8.1).

$$s = \frac{Q}{4\pi T}\left[W\left(\frac{4Tt}{Sr^2}\right) + W\left(\frac{4Tt}{Sr'^2}\right)\right]$$

When it becomes possible to use the logarithmic approximation for *both W* functions, then

$$s = \frac{Q}{4\pi T}\left[\ln\frac{2.25\,Tt}{Sr^2} + \ln\frac{2.25\,Tt}{Sr'^2}\right]$$

$$s = \frac{Q}{4\pi T}\left[2\ln\frac{2.25\,Tt}{Sr^2} + \ln\frac{r^2}{r'^2}\right]$$

If the evolution of the drawdown *s* is plotted versus the logarithm of the time (semilog paper) for a given observation point M, the slope of the line doubles as soon as the logarithmic approximation becomes valid for both the well and the image (see Fig. 8.3).

(b) *Straight-line recharge boundary (prescribed head)*.   The production is of the opposite sign (injection) at the image well:

$$s = \frac{Q}{4\pi T}\left[W\left(\frac{4Tt}{Sr^2}\right) - W\left(\frac{4Tt}{Sr'^2}\right)\right]$$

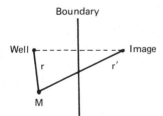

Fig. 8.1. Boundary, real and image wells.

When the logarithmic approximation can be used for *both* $W$ functions, then

$$s = \frac{Q}{4\pi T}\left[\ln\frac{2.25Tt}{Sr^2} - \ln\frac{2.25Tt}{Sr'^2}\right]$$

$$s = \frac{Q}{2\pi T}\ln\frac{r'}{r}$$

The drawdown stabilizes and does not change with time any longer (see Fig. 8.3). This is one way in which a steady state is obtained.

(c)  *Cessation of production.*  To calculate the behavior of a well after it has been stopped ("recovery curve"), an imaginary injection with the same constant flow rate is superimposed on the borehole itself, which is supposed to continue production at the same constant rate. The two flow rates thus cancel each other, and indeed represent an idle well.

Let $t_0$ be the duration of the pumping and $t_1$ the time counted from the cessation of production. The drawdown at any time after the end of production is given by

$$s = \frac{Q}{4\pi T}\left\{W\left[\frac{4T(t_0 + t_1)}{Sr^2}\right] - W\left(\frac{4Tt_1}{Sr^2}\right)\right\}$$

Three cases may arise:

(1)  The functions of $W$ must be used for one or both terms.
(2)  Jacob's approximation may be used for both of them:

$$s = \frac{Q}{4\pi T}\left\{\ln\frac{2.25T(t_0 + t_1)}{Sr^2} - \ln\frac{2.25Tt_1}{Sr^2}\right\}$$

$$s = \frac{Q}{4\pi T}\ln\frac{t_0 + t_1}{t_1} = \frac{Q}{4\pi T}\ln\left(1 + \frac{t_0}{t_1}\right)$$

If $s$ is plotted as a function of the logarithm of $(1 + t_0/t_1)$, a straight line also appears. Such a plot is called Horner's diagram.

(3)  It can be supposed that the first function $W$ is stabilized (i.e., that the pumping has gone on for long enough)—that is to say that, at least during the first part of the recovery, the drawdown $s$ only varies as a result of the second function $W$ (or its logarithmic approximation). This term is treated alone as a single pumping. This last method is known as the Houpeurt–Pouchan method.

## 8.2. Interpretation of a Pumping Test

(a)  *Jacob's method.*   If a well is pumped at a constant rate, it is possible to determine the $T$ and $S$ parameters of the aquifer: hence the frequent use of "pumping tests" in the study of an aquifer.

From an initial condition of the head in the aquifer that is as steady as possible, pumping at a constant rate is started in the well and the drawdown is observed in the well itself and, if possible, in a certain number of piezometers in the neighborhood. The rhythm of the measurements is very fast in the beginning (once or, if possible, more per minute) and slows down with time.

The pumping test is usually interpreted by graphical analysis of these measurements so that $T$ and $S$ can be deduced.

*Jacob's method* consists in plotting, on semilog paper, the drawdown $s$ at a given point (well or piezometer) versus time (Fig. 8.2).

It is also possible to plot $s/Q$ versus $t/r^2$, if the flow rate from the well has varied a little or if the aim is to plot all the piezometers at different distances on the same graph.

As soon as the logarithmic approximation holds the points must line up on one straight line, and when this is identified, the interpretation follows immediately (Fig. 8.3). The problem is that the beginning and sometimes the end of the curve deviates from Jacob's straight line, e.g., the end part, if the aquifer is not infinite or in case of leakage (see Fig. 8.3, Sections 8.1.4 and 8.3). It may thus be doubtful which is the "right" line. Theis's method, which will be discussed later, offers a way of deciding in uncertain cases.

When one line has been selected, two arbitrary points A and B on this line are chosen (Fig. 8.2):

$$s_B - s_A = \frac{Q}{4\pi T}\ln\frac{t_B}{t_A} \quad \text{i.e.,} \quad T = \frac{Q}{4\pi(s_B - s_A)}\ln\frac{t_B}{t_A}$$

The common practice is to choose

$$t_B = 10t_A$$

which gives

$$T = \frac{0.183Q}{s_B - s_A}$$

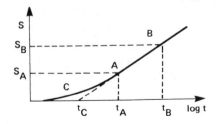

**Fig. 8.2.** Drawdown versus time on a semilog plot.

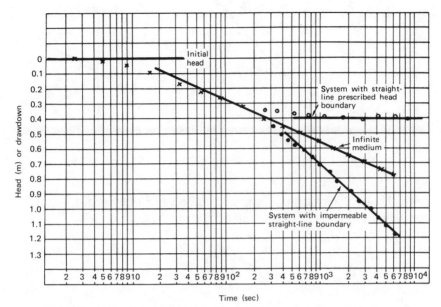

**Fig. 8.3.** Evolution of the head in a piezometer during a constant-rate pumping test.

To calculate $S$, the storage coefficient, we only need to remember that the *fictitious* point C, where the axis $s = 0$ intercepts the Jacob line (Fig. 8.2), corresponds to

$$\ln \frac{2.25 T t_C}{S r^2} = 0 \quad \text{i.e.,} \quad \frac{2.25 T t_C}{S r^2} = 1$$

whence

$$S = \frac{2.25 T t_C}{r^2}$$

*Influence of a boundary.* We have seen that an impermeable boundary doubles the slope of Jacob's straight line, as in Fig. 8.4. If the pumping test were interpreted with the second straight line, an incorrect transmissivity equal to half the true one would be found.

It is, however, possible to specify the distance from the boundary. The drawdown expression is

$$s = \frac{Q}{4\pi T} \ln \left( \frac{2.25 T t}{S r^2} + \ln \frac{2.25 T t}{S r'^2} \right)$$

Consider the fictitious intersection point I of the two straight lines. Mathematically, it is at this point that the influence of the image well is zero

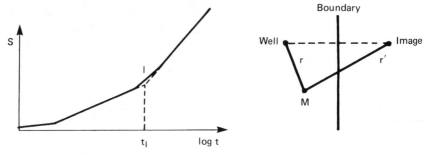

**Fig. 8.4.** Drawdown versus time for a system with a no-flow boundary.

(even though the logarithmic approximation cannot be applied until much later). In the same way as we calculated $S$ using the fictitious point C, we now write

$$\ln \frac{2.25Tt_1}{Sr'^2} = 0$$

that is,

$$r' = \sqrt{\frac{2.25Tt_1}{S}}$$

which gives an idea of the distance from the boundary. Using two piezometers and a small simple geometrical construction with two circles, it is even possible to give the exact position of the image well and, thus, of the boundary.

The procedure is precisely the same for a recharge boundary, as in Fig. 8.5.

Note that it is also possible to use Jacob's method by plotting $s$ versus $\log r$ at a given date $t$, if several piezometers are available:

$$s = \frac{-Q}{2\pi T} \ln r + \text{const}$$

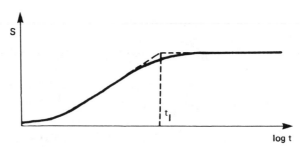

**Fig. 8.5.** Drawdown versus time for a system with a prescribed head boundary.

(b)  *Interpretation with the complete Theis formula by the curve-matching method.*  We use the complete tabulated function, valid even for small times:

$$s = \frac{Q}{4\pi T} W\left(\frac{4Tt}{Sr^2}\right)$$

A log–log paper is used for:

(1)  The "type curve" $W(u)$ versus $u$ (Fig. 8.6 or 8.7),
(2)  The experimental measurements $s$ versus $t$ (or possibly $s/Q$ versus $t/r^2$).

A transparent log–log paper (tracing paper) of the same module as the type curve* must be used. One of the graphs is drawn on tracing paper so that it can be put on top of the other.

Then, if we look at the vertical axes we can write:

$$\log s = \log\left(\frac{Q}{4\pi T} W\right) = \log\frac{Q}{4\pi T} + \log W$$

In the logarithmic graduation of the vertical axes, $s$ is deduced from $W$ through a single parallel shift ($\log Q/4\pi T$); then the corresponding sets of $s_i$ and $W_i$ points are matched.

Similarly, for the horizontal axes: one point of the horizontal axis $u$ of the type curve represents, in fact, a given value of

$$\frac{4Tt}{Sr^2} = u$$

If $u$ and $t$ are plotted in logarithmic graduation on the horizontal axes we get

$$\log u = \log\frac{4Tt}{Sr^2} = \log t + \log\frac{4T}{Sr^2}$$

Also, $t$ is deduced from $u$ through a single parallel shift ($\log 4T/Sr^2$); the set of corresponding points $t_i$ and $u_i$ are then matched.

Consequently, on log–log paper, it should be possible to match the type curve and the experimental curve through a simple parallel shift in the direction of the two axes, from one paper to the other (see Fig. 8.8), but the axes have to be kept parallel. When the two graphs are put on top of one another and matched, identification is immediate. An arbitrary point M of the plane is

---

* On Fig. 8.6, Theis' curve is drawn on a log–log paper having a module of 62.5 × 62.5 mm; this is the standard paper for the interpretation of electric soundings in geophysics. On Fig. 8.7, the same curve is drawn on a paper with a 1.85 × 1.85 in. module; this is also a very common log–log paper. Tracing paper of either module can thus be used. Figures 8.6 and 8.7 are each printed on two pages; to use them as type curves, one must first make a good photocopy of each page (without any magnification or reduction) and glue them together.

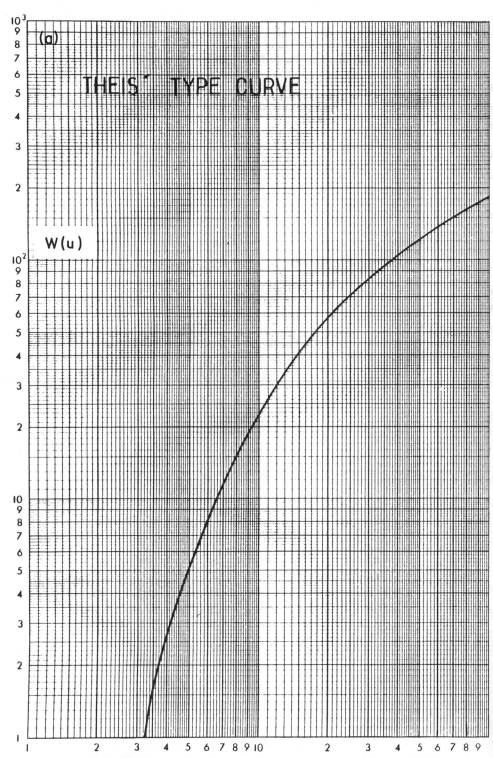

**Fig. 8.6.** Theis' type curve. $62.5 \times 62.5$ mm log module. By photocopying (without any magnification or reduction) both parts of this figure and pasting them together, it is possible to reconstruct the complete type curve.

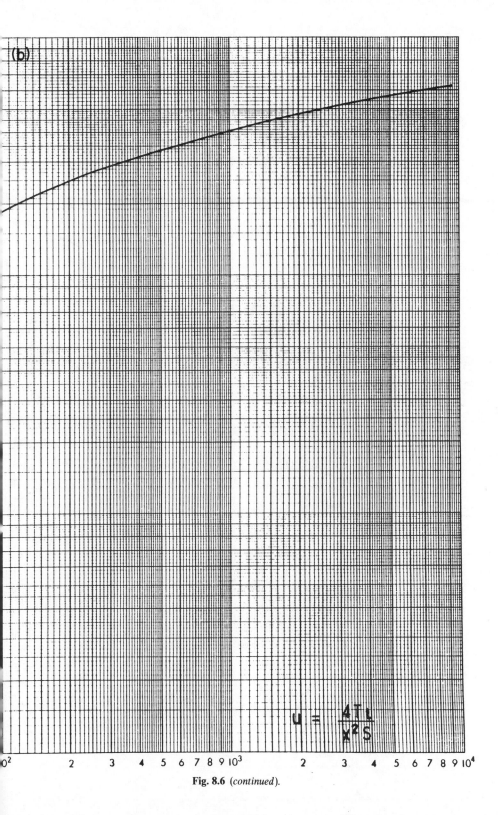

(b)

$$u = \frac{4Tt}{x^2 S}$$

10²      2      3      4    5   6  7  8  9 10³      2      3      4    5   6  7  8  9 10⁴

**Fig. 8.6** (*continued*).

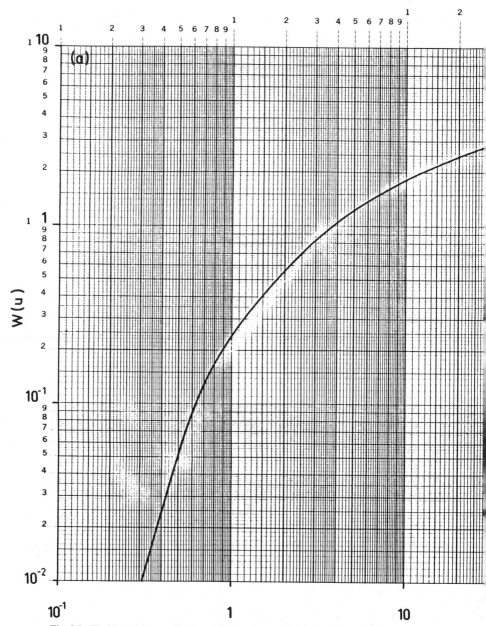

**Fig. 8.7.** Theis' type curve. 1.85 × 1.85 in. log module. By photocopying (without any magnification or reduction) both parts of this figure and pasting them together, it is possible to reconstruct the complete type curve.

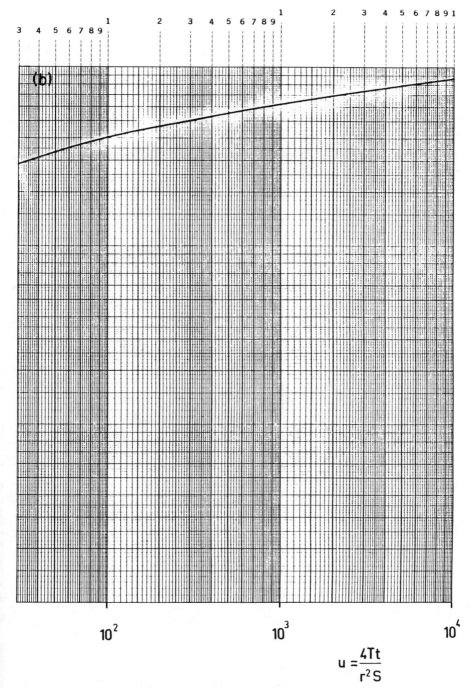

$$u = \frac{4Tt}{r^2 S}$$

**Fig. 8.7** (*continued*).

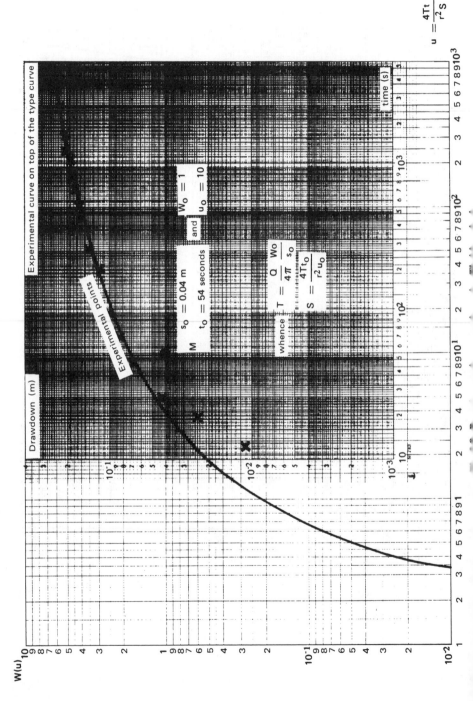

Experimental curve on top of the type curve

Drawdown (m)

Experimental points

$$M \quad s_o = 0.04 \text{ m} \qquad W_o = 1$$
$$t_o = 54 \text{ seconds} \qquad \text{and} \qquad u_o = 10$$

$$\text{whence} \quad T = \frac{Q}{4\pi} \frac{W_o}{s_o}$$

$$S = \frac{4Tt_o}{r^2 u_o}$$

$W(u)$

time (s)

$$u = \frac{4Tt}{r^2 S}$$

then chosen, not necessarily on one of the curves, and its coordinates are expressed according to the two systems:

$$M = \begin{cases} s_0 \\ t_0 \end{cases} \quad \text{and} \quad \begin{cases} W_0 \\ u_0 \end{cases}$$

By definition, we can write

$$s_0 = \frac{Q}{4\pi T} W_0 \qquad\qquad T = \frac{Q}{4\pi} \frac{W_0}{s_0}$$

$$\text{from which we get}$$

$$u_0 = \frac{4Tt_0}{Sr^2} \qquad\qquad S = \frac{4}{r^2} \frac{Tt_0}{u_0}$$

However, the influence of a boundary is more difficult to interpret in this system than in that of Jacob. The only advantages of this system are that it is not necessary to discard the first measurement points and that, for short-time tests, there is less ambiguity than when we look for a line on Jacob's graph.

Either method can give, at best, two significant figures for the parameters $T$ and $S$, never three.

A great number of computer (or hand calculator) codes have been written to adjust automatically the values of $T$ and $S$ in Theis' formula in order to match the measurements. These methods must be taken with caution. Very often, when the data are represented graphically, one realizes that only some of the measurements must be used, because there is some deviation from the hypotheses implied by the formula (e.g., influence of a boundary). A blind computer code would use all the measurements regardlessly and produce a meaningless "best fit." As the departure from the hypotheses may be due to a number of causes, it is still preferable to use graphical techniques or, at least, to check graphically the results of the computer codes.

(c) *Interpretation of recovery curves.* As we have seen in Section 8.1.3.c, there are two methods for interpreting a recovery curve:

(1) The Houpeurt–Pouchan method. Here it is assumed that the pumping has lasted long enough to allow us to suppose—at least at the beginning of the recovery—that a steady state has been attained before pumping stopped. Then, the recovery curve is interpreted as a drawdown curve with the help of either Jacob's or Theis's method.

(2) The Horner method. $\log(1 + t_p/t)$ method. Here $s$ is plotted versus $\log(1 + t_p/t)$ ($t_p$ is the duration of the pumping, $t$ the time counted from the cessation of pumping) on a semilog diagram. With the help of the straight line, which should then appear, and following Jacob's method (Section 8.2.a), we can calculate the transmissivity but not the storage coefficient.

Recovery curves are of particular importance in pumping tests where no piezometers are available and the only observation point in the aquifer is the boring itself. Indeed, during production the level in the borehole is disturbed by the losses in hydraulic head (quadratic terms), which occur as the fluid crosses the slots in the well screen and even in the first 10 or 20 cm around the well. This means that, during the pumping, the dynamic level in the well gives a poor representation of the hydraulic head in the aquifer around the well, whereas during recovery all these phenomena are cancelled, and the real level of the aquifer can be observed in the well, which makes an accurate interpretation possible.

Observe that the level in the well often fluctuates slightly because of irregularities in the running of the pump engine: the measurements in the well during production are not exact. Moreover, it must be noted that, at the boring, the radius $r$ of the well is ill-defined: of course, the radius of the borehole itself is known, as well as that of the casing, but the terrain around the boring has been disturbed during the work on the well. It is accepted that there is an effective well radius $r'$ surrounding it, which has to be taken into account in the interpretations of the level in the borehole and which is usually slightly larger than the real radius of the borehole. This is called a positive skin effect. A negative skin effect ($r' < r$) can sometimes be observed if a well is clogged or poorly developed.

(d)    *Anisotropic medium.*    The interpretation of a pumping test can be extended to the case where the medium is anisotropic in the horizontal plane (the case where the medium is anisotropic in the vertical/horizontal directions is examined in Sections 8.3.3 and 8.4.3). Let $X$ and $Y$ be the coordinate system in the horizontal plane, and $x$ and $y$ the principal directions of the anisotropy tensor of the transmissivity.

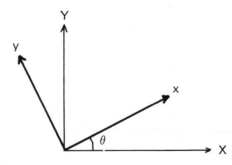

Let $\theta$ be the angle from $X$ to $x$ in the trigonometric rotation, and $T_x$ and. $T_y$ the principal components of the transmissivity **T**. As shown in Section 7.1.c, the anisotropic system can be transformed into an isotropic one by the change of coordinates

$$x' = \sqrt{T/T_x}\, x \qquad \text{and} \qquad y' = \sqrt{T/T_y}\, y \quad \text{with} \quad T = \sqrt{T_x T_y}$$

In the $(x', y')$ system, the diffusion equation is again isotropic and the Theis equation applies:

$$s = \frac{Q}{4\pi T} W\left(\frac{4Tt}{r'^2 S}\right) \qquad \text{with} \quad r'^2 = x'^2 + y'^2$$

To interpret a pumping test in an anisotropic system, let us first suppose that the principal directions $x$ and $y$ are known and that we have two piezometers a and b, in the $x$ and $y$ directions, respectively, from the well. We can write for piezometer a

$$r'^2 = x'^2 + 0 = (T/T_x)x^2$$

and for piezometer b

$$r'^2 = 0 + y'^2 = (T/T_y)y^2$$

We therefore have

$$s_a = \frac{Q}{4\pi T} W\left(\frac{4T_x t}{x^2 S}\right) \qquad \text{and} \qquad s_b = \frac{Q}{4\pi T} W\left(\frac{4T_y t}{y^2 S}\right)$$

Interpreting the drawdowns in a and b using Theis' curve-matching method will first give us directly the same $T$ for both a and b, and also $T_x/S$ for a, and $T_y/S$ for b (instead of only $S$ in the regular case). From these 3 values we can then determine $T_x$, $T_y$, and $S$. Now if the principal directions $x$ and $y$ are not known, at least 3 piezometers are needed. Let $X$ and $Y$ be the coordinates, in any system, of a piezometer; using the change of coordinates by the rotation $\theta$, we obtain

$$x = \quad X\cos\theta + Y\sin\theta$$
$$y = -X\sin\theta + Y\cos\theta$$

Then

$$r'^2 = T/T_x(X\cos\theta + Y\sin\theta)^2 + T/T_y(-X\sin\theta + Y\cos\theta)^2$$

The interpretation using Theis' curve matching method still gives us first $T = \sqrt{T_x T_y}$ for all piezometers, but then the second parameter obtained for each piezometer is a rather complex expression of $T_x$, $T_y$, $S$, and $\theta$. One can solve it by trial and error, graphically, or mathematically. See Hantush (1966), Neuman et al. (1984).

## 8.3. Leakage in Radial Coordinate Systems

We have defined leakage in Section 5.3.g as the flux $F_t$ and $F_b$ exchanged at the upper and lower boundaries of a confined aquifer with its confining beds. We shall study three analytical solutions of this problem for a well pumping in an aquifer, where at least one of the confining layers is an aquitard through which leakage occurs. These solutions are by Hantush, Boulton, and Streltsova.

### 8.3.1. Hantush's Solution

Hantush (1956) assumed that the confined aquifer is recharged from an overlying unconfined aquifer, which percolates through the aquitard separating them, as in Fig. 8.9.

The leakage flux $F_t^0$, in the steady state, is given according to Darcy's law by the hydraulic head gradient in the aquitard between the two aquifers:

$$F_t^0 = -K' \frac{h_2^0 - h_1^0}{e'} \tag{8.3.1.1}$$

where $K'$ is the hydraulic conductivity of the aquitard, $e'$ the thickness of the aquitard, $h_2$ the hydraulic head in the confined aquifer, $h_1$ the hydraulic head in the unconfined aquifer, and the superscript 0 means steady state.

Hantush examined the reaction of such a system, when pumping at a constant rate is started in the confined aquifer. He then made two assumptions:

(1) The hydraulic head $h_1$ in the unconfined aquifer is not going to change even if the leakage flux $F_t$ increases. This is true if the unconfined aquifer is well recharged (e.g., by rainfall) or if the pumping does not last too long.

(2) The increase in the leakage flux is assumed to take place instantly and to be always given by Darcy's law. If the drawdown in the confined aquifer is denoted by $s$, then

$$F_t = -K' \frac{(h_2^0 - s) - h_1^0}{e'} \tag{8.3.1.2}$$

This disregards the existence of a transient state in the aquitard (see Section 8.5). Thus, the leakage flux is given by

$$F_t = F_t^0 + \frac{K'}{e'} s \tag{8.3.1.3}$$

If the initial steady state $h_2^0$ satisfies the equation $\nabla^2 h_2^0 = -F_t^0/T$, the drawdown $s$ then satisfies the following diffusion equation:

$$\nabla^2 s = \frac{S}{T} \frac{\partial s}{\partial t} + \frac{K'}{Te'} s \tag{8.3.1.4}$$

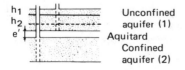

| $h_1$ | | Unconfined |
| $h_2$ | | aquifer (1) |
| $e'$ | | Aquitard |
| | | Confined |
| | | aquifer (2) |

**Fig. 8.9.** Leaky system.

## Table 8.2

Values of $W'(4Tt/r^2S, r/B)$[a]

| $\dfrac{r^2S}{4Tt}$ \ $r/B$ | 0.01 | 0.015 | 0.03 | 0.05 | 0.075 | 0.10 | 0.15 | 0.2 | 0.3 | 0.4 | 0.5 | 0.6 | 0.7 | 0.8 | 0.9 | 1.0 | 1.5 | 2.0 | 2.5 |
|---|---|---|---|---|---|---|---|---|---|---|---|---|---|---|---|---|---|---|---|
| 0.000001 | | | | | | | | | | | | | | | | | | | |
| 0.000005 | 9.4413 | | | | | | | | | | | | | | | | | | |
| 0.00001 | 9.4176 | 8.6313 | | | | | | | | | | | | | | | | | |
| 0.00005 | 8.8827 | 8.4533 | 7.2450 | | | | | | | | | | | | | | | | |
| 0.0001 | 8.3983 | 8.1414 | 7.2122 | 6.2282 | 5.4228 | | | | | | | | | | | | | | |
| 0.0005 | 6.9750 | 6.9152 | 6.6219 | 6.0821 | 5.4062 | 4.8530 | | | | | | | | | | | | | |
| 0.001 | 6.3069 | 6.2765 | 6.1202 | 5.7965 | 5.3078 | 4.8292 | 4.0595 | 3.5054 | | | | | | | | | | | |
| 0.005 | 4.7212 | 4.7152 | 4.6829 | 4.6084 | 4.4713 | 4.2960 | 3.8821 | 3.4567 | 2.7428 | 2.2290 | | | | | | | | | |
| 0.01 | 4.0356 | 4.0326 | 4.0167 | 3.9795 | 3.9091 | 3.8150 | 3.5725 | 3.2875 | 2.7104 | 2.2253 | 1.8486 | 1.5550 | 1.3210 | 1.1307 | | | | | |
| 0.05 | 2.4675 | 2.4670 | 2.4642 | 2.4576 | 2.4448 | 2.4271 | 2.3776 | 2.3110 | 2.1371 | 1.9283 | 1.7075 | 1.4927 | 1.2955 | 1.1210 | 0.9700 | 0.8409 | | | |
| 0.1 | 1.8227 | 1.8225 | 1.8213 | 1.8184 | 1.8128 | 1.8050 | 1.7829 | 1.7527 | 1.6704 | 1.5644 | 1.4422 | 1.3115 | 1.1791 | 1.0505 | 0.9297 | 0.8190 | 0.4271 | 0.2278 | |
| 0.5 | 0.5598 | 0.5597 | 0.5596 | 0.5594 | 0.5588 | 0.5581 | 0.5561 | 0.5532 | 0.5453 | 0.5344 | 0.5206 | 0.5044 | 0.4860 | 0.4658 | 0.4440 | 0.4210 | 0.3007 | 0.1944 | 0.1174 |
| 1.0 | 0.2194 | 0.2194 | 0.2193 | 0.2193 | 0.2191 | 0.2190 | 0.2186 | 0.2179 | 0.2161 | 0.2135 | 0.2103 | 0.2065 | 0.2020 | 0.1970 | 0.1914 | 0.1855 | 0.1509 | 0.1139 | 0.0803 |
| 5.0 | 0.0011 | 0.0011 | 0.0011 | 0.0011 | 0.0011 | 0.0011 | 0.0011 | 0.0011 | 0.0011 | 0.0011 | 0.0011 | 0.0011 | 0.0011 | 0.0011 | 0.0011 | 0.0011 | 0.0010 | 0.0010 | 0.0009 |

[a] After Hantush (1956) and Walton (1970).

We define the Hantush leakage factor $B = \sqrt{Te'/K'}$ of dimension (length). The Hantush radial solution of this equation then becomes

$$s = \frac{Q}{4\pi T} \int_{r^2S/4Tt}^{\infty} \frac{\exp(-\tau - r^2/4B^2\tau)}{\tau}\, d\tau = \frac{Q}{4\pi T} W'\left(\frac{4Tt}{r^2S}, \frac{r}{B}\right)$$

This solution depends on two parameters ($u = 4Tt/r^2S$ and $r/B$) and takes the following form:

(1) The envelope curve is that of Theis (corresponding to negligible $r/B$);

(2) For a given value of $r/B$ (i.e., for a given hydraulic conductivity $K'$ of the aquitard and a distance $r$ to the pumping well), the response curve stabilizes with time: a steady state is reached. Table 8.2 gives the function $W'$ from Walton (1970). This function is drawn as a type curve on a log–log paper in Figs. 8.10 and 8.11.*

This explains why, in certain cases, we obtain a stabilization in a pumping test that is due to a leakage phenomenon but that may be wrongly interpreted as a steady state because of the existence of a fictitious "radius of action" $R$ around the borehole (see Section 7.3.a, Dupuit's formula).

This stabilization of the drawdown in a piezometer close to the borehole after a certain lapse of time occurs at the same time in all the borings at the same distance from the well. This would not be true for a stabilization due to a recharge boundary (stream), where the piezometers closest to the fictitious image well would be the first to stabilize. To identify this type of leakage, the Hantush type curve must be used (Fig. 8.10 or 8.11) and the procedure is the same as for the Theis curve in Section 8.2.b, but the best fit using the Hantush type curve has to be found in order to achieve a satisfactory matching, which also gives $r/B$.

---

* See the footnote on p. 171. The same comments apply for Figs. 8.10 and 8.11 as for 8.6 and 8.7, respectively.

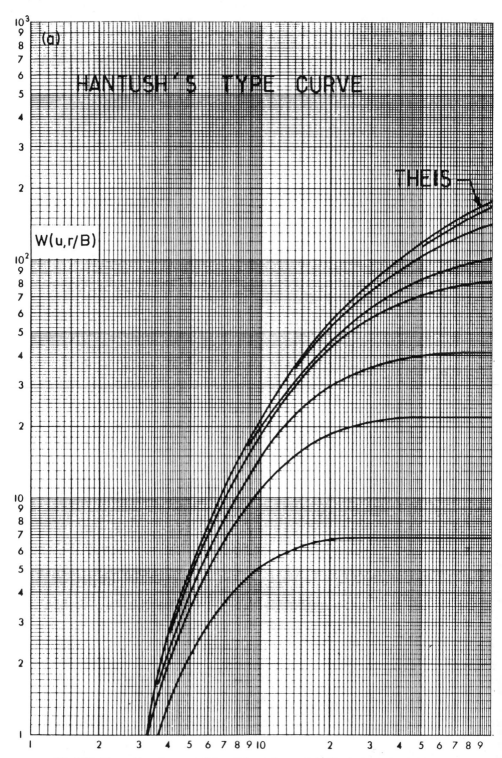

**Fig. 8.10.** Hantush's type curve. $62.5 \times 62.5$ mm log module. By photocopying (without any magnification or reduction) both parts of this figure, and pasting them together, it is possible to reconstruct the complete type curve.

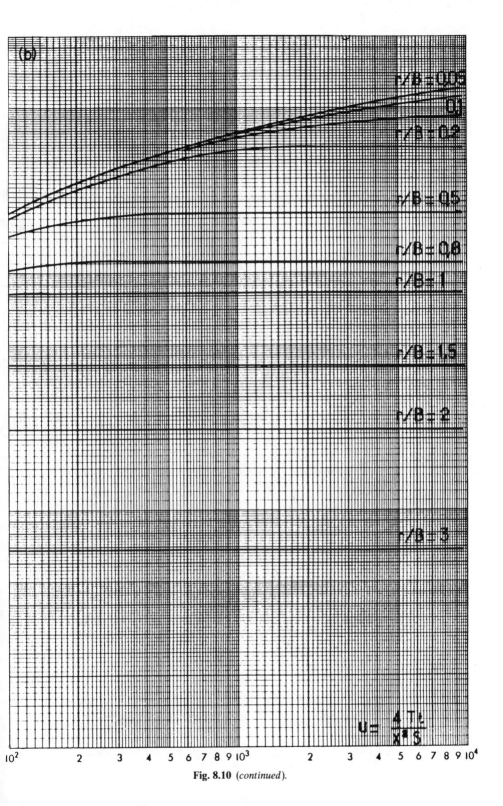

**Fig. 8.10** (*continued*).

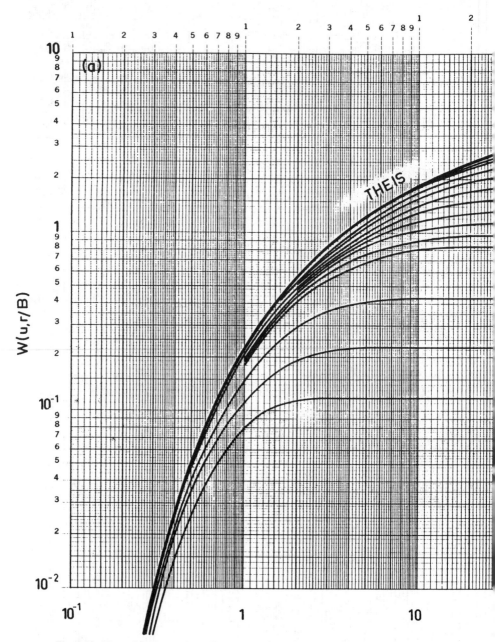

**Fig. 8.11.** Hantush's type curve. 1.85 × 1.85 in. log module. By photocopying (without any magnification or reduction) both parts of this figure, and pasting them together, it is possible to reconstruct the complete type curve.

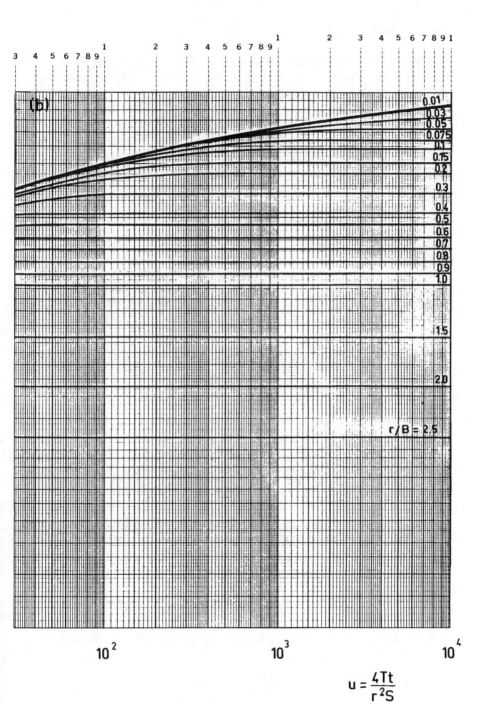

$$u = \frac{4Tt}{r^2 S}$$

**Fig. 8.11** (*continued*).

185

Neuman and Witherspoon (1968, 1969a,b, 1972) have shown, however, that this solution, which disregards the storage in the confining beds, may sometimes lead to considerable errors. They suggest other methods of interpretation, which take this storage into account as well as the variations of hydraulic head in the overlying aquifer (see also Section 8.3.3).

We must remember that the identification of a leakage phenomenon during a pumping test does not in the least affect the direction of the exchanges: the leakage may stem from an overlying or an underlying aquifer and may be, in a steady state (before the pumping has begun), a recharge of or a withdrawal from the studied aquifer; the flow $F_t^0$ of Eq. (8.3.1.1) is algebraic and Eqs. (8.3.1.1)–(8.3.1.4) are valid whatever its sign may be.

### 8.3.2. Boulton's Solution

Boulton (1963) made another assumption concerning the leakage flux caused by the drawdown $s$: he assumed semiempirically that an increase in the drawdown $\Delta s$ at time $t$ gives rise to a leakage flux $\Delta q$, which decreases exponentially with time:

$$\Delta q(\tau) = S'f \exp[-f(\tau - t)] \Delta s$$

where $f$ is a parameter of dimension (time$^{-1}$).

The integral of this flux between $t$ and infinity is

$$q = \int_t^\infty S'f \exp[-f(\tau - t)] \Delta s \, d\tau$$

$$q = S' \Delta s$$

The term $S'$ is the storage coefficient of the overlying (or underlying) aquifer, which recharges the confined aquifer through leakage, since a drawdown $\Delta s$ causes accumulated flux $S' \Delta s$. However, this flow is not instantly released: the suggested solution corresponds to an exponential decay of the leakage flux.

The diffusion equation is obtained by calculating the leakage flux $F_t$ at every instant by convolution, i.e., by adding the elementary fluxes produced by all the drawdowns from the beginning of the pumping:

$$\nabla^2 s = \frac{S}{T} \frac{\partial s}{\partial t} + \frac{S'}{T} \int_0^t f \exp[-f(t - \tau)] \left(\frac{\partial s}{\partial t}\right)_\tau d\tau$$

Boulton gives a radial solution to this equation, which takes the following form for small $r$:

$$s = \frac{Q}{4\pi T} W''(u, S', f)$$

where $u = 4Tt/r^2 S$. Figure 8.12 illustrates this solution.

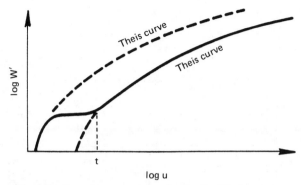

**Fig. 8.12.** Drawdown versus time for Boulton's leaky systems.

The evolution of the drawdown conforms initially to Theis's solution, which corresponds to the parameter couple $(T, S)$. Then comes a level stage at which it might be possible to identify $f$, and, finally, the drawdown again takes the form of a Theis function but this time displaced in relation to the first by a shift parallel to the $u$ axis (no vertical parallel shift) and corresponding to the parameters $(T, S + S')$. This type of leakage is therefore easy to recognize and identify with the help of a Theis type curve, and we can then calculate $S'$.

If $t$ is the time when the leveling off of the withdrawal intercepts the second Theis curve (see Fig. 8.12), Berkaloff (1966) shows that

$$f = \frac{0.561}{t}$$

This type of behavior is rather frequent in unconfined aquifers, where the delayed flow is simply due to the draining of the unsaturated medium when the free surface is drawn down (see Sections 6.2.a and 8.4.3).

### 8.3.3. Streltsova's Solution

A more elaborate solution of delayed leakage has been proposed by Boulton and Streltsova (1975) and Streltsova (1976b). These authors consider a confined aquifer with a producing well overlain by an aquitard containing a free surface. The following assumptions are made:

(1)  The aquitard containing the water table is homogeneous and the flow through it is only vertical downwards. Both the water and the aquitard itself are supposed incompressible: the production of water in the aquitard is only by drainage and lowering of the water table. However, this lowering of the water table is small enough so that the saturated thickness of the aquitard is

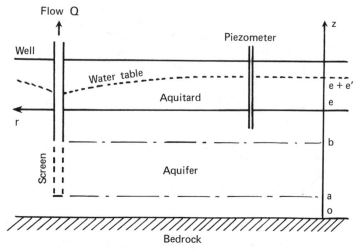

**Fig. 8.13.** Aquifer and aquitard in Streltsova's solution.

taken as constant and the position of the water table is considered as fixed (see Fig. 8.13). There is no vertical recharge in the aquitard.

(2) The confined aquifer is compressible and in general anisotropic, the horizontal and vertical hydraulic conductivities being constant. It is underlain by a horizontal impermeable bed. In this aquifer a production well is pumped at a constant rate from the instant $t = 0$. This well is only screened over a portion of the thickness of the aquifer (partially penetrating well). The discharge per unit length of the unlined part is constant, and the radius of the well is vanishingly small.

The equations and boundary conditions used are, (see Fig. 8.13) in the aquifer,

$$K_r\left(\frac{\partial^2 s}{\partial r^2} + \frac{1}{r}\frac{\partial s}{\partial r}\right) + K_z\frac{\partial^2 s}{\partial z^2} = S_s\frac{\partial s}{\partial t}, \qquad 0 < z < e$$

drawdown at the water table (see Section 6.3.d)

$$K'\frac{\partial s'}{\partial z} = -\omega'_d\frac{\partial s'}{\partial t}, \qquad z = e + e'$$

at the interface between aquitard and aquifer

$$K_z\frac{\partial s}{\partial z} = K'\frac{\partial s'}{\partial z}, \qquad z = e, \quad \forall r, t$$

at the impervious bottom

$$\frac{\partial s}{\partial z} = 0, \qquad z = 0, \quad \forall r, t$$

along the well

$$\lim_{r \to 0} r \frac{\partial s}{\partial r} = \frac{Q}{2\pi K_r(b-a)}, \qquad a < z < b, \quad \forall t$$

along the nonscreened section of the axis of the well

$$\frac{\partial s}{\partial r} = 0, \qquad 0 < z < a \quad \text{and} \quad b < z < e, \quad r = 0, \forall t$$

at infinity

$$s = 0, \qquad r \to \infty, \quad \forall z, t$$

initial condition

$$s = 0, \qquad t = 0, \qquad \forall r, z$$

with $s, s'$ the drawdown in aquifer and aquitard, respectively; $K_r, K_z$ the horizontal and vertical hydraulic conductivity in the aquifer; $S_s^-$ the specific storage coefficient in the aquifer; where $S = S_s e$ is the storage coefficient; $K'$, $\omega_d'$ the vertical hydraulic conductivity and specific yield of the aquitard; $e, e'$ the thickness of aquifer and aquitard, respectively; $a, b$ the position of the well screen (see Fig. 8.13); and $Q$ the flow rate of the well.

The drawdown in the aquifer is given by

$$s(r, z, t) = \frac{Q}{4\pi T} \int_0^\infty 4y J_0(y\beta^{1/2})\left[u_0(y) + \sum_1^\infty u_n(y)\right] dy$$

with the transmissivity of the aquifer given by

$$T = K_r e$$

and

$$\beta = (K_z/K_r)(r^2/e^2)$$

$$\sigma = \frac{S}{\omega_d'}$$

$$C = K'e/K_z e'$$

and

$$u_0(y) = [\{1 - \exp[-t_s\beta(y^2 - \gamma_0^2)]\}\{\sinh(\gamma_0 b/e) - \sinh(\gamma_0 a/e)\}$$
$$\times \cosh(\gamma_0 z/e)][\{[(b-a)/e](y^2 - \gamma_0^2)x_0\cosh(\gamma_0)\}^{-1}$$

where $\gamma_0$ is the positive root of the equation

$$(y^2 - \gamma_0^2 - C\sigma)\gamma_0 \sinh(\beta_0) + C(y^2 - \gamma_0^2)\cosh(\beta_0) = 0$$

and

$$\chi_0 = \left[1 + \frac{C(y^2 - \gamma_0^2)}{y^2 - \gamma_0^2 - C\sigma}\right]\sinh(\gamma_0) + \left[1 + \frac{2C^2\sigma}{(y^2 - \gamma_0^2 - C\sigma)^2}\right]\gamma_0\cosh(\gamma_0)$$

$$u_n(y) = \left[\{1 - \exp[-t_s\beta(y^2 + \gamma_n^2)]\}\{\sinh)(\gamma_n b/e) - \sinh(\gamma_n a/e)\}\right.$$
$$\times \cosh(\gamma_n z/e)]\{[(b - a)/e](y^2 + \gamma_n^2)x_n\cosh(\gamma_n)\}^{-1}$$

where $\gamma_n$ is the $n$th positive root of:

$$(y^2 + \gamma_n^2 - C\sigma)\gamma_n\sin(\gamma_n) - C(y^2 + \gamma_n^2)\cos(\gamma_n) = 0$$

and

$$\chi_n = \left[1 + \frac{C(y^2 + \gamma_n^2)}{y^2 + \gamma_n^2 - C\sigma}\right]\sin(\gamma_n) + \left[1 + \frac{2C^2\sigma}{(y^2 + \gamma_n^2 - C\sigma)^2}\right]\gamma_n\cos(\gamma_n)$$

This solution can be computed numerically although it is not very easy when accurate results are required. $J_0$ is the Bessel function of the first kind and zero order.

The drawdown $s$ is calculated at elevation $z$ in the aquifer, i.e., for a piezometer open only over a short distance at elevation $z$. If a fully screened piezometer is used, the average of $s$ from $z = 0$ to $e$ must be calculated.

This solution is very close to that developed by Neuman for delayed yield in an unconfined aquifer (see Section 8.4.3). Additional development can be found in Streltsova (1984).

## 8.4. Additional Analytical Solutions for the Flow toward a Well

The interpretation of pumping tests is a science in itself. This kind of test is very useful in hydrogeology, because it is one of the most widely used means of measuring *in situ* the values of the parameters $T$ and $S$. Whole books are devoted to the subject: see, for example Kruseman and de Ridder (1970).

In the following, some examples of particular importance will be given.

### 8.4.1. *Effect of the Well Capacity*

At the beginning of pumping, if the production rate is $Q$, the flow withdrawn from the formation is not $Q$, since the well is starting to empty. Papadopoulos and Cooper (1967) have given the following solution to this problem for confined aquifers:

$$s_p = \frac{Q}{4\pi T}F(u_p, \alpha)$$

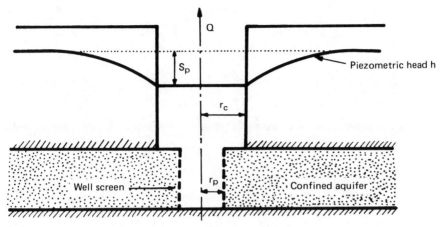

**Fig. 8.14.** Well capacity effect.

with

$$F(u_p, \alpha) = \frac{32\alpha^2}{\pi^2} \int_0^\infty \frac{1 - \exp(-\beta^2/4u_p)}{\beta^3 \, \Delta(\beta)} \, d\beta$$

where $s_p$ is the drawdown at the well, $u_p = r_p^2 S/4Tt$, $r_p$ is the well radius at the well screen, $\alpha = r_p^2 S/r_c^2$, $r_c$ is the well radius at the casing, $Q$ the production rate, $S$ the storage coefficient, $T$ the transmissivity, and

$$\Delta(\beta) = [\beta J_0(\beta) - 2\alpha J_1(\beta)]^2 + [\beta Y_0(\beta) - 2\alpha Y_1(\beta)]^2$$

where $J_n$ is the Bessel function of the first kind and order $n$ and $Y_n$ the Bessel function of the second kind and order $n$.

The difference between $r_p$ and $r_c$ is illustrated in Fig. 8.14. This solution is particularly useful for pumping tests in dug wells with large diameters, frequently found in developing countries. The function $F(u_p, \alpha)$ is given by the Table 8.3 and the type curve of Fig. 8.15, taken from Papadopoulos and Cooper. Note that the horizontal axis is graduated in $u_p$, not in $1/u_p$. Finally, the expression for the drawdown in a piezometer at some distance from the well is given by Carslaw and Jaeger (1959).

### 8.4.2. Artesian Tests

In an artesian boring, when the well is opened the water flows out naturally at a rate that decreases with time. Instead of a constant flow rate, a constant drawdown is imposed ($h = z$ at the well head). Jacob and Lohman (1952) have given the expression for artesian flow versus time:

$$Q = 2\pi T(h_0 - h)G(\alpha)$$

### Table 8.3

Values of the Well Capacity Function $F(u_p, \alpha)^a$

| $u_p$ \ $\alpha$ | $10^{-1}$ | $10^{-2}$ | $10^{-3}$ | $10^{-4}$ | $10^{-5}$ |
|---|---|---|---|---|---|
| 10 | $9.755 \times 10^{-3}$ | $9.976 \times 10^{-4}$ | $9.998 \times 10^{-5}$ | $1.000 \times 10^{-5}$ | $1.000 \times 10^{-6}$ |
| 1 | $9.192 \times 10^{-2}$ | $9.914 \times 10^{-3}$ | $9.991 \times 10^{-4}$ | $1.000 \times 10^{-4}$ | $1.000 \times 10^{-5}$ |
| $5 \times 10^{-1}$ | $1.767 \times 10^{-1}$ | $1.974 \times 10^{-2}$ | $1.997 \times 10^{-3}$ | 2.000 | 2.000 |
| 2 | 4.062 | 4.890 | 4.989 | 4.999 | 5.000 |
| 1 | 7.336 | 9.665 | 9.966 | 9.997 | $1.000 \times 10^{-4}$ |
| $5 \times 10^{-2}$ | $1.260 \times 10^{0}$ | $1.896 \times 10^{-1}$ | $1.989 \times 10^{-2}$ | $1.999 \times 10^{-3}$ | 2.000 |
| 2 | 2.303 | 4.529 | 4.949 | 4.995 | 5.000 |
| 1 | 3.276 | 8.520 | 9.834 | 9.984 | $1.000 \times 10^{-3}$ |
| $5 \times 10^{-3}$ | 4.255 | $1.540 \times 10^{0}$ | $1.945 \times 10^{-1}$ | $1.994 \times 10^{-2}$ | 2.000 |
| 2 | 5.420 | 3.043 | 4.725 | 4.972 | 4.998 |
| 1 | 6.212 | 4.545 | 9.069 | 9.901 | 9.992 |
| $5 \times 10^{-4}$ | 6.960 | 6.031 | $1.688 \times 10^{0}$ | $1.965 \times 10^{-1}$ | $1.997 \times 10^{-2}$ |
| 2 | 7.886 | 7.557 | 3.523 | 4.814 | 4.982 |
| 1 | 8.572 | 8.443 | 5.526 | 9.340 | 9.932 |
| $5 \times 10^{-5}$ | 9.318 | 9.229 | 7.631 | $1.768 \times 10^{0}$ | $1.975 \times 10^{-1}$ |
| 2 | $1.024 \times 10^{1}$ | $1.020 \times 10^{1}$ | 9.676 | 3.828 | 4.861 |
| 1 | 1.093 | 1.087 | $1.068 \times 10^{1}$ | 6.245 | 9.493 |
| $5 \times 10^{-6}$ | 1.163 | 1.162 | 1.150 | 8.991 | $1.817 \times 10^{0}$ |
| 2 | 1.255 | 1.254 | 1.249 | $1.174 \times 10^{1}$ | 4.033 |
| 1 | 1.324 | 1.324 | 1.321 | 1.291 | 6.779 |
| $5 \times 10^{-7}$ | 1.393 | 1.393 | 1.392 | 1.378 | $1.013 \times 10^{1}$ |
| 2 | 1.485 | 1.485 | 1.484 | 1.479 | 1.371 |
| 1 | 1.554 | 1.554 | 1.554 | 1.551 | 1.513 |
| $5 \times 10^{-8}$ | 1.623 | 1.623 | 1.623 | 1.622 | 1.605 |
| 2 | 1.705 | 1.705 | 1.705 | 1.714 | 1.708 |
| 1 | 1.784 | 1.784 | 1.784 | 1.784 | 1.781 |
| $5 \times 10^{-9}$ | 1.854 | 1.854 | 1.854 | 1.854 | 1.851 |
| 2 | 1.945 | 1.945 | 1.945 | 1.945 | 1.940 |
| 1 | 2.015 | 2.015 | 2.015 | 2.015 | 2.015 |

[a] From Papadopoulos and Cooper (1967).

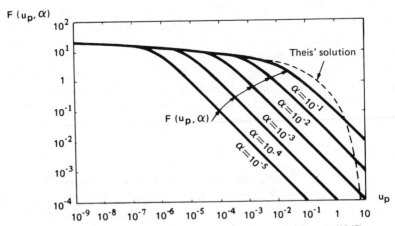

**Fig. 8.15.** Well capacity function. From Papadopoulos and Cooper (1967).

where $Q$ is the artesian flow rate, $T$ is the transmissivity, and $h_0 - h$ is the imposed drawdown (hydraulic head in the aquifer before the test minus the head imposed at the opening by the elevation of the top of the well). If the boring is very deep, losses of hydraulic head in the casing have to be taken into account, as the hydraulic head $h$ required is the one at the level of the formation

$$\alpha = \frac{Tt}{r_p^2 S}$$

where $S$ is the storage coefficient and $r_p$ the borehole radius, and

$$G(\alpha) = \frac{4\alpha}{\pi} \int_0^\infty x e^{-\alpha x^2} \left\{ \frac{\pi}{2} + \text{arctg}[Y_0(x)/J_0(x)] \right\} dx$$

Here $J_0$ and $Y_0$ are the zero-order Bessel functions of the first and second kinds, respectively.

The function $G$ and the corresponding type curve are given in Table 8.4 and Fig. 8.16 respectively.

**Table 8.4**

Well Function for Artesian Conditions[a]

| $\alpha$ | $10^{-4}$ | $10^{-3}$ | $10^{-2}$ | $10^{-1}$ | 1 | 10 | $10^2$ | $10^3$ |
|---|---|---|---|---|---|---|---|---|
| 1 | 56.9 | 18.34 | 6.13 | 2.249 | 0.985 | 0.534 | 0.346 | 0.251 |
| 2 | 40.4 | 13.11 | 4.47 | 1.716 | 0.803 | 0.461 | 0.311 | 0.232 |
| 3 | 33.1 | 10.79 | 3.74 | 1.477 | 0.719 | 0.427 | 0.294 | 0.222 |
| 4 | 28.7 | 9.41 | 3.30 | 1.333 | 0.667 | 0.405 | 0.283 | 0.215 |
| 5 | 25.7 | 8.47 | 3.00 | 1.234 | 0.630 | 0.389 | 0.274 | 0.210 |
| 6 | 23.5 | 7.77 | 2.78 | 1.160 | 0.602 | 0.377 | 0.268 | 0.206 |
| 7 | 21.8 | 7.23 | 2.60 | 1.103 | 0.580 | 0.367 | 0.263 | 0.203 |
| 8 | 20.4 | 6.79 | 2.46 | 1.057 | 0.562 | 0.359 | 0.258 | 0.200 |
| 9 | 19.3 | 6.43 | 2.35 | 1.018 | 0.547 | 0.352 | 0.254 | 0.198 |
| 10 | 18.3 | 6.13 | 2.25 | 0.985 | 0.534 | 0.346 | 0.251 | 0.196 |

| $\alpha$ | $10^4$ | $10^5$ | 10 | $10^7$ | $10^8$ | $10^9$ | $10^{10}$ | $10^{11}$ |
|---|---|---|---|---|---|---|---|---|
| 1 | 0.1964 | 0.1608 | 0.1360 | 0.1177 | 0.1037 | 0.0927 | 0.0838 | 0.0764 |
| 2 | 0.1841 | 0.1524 | 0.1299 | 0.1131 | 0.1002 | 0.0899 | 0.0814 | 0.0744 |
| 3 | 0.1777 | 0.1479 | 0.1266 | 0.1106 | 0.0982 | 0.0883 | 0.0801 | 0.0733 |
| 4 | 0.1733 | 0.1449 | 0.1244 | 0.1089 | 0.0968 | 0.0872 | 0.0792 | 0.0726 |
| 5 | 0.1701 | 0.1426 | 0.1227 | 0.1076 | 0.0958 | 0.0864 | 0.0785 | 0.0720 |
| 6 | 0.1675 | 0.1408 | 0.1213 | 0.1066 | 0.0950 | 0.0857 | 0.0779 | 0.0716 |
| 7 | 0.1654 | 0.1393 | 0.1202 | 0.1057 | 0.0943 | 0.0851 | 0.0774 | 0.0712 |
| 8 | 0.1636 | 0.1380 | 0.1192 | 0.1049 | 0.0937 | 0.0846 | 0.0770 | 0.0709 |
| 9 | 0.1621 | 0.1369 | 0.1184 | 0.1043 | 0.0932 | 0.0842 | 0.0767 | 0.0706 |
| 10 | 0.1608 | 0.1360 | 0.1177 | 0.1037 | 0.0927 | 0.0838 | 0.0764 | 0.0704 |

[a] From Jacob and Lohman (1952).

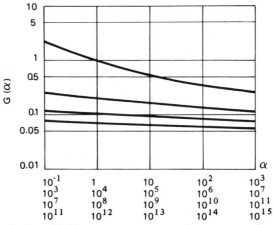

**Fig. 8.16.** Well function for artesian conditions. $\alpha = Tt/r_p^2 S$.

### 8.4.3. Anisotropic Unconfined Aquifer

Neuman (1972, 1973b, 1974, 1975a,b) has studied the problem of a fully or partially penetrating well pumping in an anisotropic unconfined aquifer, taking into account the delayed drainage of the unsaturated zone by gravity. The anisotropy is understood to be that of the vertical/horizontal hydraulic conductivity. His solution has a very similar expression to that of Streltsova (Section 8.3.3): he only assumes that the free surface remains always at $z = e$ (see Fig. 8.13) and that the boundary condition at this surface is

$$K_z \, \partial s/\partial z = -\omega_d \, \partial s/\partial t, \qquad z = e, \quad \forall r, t$$

Otherwise he uses all the equations and boundary conditions given for the aquifer in Section 8.3.3. Note however that his solution was published prior to that of Streltsova.

If the well is fully penetrating and screened along its entire length, the drawdown in a piezometer, which is also entirely screened, is given by

$$s(r, t) = \frac{Q}{4\pi T} \int_0^\infty 4y J_0(y\beta^{1/2})[u_0(y) + \sum_{n=1}^\infty u_n(y)] \, dy$$

$$u_0(y) = \frac{\{1 - \exp[-t_s\beta(y^2 - \gamma_0^2)]\} \tanh(\gamma_0)}{y^2 + (1 + \sigma)\gamma_0^2 - (y^2 - \gamma_0^2)^2/\sigma\}\gamma_0}$$

$$u_n(y) = \frac{\{1 - \exp[-t_s\beta(y^2 + \gamma_n^2)]\} \tan(\gamma_n)}{[y^2 - (1 + \sigma)\gamma_n^2 - (y^2 + \gamma_n^2)^2/\sigma]\gamma_n}$$

where $\gamma_0$ and $\gamma_n$ are the roots of

$$\sigma\gamma_0 \sinh(\gamma_0) - (y^2 - \gamma_0^2)\cosh(\gamma_0) = 0 \qquad \gamma_0^2 < y^2$$

$$\sigma\gamma_n \sin(\gamma_n) + (y^2 + \gamma_n^2)\cos(\gamma_n) = 0$$

with $\qquad\qquad (2n-1)(\pi/2) < \gamma_n < n\pi \qquad n \geq 1$

where $r$ is the distance from the piezometer to the well, $Q$ the flow rate of the well (constant), $T$ the transmissivity, $J_0$ the Bessel function of the first kind and zero order, $t_s = Tt/Sr^2$, the dimensionless "elastic" time, and $S$ the storage coefficient of the formation (indeed, in the same way as in a confined aquifer, owing to the elasticity, the pressures are transmitted up to the free surface, where the drainage comes into play. Hence the notion of delayed drainage), $t_y = Tt/\omega_d r^2$, the dimensionless "drainage" time, $\omega_d$ is the specific yield of the formation, $\sigma = S/\omega_d = y/t_s$, $\beta = (K_z/K_r)(r^2/e^2)$, $K_z$ and $K_r$ are the anisotropic hydraulic conductivity in the directions $z$ and $r$, and $e$ is the initial saturated thickness of the aquifer, assumed constant through time.

This function is given in Table 8.5 and Fig. 8.17 (Neuman, 1975a). The curves are drawn for $\sigma$ close to zero; thus we obtain two families of curves (type A and type B), which are united by a horizontal line. The length of this horizontal stretch is directly dependent on the value of $\sigma$. To avoid introducing this parameter into the type curves, the curves A are shown as functions of dimensionless time $t_s$ (upper scale) and the curves B as functions of $t_y$ (lower scale).

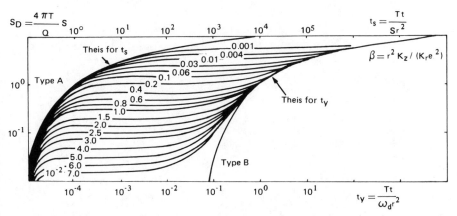

**Fig. 8.17** Neuman's well function for unconfined aquifers. [From Neuman (1975a).]

**Table 8.5**

Neuman Function[a]

Type A curve

| $t_s$ | $\beta=0.001$ | $\beta=0.004$ | $\beta=0.01$ | $\beta=0.03$ | $\beta=0.06$ | $\beta=0.1$ | $\beta=0.2$ | $\beta=0.4$ | $\beta=0.6$ | $\beta=0.8$ | $\beta=1.0$ | $\beta=1.5$ | $\beta=2.0$ | $\beta=2.5$ | $\beta=3.0$ | $\beta=4.0$ | $\beta=5.0$ | $\beta=6.0$ | $\beta=7.0$ |
|---|---|---|---|---|---|---|---|---|---|---|---|---|---|---|---|---|---|---|---|
| $1\times10^{-1}$ | $2.48\times10^{-2}$ | $2.43\times10^{-2}$ | $2.41\times10^{-2}$ | $2.35\times10^{-2}$ | $2.30\times10^{-2}$ | $2.24\times10^{-2}$ | $2.14\times10^{-2}$ | $1.99\times10^{-2}$ | $1.88\times10^{-2}$ | $1.79\times10^{-2}$ | $1.70\times10^{-2}$ | $1.53\times10^{-2}$ | $1.38\times10^{-2}$ | $1.25\times10^{-2}$ | $1.13\times10^{-2}$ | $9.33\times10^{-3}$ | $7.72\times10^{-3}$ | $6.39\times10^{-3}$ | $5.30\times10^{-3}$ |
| $2\times10^{-1}$ | $1.45\times10^{-1}$ | $1.42\times10^{-1}$ | $1.40\times10^{-1}$ | $1.36\times10^{-1}$ | $1.31\times10^{-1}$ | $1.27\times10^{-1}$ | $1.19\times10^{-1}$ | $1.08\times10^{-1}$ | $9.88\times10^{-2}$ | $9.15\times10^{-2}$ | $8.49\times10^{-2}$ | $7.13\times10^{-2}$ | $6.03\times10^{-2}$ | $5.11\times10^{-2}$ | $4.35\times10^{-2}$ | $3.17\times10^{-2}$ | $2.34\times10^{-2}$ | $1.74\times10^{-2}$ | $1.31\times10^{-2}$ |
| $3.5\times10^{-1}$ | $3.58\times10^{-1}$ | $3.52\times10^{-1}$ | $3.45\times10^{-1}$ | $3.31\times10^{-1}$ | $3.18\times10^{-1}$ | $3.04\times10^{-1}$ | $2.79\times10^{-1}$ | $2.44\times10^{-1}$ | $2.17\times10^{-1}$ | $1.94\times10^{-1}$ | $1.75\times10^{-1}$ | $1.36\times10^{-1}$ | $1.07\times10^{-1}$ | $8.46\times10^{-2}$ | $6.78\times10^{-2}$ | $4.45\times10^{-2}$ | $3.02\times10^{-2}$ | $2.10\times10^{-2}$ | $1.51\times10^{-2}$ |
| $6\times10^{-1}$ | $6.62\times10^{-1}$ | $6.48\times10^{-1}$ | $6.33\times10^{-1}$ | $6.01\times10^{-1}$ | $5.70\times10^{-1}$ | $5.40\times10^{-1}$ | $4.83\times10^{-1}$ | $4.03\times10^{-1}$ | $3.43\times10^{-1}$ | $2.96\times10^{-1}$ | $2.56\times10^{-1}$ | $1.82\times10^{-1}$ | $1.33\times10^{-1}$ | $1.01\times10^{-1}$ | $7.67\times10^{-2}$ | $4.76\times10^{-2}$ | $3.13\times10^{-2}$ | $2.14\times10^{-2}$ | $1.52\times10^{-2}$ |
| $1\times10^{0}$ | $1.02\times10^{0}$ | $9.92\times10^{-1}$ | $9.63\times10^{-1}$ | $9.05\times10^{-1}$ | $8.49\times10^{-1}$ | $7.92\times10^{-1}$ | $6.88\times10^{-1}$ | $5.42\times10^{-1}$ | $4.38\times10^{-1}$ | $3.60\times10^{-1}$ | $3.00\times10^{-1}$ | $1.99\times10^{-1}$ | $1.40\times10^{-1}$ | $1.03\times10^{-1}$ | $7.79\times10^{-2}$ | $4.78\times10^{-2}$ | | $2.15\times10^{-2}$ | |
| $2\times10^{0}$ | $1.57\times10^{0}$ | $1.52\times10^{0}$ | $1.46\times10^{0}$ | $1.35\times10^{0}$ | $1.23\times10^{0}$ | $1.12\times10^{0}$ | $9.18\times10^{-1}$ | $6.59\times10^{-1}$ | $4.97\times10^{-1}$ | $3.91\times10^{-1}$ | $3.17\times10^{-1}$ | $2.03\times10^{-1}$ | $1.41\times10^{-1}$ | | | | | | |
| $3.5\times10^{0}$ | $2.05\times10^{0}$ | $1.97\times10^{0}$ | $1.88\times10^{0}$ | $1.70\times10^{0}$ | $1.51\times10^{0}$ | $1.34\times10^{0}$ | $1.03\times10^{0}$ | $6.90\times10^{-1}$ | $5.07\times10^{-1}$ | $3.94\times10^{-1}$ | | | | | | | | | |
| $6\times10^{0}$ | $2.52\times10^{0}$ | $2.41\times10^{0}$ | $2.27\times10^{0}$ | $1.99\times10^{0}$ | $1.73\times10^{0}$ | $1.47\times10^{0}$ | $1.07\times10^{0}$ | $6.96\times10^{-1}$ | | | | | | | | | | | |
| $1\times10^{1}$ | $2.97\times10^{0}$ | $2.80\times10^{0}$ | $2.61\times10^{0}$ | $2.22\times10^{0}$ | $1.85\times10^{0}$ | $1.53\times10^{0}$ | $1.08\times10^{0}$ | | | | | | | | | | | | |
| $2\times10^{1}$ | $3.56\times10^{0}$ | $3.30\times10^{0}$ | $3.00\times10^{0}$ | $2.41\times10^{0}$ | $1.92\times10^{0}$ | $1.55\times10^{0}$ | | | | | | | | | | | | | |
| $3.5\times10^{1}$ | $4.01\times10^{0}$ | $3.65\times10^{0}$ | $3.23\times10^{0}$ | $2.48\times10^{0}$ | $1.93\times10^{0}$ | | | | | | | | | | | | | | |
| $6\times10^{1}$ | $4.42\times10^{0}$ | $3.93\times10^{0}$ | $3.37\times10^{0}$ | $2.49\times10^{0}$ | $1.94\times10^{0}$ | | | | | | | | | | | | | | |
| $1\times10^{2}$ | $4.77\times10^{0}$ | $4.12\times10^{0}$ | $3.43\times10^{0}$ | $2.50\times10^{0}$ | | | | | | | | | | | | | | | |
| $2\times10^{2}$ | $5.16\times10^{0}$ | $4.26\times10^{0}$ | $3.45\times10^{0}$ | | | | | | | | | | | | | | | | |
| $3.5\times10^{2}$ | $5.40\times10^{0}$ | $4.29\times10^{0}$ | $3.46\times10^{0}$ | | | | | | | | | | | | | | | | |
| $6\times10^{2}$ | $5.54\times10^{0}$ | $4.30\times10^{0}$ | | | | | | | | | | | | | | | | | |
| $1\times10^{3}$ | $5.59\times10^{0}$ | | | | | | | | | | | | | | | | | | |
| $2\times10^{3}$ | $5.62\times10^{0}$ | | | | | | | | | | | | | | | | | | |
| $3.5\times10^{3}$ | $5.62\times10^{0}$ | $4.30\times10^{0}$ | $3.46\times10^{0}$ | $2.50\times10^{0}$ | $1.94\times10^{0}$ | $1.55\times10^{0}$ | $1.08\times10^{0}$ | $6.96\times10^{-1}$ | $5.07\times10^{-1}$ | $3.94\times10^{-1}$ | $3.17\times10^{-1}$ | $2.03\times10^{-1}$ | $1.41\times10^{-1}$ | $1.03\times10^{-1}$ | $7.79\times10^{-2}$ | $4.78\times10^{-2}$ | $3.13\times10^{-2}$ | $2.15\times10^{-2}$ | $1.52\times10^{-2}$ |

Type B curve

| $t_y$ | $\beta=0.001$ | $\beta=0.004$ | $\beta=0.01$ | $\beta=0.03$ | $\beta=0.06$ | $\beta=0.1$ | $\beta=0.2$ | $\beta=0.4$ | $\beta=0.6$ | $\beta=0.8$ | $\beta=1.0$ | $\beta=1.5$ | $\beta=2.0$ | $\beta=2.5$ | $\beta=3.0$ | $\beta=4.0$ | $\beta=5.0$ | $\beta=6.0$ | $\beta=7.0$ |
|---|---|---|---|---|---|---|---|---|---|---|---|---|---|---|---|---|---|---|---|
| $1\times10^{-4}$ | $5.62\times10^{0}$ | $4.30\times10^{0}$ | $3.46\times10^{0}$ | $2.50\times10^{0}$ | $1.94\times10^{0}$ | $1.56\times10^{0}$ | $1.09\times10^{0}$ | $6.97\times10^{-1}$ | $5.08\times10^{-1}$ | $3.95\times10^{-1}$ | $3.18\times10^{-1}$ | $2.04\times10^{-1}$ | $1.42\times10^{-1}$ | $1.03\times10^{-1}$ | $7.80\times10^{-2}$ | $4.79\times10^{-2}$ | $3.14\times10^{-2}$ | $2.15\times10^{-2}$ | $1.53\times10^{-2}$ |
| $2\times10^{-4}$ | | | | | | | | | | | | | | | $7.81\times10^{-2}$ | $4.80\times10^{-2}$ | $3.15\times10^{-2}$ | $2.16\times10^{-2}$ | $1.53\times10^{-2}$ |
| $3.5\times10^{-4}$ | | | | | | | | | | | | | | $1.03\times10^{-1}$ | $7.83\times10^{-2}$ | $4.81\times10^{-2}$ | $3.16\times10^{-2}$ | $2.17\times10^{-2}$ | $1.54\times10^{-2}$ |
| $6\times10^{-4}$ | | | | | | | | | | | | | | $1.04\times10^{-1}$ | $7.85\times10^{-2}$ | $4.84\times10^{-2}$ | $3.18\times10^{-2}$ | $2.19\times10^{-2}$ | $1.56\times10^{-2}$ |
| $1\times10^{-3}$ | | | | | | | | $6.97\times10^{-1}$ | $5.08\times10^{-1}$ | $3.95\times10^{-1}$ | $3.18\times10^{-1}$ | $2.04\times10^{-1}$ | $1.42\times10^{-1}$ | $1.04\times10^{-1}$ | $7.89\times10^{-2}$ | $4.88\times10^{-2}$ | $3.21\times10^{-2}$ | $2.21\times10^{-2}$ | $1.58\times10^{-2}$ |
| $2\times10^{-3}$ | | | | | | | | $6.97\times10^{-1}$ | $5.09\times10^{-1}$ | $3.96\times10^{-1}$ | $3.19\times10^{-1}$ | $2.05\times10^{-1}$ | $1.43\times10^{-1}$ | $1.05\times10^{-1}$ | $7.99\times10^{-2}$ | $4.96\times10^{-2}$ | $3.29\times10^{-2}$ | $2.28\times10^{-2}$ | $1.64\times10^{-2}$ |
| $3.5\times10^{-3}$ | | | | | | | | $6.98\times10^{-1}$ | $5.10\times10^{-1}$ | $3.97\times10^{-1}$ | $3.21\times10^{-1}$ | $2.07\times10^{-1}$ | $1.45\times10^{-1}$ | $1.07\times10^{-1}$ | $8.14\times10^{-2}$ | $5.09\times10^{-2}$ | $3.41\times10^{-2}$ | $2.39\times10^{-2}$ | $1.73\times10^{-2}$ |
| $6\times10^{-3}$ | | | | | | | | $7.00\times10^{-1}$ | $5.12\times10^{-1}$ | $3.99\times10^{-1}$ | $3.23\times10^{-1}$ | $2.09\times10^{-1}$ | $1.47\times10^{-1}$ | $1.09\times10^{-1}$ | $8.38\times10^{-2}$ | $5.32\times10^{-2}$ | $3.61\times10^{-2}$ | $2.57\times10^{-2}$ | $1.89\times10^{-2}$ |
| $1\times10^{-2}$ | | | | | | | | $7.03\times10^{-1}$ | $5.16\times10^{-1}$ | $4.03\times10^{-1}$ | $3.27\times10^{-1}$ | $2.13\times10^{-1}$ | $1.52\times10^{-1}$ | $1.13\times10^{-1}$ | $8.79\times10^{-2}$ | $5.68\times10^{-2}$ | $3.93\times10^{-2}$ | $2.86\times10^{-2}$ | $2.15\times10^{-2}$ |
| $2\times10^{-2}$ | | | | | | $1.56\times10^{0}$ | $1.09\times10^{0}$ | $7.10\times10^{-1}$ | $5.24\times10^{-1}$ | $4.12\times10^{-1}$ | $3.37\times10^{-1}$ | $2.24\times10^{-1}$ | $1.62\times10^{-1}$ | $1.24\times10^{-1}$ | $9.80\times10^{-2}$ | $6.61\times10^{-2}$ | $4.78\times10^{-2}$ | $3.62\times10^{-2}$ | $2.84\times10^{-2}$ |
| $3.5\times10^{-2}$ | | | | | $1.94\times10^{0}$ | $1.56\times10^{0}$ | $1.10\times10^{0}$ | $7.20\times10^{-1}$ | $5.37\times10^{-1}$ | $4.25\times10^{-1}$ | $3.50\times10^{-1}$ | $2.39\times10^{-1}$ | $1.78\times10^{-1}$ | $1.39\times10^{-1}$ | $1.13\times10^{-1}$ | $8.06\times10^{-2}$ | $6.12\times10^{-2}$ | $4.86\times10^{-2}$ | $3.98\times10^{-2}$ |
| $6\times10^{-2}$ | | | | $2.50\times10^{0}$ | $1.95\times10^{0}$ | $1.57\times10^{0}$ | $1.11\times10^{0}$ | $7.37\times10^{-1}$ | $5.57\times10^{-1}$ | $4.47\times10^{-1}$ | $3.74\times10^{-1}$ | $2.65\times10^{-1}$ | $2.05\times10^{-1}$ | $1.66\times10^{-1}$ | $1.40\times10^{-1}$ | $1.06\times10^{-1}$ | $8.53\times10^{-2}$ | $7.14\times10^{-2}$ | $6.14\times10^{-2}$ |
| $1\times10^{-1}$ | | | | $2.51\times10^{0}$ | $1.96\times10^{0}$ | $1.58\times10^{0}$ | $1.13\times10^{0}$ | $7.63\times10^{-1}$ | $5.89\times10^{-1}$ | $4.83\times10^{-1}$ | $4.12\times10^{-1}$ | $3.07\times10^{-1}$ | $2.48\times10^{-1}$ | $2.10\times10^{-1}$ | $1.84\times10^{-1}$ | $1.49\times10^{-1}$ | $1.28\times10^{-1}$ | $1.13\times10^{-1}$ | $1.02\times10^{-1}$ |
| $2\times10^{-1}$ | $5.62\times10^{0}$ | $4.30\times10^{0}$ | $3.46\times10^{0}$ | $2.52\times10^{0}$ | $1.98\times10^{0}$ | $1.61\times10^{0}$ | $1.18\times10^{0}$ | $8.29\times10^{-1}$ | $6.67\times10^{-1}$ | $5.71\times10^{-1}$ | $5.06\times10^{-1}$ | $4.10\times10^{-1}$ | $3.57\times10^{-1}$ | $3.23\times10^{-1}$ | $2.98\times10^{-1}$ | $2.66\times10^{-1}$ | $2.45\times10^{-1}$ | $2.31\times10^{-1}$ | $2.20\times10^{-1}$ |
| $3.5\times10^{-1}$ | $5.63\times10^{0}$ | $4.31\times10^{0}$ | $3.47\times10^{0}$ | $2.54\times10^{0}$ | $2.01\times10^{0}$ | $1.66\times10^{0}$ | $1.24\times10^{0}$ | $9.22\times10^{-1}$ | $7.80\times10^{-1}$ | $6.97\times10^{-1}$ | $6.42\times10^{-1}$ | $5.62\times10^{-1}$ | $5.17\times10^{-1}$ | $4.89\times10^{-1}$ | $4.70\times10^{-1}$ | $4.45\times10^{-1}$ | $4.30\times10^{-1}$ | $4.19\times10^{-1}$ | $4.11\times10^{-1}$ |
| $6\times10^{-1}$ | $5.63\times10^{0}$ | $4.31\times10^{0}$ | $3.49\times10^{0}$ | $2.57\times10^{0}$ | $2.06\times10^{0}$ | $1.73\times10^{0}$ | $1.35\times10^{0}$ | $1.07\times10^{0}$ | $9.54\times10^{-1}$ | $8.89\times10^{-1}$ | $8.50\times10^{-1}$ | $7.92\times10^{-1}$ | $7.63\times10^{-1}$ | $7.45\times10^{-1}$ | $7.33\times10^{-1}$ | $7.18\times10^{-1}$ | $7.09\times10^{-1}$ | $7.03\times10^{-1}$ | $6.99\times10^{-1}$ |
| $1\times10^{0}$ | $5.63\times10^{0}$ | $4.32\times10^{0}$ | $3.51\times10^{0}$ | $2.62\times10^{0}$ | $2.13\times10^{0}$ | $1.83\times10^{0}$ | $1.50\times10^{0}$ | $1.29\times10^{0}$ | $1.20\times10^{0}$ | $1.16\times10^{0}$ | $1.13\times10^{0}$ | $1.10\times10^{0}$ | $1.08\times10^{0}$ | $1.07\times10^{0}$ | $1.07\times10^{0}$ | $1.06\times10^{0}$ | $1.06\times10^{0}$ | $1.05\times10^{0}$ | $1.05\times10^{0}$ |
| $2\times10^{0}$ | $5.64\times10^{0}$ | $4.35\times10^{0}$ | $3.56\times10^{0}$ | $2.73\times10^{0}$ | $2.31\times10^{0}$ | $2.07\times10^{0}$ | $1.85\times10^{0}$ | $1.72\times10^{0}$ | $1.68\times10^{0}$ | $1.66\times10^{0}$ | $1.65\times10^{0}$ | $1.64\times10^{0}$ | $1.63\times10^{0}$ | $1.63\times10^{0}$ | $1.63\times10^{0}$ | $1.63\times10^{0}$ | $1.63\times10^{0}$ | $1.63\times10^{0}$ | $1.63\times10^{0}$ |
| $3.5\times10^{0}$ | $5.65\times10^{0}$ | $4.38\times10^{0}$ | $3.63\times10^{0}$ | $2.88\times10^{0}$ | $2.55\times10^{0}$ | $2.37\times10^{0}$ | $2.23\times10^{0}$ | $2.17\times10^{0}$ | $2.15\times10^{0}$ | $2.15\times10^{0}$ | $2.14\times10^{0}$ | $2.14\times10^{0}$ | $2.14\times10^{0}$ | $2.14\times10^{0}$ | $2.14\times10^{0}$ | $2.14\times10^{0}$ | $2.14\times10^{0}$ | $2.14\times10^{0}$ | $2.14\times10^{0}$ |
| $6\times10^{0}$ | $5.67\times10^{0}$ | $4.44\times10^{0}$ | $3.74\times10^{0}$ | $3.11\times10^{0}$ | $2.86\times10^{0}$ | $2.75\times10^{0}$ | $2.68\times10^{0}$ | $2.66\times10^{0}$ | $2.65\times10^{0}$ | $2.65\times10^{0}$ | $2.65\times10^{0}$ | $2.65\times10^{0}$ | $2.64\times10^{0}$ | $2.64\times10^{0}$ | $2.64\times10^{0}$ | $2.64\times10^{0}$ | $2.64\times10^{0}$ | $2.64\times10^{0}$ | $2.64\times10^{0}$ |
| $1\times10^{1}$ | $5.70\times10^{0}$ | $4.52\times10^{0}$ | $3.90\times10^{0}$ | $3.40\times10^{0}$ | $3.24\times10^{0}$ | $3.18\times10^{0}$ | $3.15\times10^{0}$ | $3.14\times10^{0}$ | $3.14\times10^{0}$ | $3.14\times10^{0}$ | $3.14\times10^{0}$ | $3.14\times10^{0}$ | $3.14\times10^{0}$ | $3.14\times10^{0}$ | $3.14\times10^{0}$ | $3.14\times10^{0}$ | $3.14\times10^{0}$ | $3.14\times10^{0}$ | $3.14\times10^{0}$ |
| $2\times10^{1}$ | $5.76\times10^{0}$ | $4.71\times10^{0}$ | $4.22\times10^{0}$ | $3.92\times10^{0}$ | $3.85\times10^{0}$ | $3.83\times10^{0}$ | $3.82\times10^{0}$ | $3.82\times10^{0}$ | $3.82\times10^{0}$ | $3.82\times10^{0}$ | $3.82\times10^{0}$ | $3.82\times10^{0}$ | $3.82\times10^{0}$ | $3.82\times10^{0}$ | $3.82\times10^{0}$ | $3.82\times10^{0}$ | $3.82\times10^{0}$ | $3.82\times10^{0}$ | $3.82\times10^{0}$ |
| $3.5\times10^{1}$ | $5.85\times10^{0}$ | $4.94\times10^{0}$ | $4.58\times10^{0}$ | $4.40\times10^{0}$ | $4.38\times10^{0}$ | $4.38\times10^{0}$ | $4.37\times10^{0}$ | $4.37\times10^{0}$ | $4.37\times10^{0}$ | $4.37\times10^{0}$ | $4.37\times10^{0}$ | $4.37\times10^{0}$ | $4.37\times10^{0}$ | $4.37\times10^{0}$ | $4.37\times10^{0}$ | $4.37\times10^{0}$ | $4.37\times10^{0}$ | $4.37\times10^{0}$ | $4.37\times10^{0}$ |
| $6\times10^{1}$ | $5.99\times10^{0}$ | $5.23\times10^{0}$ | $5.00\times10^{0}$ | $4.92\times10^{0}$ | $4.91\times10^{0}$ | $4.91\times10^{0}$ | $4.91\times10^{0}$ | $4.91\times10^{0}$ | $4.91\times10^{0}$ | $4.91\times10^{0}$ | $4.91\times10^{0}$ | $4.91\times10^{0}$ | $4.91\times10^{0}$ | $4.91\times10^{0}$ | $4.91\times10^{0}$ | $4.91\times10^{0}$ | $4.91\times10^{0}$ | $4.91\times10^{0}$ | $4.91\times10^{0}$ |
| $1\times10^{2}$ | $6.16\times10^{0}$ | $5.59\times10^{0}$ | $5.46\times10^{0}$ | $5.42\times10^{0}$ | $5.42\times10^{0}$ | $5.42\times10^{0}$ | $5.42\times10^{0}$ | $5.42\times10^{0}$ | $5.42\times10^{0}$ | $5.42\times10^{0}$ | $5.42\times10^{0}$ | $5.42\times10^{0}$ | $5.42\times10^{0}$ | $5.42\times10^{0}$ | $5.42\times10^{0}$ | $5.42\times10^{0}$ | $5.42\times10^{0}$ | $5.42\times10^{0}$ | $5.42\times10^{0}$ |

[a] From Neuman (1975a).

The interpretation is done on log–log paper as follows:

(1)   The end of the test is fitted on the B-type curves by the same method as for the Theis curve, and $\beta$, $T$, and $\omega_d$ are identified.

(2)   The beginning of the test is fitted without vertical parallel shift of the curves, so that the same $T$ and $\beta$ are retained. Then $S$ is calculated on the A-type curves.

(3)   From $T$ we calculate $K_r = T/e$, and with the help of $\beta$ we can calculate $K_z$. Therefore, this is one of the few methods that allows us to estimate the anisotropy of the formation.

The above solution is, however, an approximation, obtained by linearization. It omits, for example, the reduction in the saturated thickness with time. The solution may also be calculated for any piezometer open at a given elevation $z$, but not screened along its entire length.

*In the case of a partially penetrating well*, Neuman (1975a) gives the changes which should be made in the previous function. However, here the number of parameters becomes too large to be shown on type curves: a program for calculating the type curve for a given geometry (well depth, aquifer thickness, position and intake of the piezometer) is available from the author (Neuman, unpublished, 1975b). Additional developments on this approach of delayed yield can be found in Neuman (1979).

### 8.4.4. *Variations in Flow during the Test*

All the solutions shown are linear in $Q$. If the flow varies in time, the response is obtained by convolution of the elementary solution with the flow variation. For example, for the Theis solution,

$$s = \frac{1}{4\pi T} \left\{ Q(0) W\left(\frac{4Tt}{r^2 S}\right) + \int_0^t \left(\frac{\partial Q}{\partial t}\right)_\tau W\left[\frac{4T(t-\tau)}{r^2 S}\right] d\tau \right\}$$

Inversely, the step-function response of the system may be calculated by deconvolution of the observations, i.e., the drawdown $s(t)$ that would have been observed if the flow $Q$ had been kept constant. This is the one that will be used for the interpretation.

### 8.4.5. *Flow in Fractured Systems*

Pumping tests in porous fractured reservoirs are very common in the petroleum industry, and a large number of particular solutions have been developed for such systems, assuming particular geometries for the fractures. [See Boulton and Streltsova (1977), DeSwann (1976), Gringarten *et al.* (1974a,b; 1975), Gringarten (1982), Hartsock and Warren (1961), and Streltsova (1976a).]

## 8.5.  Other (One-Dimensional) Solutions to the Diffusion Equation

We shall only give two analytical solutions for one-dimensional flow that are especially useful for the interpretation of natural variations of the piezometric head in aquifers.

(a)  *Semiinfinite domain, sudden variation of hydraulic head.*

Equation:  $$\frac{\partial^2 h}{\partial x^2} = \frac{S}{T}\frac{\partial h}{\partial t}, \quad x \text{ and } t \geq 0$$

Conditions:   $h(x,0) = h_0, \quad \forall\, x > 0 \quad$ (initial)

$\qquad\qquad\quad h(0,t) = 0 \qquad t \geq 0 \quad$ (at the boundary)

Solution:   $$h(x,t) = h_0 \operatorname{erf}\left(x\sqrt{\frac{S}{4Tt}}\right)$$

where $\operatorname{erf}(u)$ is the error function, known and tabulated (Table 8.6) as

$$\operatorname{erf}(u) = \frac{2}{\sqrt{\pi}}\int_0^u e^{-v^2}\,dv$$

This solution fits the case of a semi-infinite confined aquifer, initially in equilibrium at the hydraulic head $h_0$ with a stream as one boundary. At the time $t = 0$, the level of the stream suddenly drops to the elevation $h = 0$ as in Fig. 8.18.

This solution is also applicable to the study of recharge in an aquifer. Suppose that the aquifer is initially in equilibrium with a river at $h = 0$, and that at time $t = 0$ it receives a uniform and instantaneous recharge $h_0$ all over its surface. The head $h(x,t)$ will be given by the same expression as above.

On the other hand, a sudden raising $h_0$ of the level of a stream, initially in equilibrium with the aquifer at the elevation 0, causes a variation of the

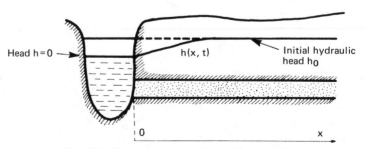

**Fig. 8.18.** Stream in contact with a confined aquifer.

**Table 8.6**

The Error Function[a]

| u | erf u | erfc u | | u | erf u | erfc u |
|---|---|---|---|---|---|---|
| 0 | 0 | 1.0 | | 1.0 | 0.842701 | 0.157299 |
| 0.05 | 0.056372 | 0.943628 | | 1.1 | 0.880205 | 0.119795 |
| 0.1 | 0.112463 | 0.887537 | | 1.2 | 0.910314 | 0.089686 |
| 0.15 | 0.167996 | 0.832004 | | 1.3 | 0.934008 | 0.065992 |
| 0.2 | 0.222703 | 0.777297 | | 1.4 | 0.952285 | 0.047715 |
| 0.25 | 0.276326 | 0.723674 | | 1.5 | 0.966105 | 0.033895 |
| 0.3 | 0.328627 | 0.671373 | | 1.6 | 0.976348 | 0.023652 |
| 0.35 | 0.379382 | 0.620618 | | 1.7 | 0.983790 | 0.016210 |
| 0.4 | 0.428392 | 0.571608 | | 1.8 | 0.989091 | 0.010909 |
| 0.45 | 0.475482 | 0.524518 | | 1.9 | 0.992790 | 0.007210 |
| 0.5 | 0.520500 | 0.479500 | | 2.0 | 0.995322 | 0.004678 |
| 0.55 | 0.563323 | 0.436677 | | 2.1 | 0.997021 | 0.002979 |
| 0.6 | 0.603856 | 0.396144 | | 2.2 | 0.998137 | 0.001863 |
| 0.65 | 0.642029 | 0.357971 | | 2.3 | 0.998857 | 0.001143 |
| 0.7 | 0.677801 | 0.322199 | | 2.4 | 0.999311 | 0.000689 |
| 0.75 | 0.711156 | 0.288844 | | 2.5 | 0.999593 | 0.000407 |
| 0.8 | 0.742101 | 0.257899 | | 2.6 | 0.999764 | 0.000236 |
| 0.85 | 0.770668 | 0.229332 | | 2.7 | 0.999866 | 0.000134 |
| 0.9 | 0.796908 | 0.203092 | | 2.8 | 0.999925 | 0.000075 |
| 0.95 | 0.820891 | 0.179109 | | 2.9 | 0.999959 | 0.000041 |
| | | | | 3.0 | 0.999978 | 0.000022 |

[a] After Carslaw and Jaeger (1959).

hydraulic head:

$$h(x,t) = h_0 \left[ 1 - \text{erf}\left( x \sqrt{\frac{S}{4Tt}} \right) \right] = h_0 \, \text{erfc}\left( x \sqrt{\frac{S}{4Tt}} \right)$$

where erfc $= 1 - $ erf is the complementary error function (see Table 8.6).

By convolution it is also possible to calculate the response of an aquifer to continuous variations in stream level.

In practice, these solutions are always used for the variation of the head with respect to an initial steady state, and not for the head itself (see Section 7.1.b). They are also applied to unconfined aquifers if the variation of the head is small enough that the saturated thickness of the aquifer can be considered constant.

Figure 8.19. gives the error function (erf, curve I) and the derivative of $h$ versus time or space (curve II).

$$-\frac{t\sqrt{\pi}}{h_0} \frac{\partial h}{\partial t} = \frac{x\sqrt{\pi}}{2h_0} \frac{\partial h}{\partial x} = x \sqrt{\frac{S}{4Tt}} \exp\left( -\frac{x^2 S}{4Tt} \right)$$

(b) *Bounded domains.*

Equation: $$\frac{\partial^2 h}{\partial x^2} = \frac{S}{T} \frac{\partial h}{\partial t}, \qquad 0 \le x \le l \qquad t \ge 0$$

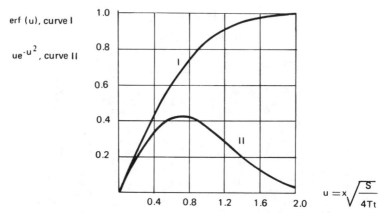

**Fig. 8.19.** The error function and its derivative. [From Carslaw and Jaeger (1959).]

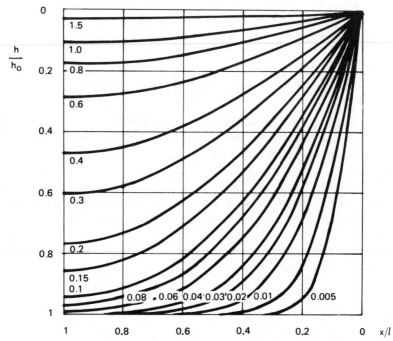

**Fig. 8.20** Type curve for a bounded domain. The parameter given on the curves is $Tt/Sl^2$. [From Carslaw and Jaeger (1959).]

Conditions:

$$h(x, 0) = h_0, \qquad 0 < x \le l, \qquad \text{(initial)}$$

$$h(0, t) = 0, \qquad t \ge 0, \qquad \text{(first boundary with prescribed head)}$$

$$\left(\frac{\partial h}{\partial x}\right)_{x=l} = 0, \qquad t \ge 0, \qquad \text{(second no-flow boundary)}$$

Solution:

$$h = h_0 \left\{ 1 - \sum_{n=0}^{\infty} (-1)^n \left[ \text{erfc} \frac{2nl + x}{2} \sqrt{\frac{S}{Tt}} + \text{erfc}\left( \frac{(2n+1)l - x}{2} \sqrt{\frac{S}{Tt}} \right) \right] \right\}$$

or again:

$$h = h_0 \frac{4}{\pi} \sum_{n=0}^{\infty} \frac{(-1)^n}{(2n+1)} \exp\left[ -\frac{(2n+1)^2 \pi^2 Tt}{4Sl^2} \right] \cos \frac{(2n+1)\pi x}{2l}$$

This solution corresponds to the same conditions as in Fig. 8.18, but with an aquifer limited by a no-flow boundary at the distance $x = l$. Figure 8.20 is the corresponding type curve, where the curves are indexed on the parameter $Tt/Sl^2$.

## 8.6. *In Situ* Point Measurements of Permeability

The pumping tests described in Sections 8.2 and 8.4 offer the best estimates of average hydraulic conductivities in a medium. However, since they are quite arduous to work out, it has been suggested that more basic and localized methods should be used to estimate the hydraulic conductivities. We shall briefly describe three of these, which are chiefly used in civil engineering.

(a) *Pocket or Lefranc test.* A "pocket" of length $l$ facing the terrain to be explored is made in a boring (open or screened) of diameter $D$. This pocket is created either by putting a casing into the remainder of the borehole or by isolating the section by means of an inflatable packer (rubber sleeve tightly fitted to the terrain).

The test consists in injecting (or pumping) at a constant flow rate $Q$ and waiting until the hydraulic head (or the pressure) is approximately stabilized (quasi-steady state; one only waits for a few minutes). The hydraulic conductivity is given by the relation (Schneebeli, 1966):

$$K = \frac{\alpha}{D} \frac{Q}{\Delta h}$$

where $\Delta h$ is the change in hydraulic head between the initial condition and the quasi-steady state, $D$ is the diameter of the pocket, $Q$ is the injected or withdrawn constant flow rate, and $\alpha$ is a dimensionless coefficient that depends on the shape of the pocket:

(1)    More or less spherical pocket:

$$\alpha = \frac{1}{2\pi\sqrt{l/D + 1/4}}$$

(2)    Ellipsoidal pocket:

$$\alpha = \frac{\ln(l/D + \sqrt{(l/D)^2 + 1})}{2\pi l/D}$$

(3)    Very elongated pocket ($l/D > 4$):

$$\alpha = \frac{\ln(2l/D)}{2\pi l/D}$$

Here, $l$ is the height of the pocket. When the pocket is close to a no-flow boundary (free surface or impervious bedrock), the boundary effect creates an image, which is accounted for by multiplying $\alpha$ by $D/8\pi z$, where $z$ is the distance from the center of the pocket to the boundary, assumed to be large compared to $l$ and $D$.

The test is made with different flow rates in order to verify the linearity of the relation $Q - \Delta h$, since any nonlinearity may indicate a leak in the casing or the packer or hydraulic fracturing of the terrain.

(b)    *Lugeon's test in fractured rocks.*    This is a very well-known empirical test using a boring in fractured rocks. A portion of the boring, usually 5 m long, is isolated by a packer. Quite often, the test is made as the boring progresses: every time a 5-m-long section of boring is finished, it is sealed off with a single packer and the test made. The packer is then removed and the boring resumed. Sometimes the test is made when the boring is already completed. The procedure is then to isolate 5-m sections using two packers, inject water under pressure, and measure the stabilized flow (after 5–10 min) versus the pressure. The measurement program proceeds as follows: the pressure is made to increase gradually from 0 to 10 bars, then to decrease from 10 to 0 bars. Then, the flow rate in liters per minute versus pressure in bars is plotted as in Fig. 8.21.

The flow is generally stronger when the pressure decreases than when it increases. This also conveys information concerning the behavior of fractured rocks (unclogging of fractures, hydraulic fracturing of the terrain, etc.).

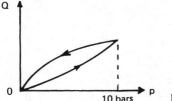

**Fig. 8.21.** Lugeon's diagram.

In this test, the hydraulic conductivity of the terrain is defined in "Lugeon units," i.e., the flow injected in liters per minute under a pressure of 10 bars and per linear meter of boring for a test time of 10 min at constant pressure.

It is admitted that, if the hydraulic conductivity in Lugeon units is small (a few to some tens of units), then the Lugeon unit is *very approximately* equal to 1- or $2 \times 10^{-7}$ m/s (Cambefort, 1966).

(c) *Slug tests.* Whereas the preceding tests are usually interpreted in a steady state, the slug test consists in creating a very brief pressure pulse at one point in the aquifer and observing the transient response at the same point. The interpretation varies depending on the shape of the cavity where the impulse occurs (cylindrical or spherical symmetry). This test measures chiefly the transmissivity (or the hydraulic conductivity) and, with a lesser degree of precision, the storage coefficient (Papadopoulos *et al.*, 1973).

In cylindrical symmetry (Fig. 8.22), a slug test may be interpreted in a fully penetrating well or piezometer, i.e., one that penetrates the entire thickness of the aquifer. Let $T$ be the transmissivity and $S$ the storage coefficient of the aquifer, and $R$ the radius of the boring at the level of the aquifer (radius of the borehole). Let $R'$ be the radius of the boring at the static level of the water, assumed to be at equilibrium before the test: $R'$ is usually the interior radius of the casing. At time $t = 0$, a sudden variation $\Delta h_0$ of the hydraulic head in the borehole of radius $R'$ is caused by injection or withdrawal of a volume of

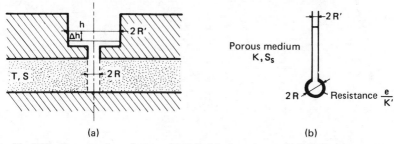

**Fig. 8.22.** Geometry for a slug test. (a) Cylindrical symmetry. (b) Spherical symmetry.

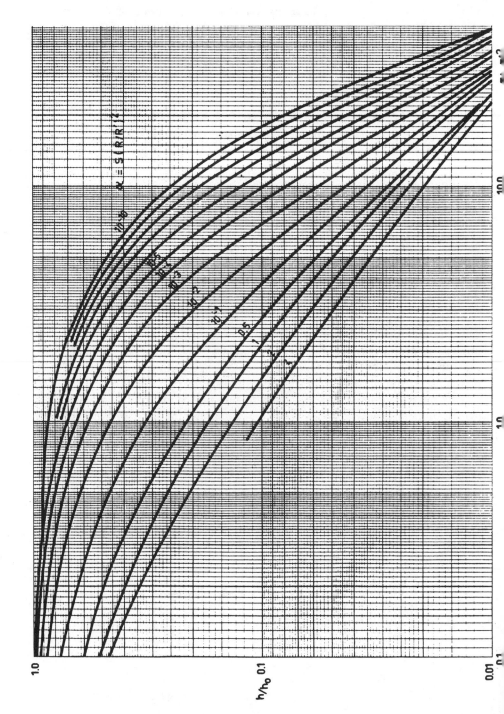

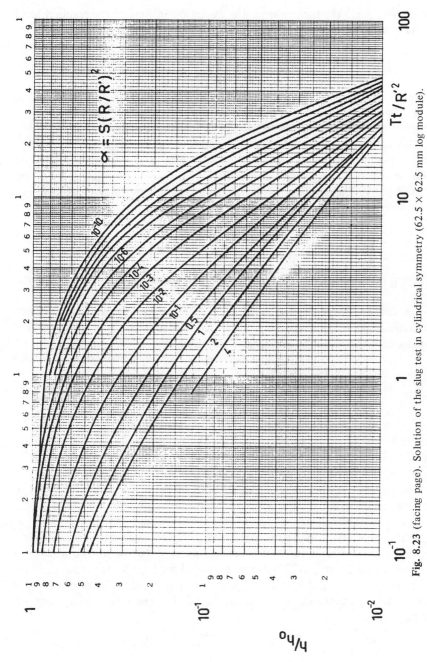

**Fig. 8.23** (facing page). Solution of the slug test in cylindrical symmetry (62.5 × 62.5 mm log module).

**Fig. 8.24** (this page). Solution of the slug test in cylindrical symmetry (1.85 × 1.85 in. log module).

water $\pi R'^2 \Delta h_0$. We then observe the evolution in time of the residual head $\Delta h(t)$ in the casing. It is given by

$$\frac{\Delta h(t)}{\Delta h_0} = \frac{4\alpha}{\pi^2} \int_0^\infty \frac{\exp(-Ttu^2/R^2 S)}{uF(u)} du$$

where $\alpha = S(R/R')^2$, $F(u) = [uJ_0(u) - \alpha J_1(u)]^2 + [uY_0(u) - \alpha Y_1(u)]^2$, and $J_n$ and $Y_n$ are Bessel functions of the first and second kind, respectively, and order $n$.

This solution is given on Figs. 8.23 and 8.24 on log–log paper as $\Delta h/\Delta h_0$ versus the dimensionless time $Tt/R'^2$ (Degallier and Marsily, 1977). The measurements $\Delta h/\Delta h_0$ versus time are drawn on log–log tracing paper of the same module. The matching of the measured curve with the type curve gives $\alpha$ and the correspondence between dimensionless time and real time, whence

$$T = R'^2/t_i$$

$$S = \alpha(R'/R)^2$$

where $t_i$ is real time coinciding with dimensionless time of the value 1 on the type curve.

Carslaw and Jaeger also consider the case where the observations are made at a distance $r > R$ from the boring or where there is a zone of low conductivity between the boring and the aquifer.

If the cavity on which the slug test is made has a spherical shape, the same authors give the following solution:

$$\frac{\Delta h(t)}{\Delta h_0} = \frac{2\alpha\gamma^2}{\pi} \int_0^\infty \frac{\exp(-Ktu^2/S_s R^2)u^2}{[u^2(1 + \gamma) - \alpha\gamma]^2 + (u^3 - \alpha\gamma u)} du$$

where $K$ and $S_s$ are the hydraulic conductivity and specific storage coefficient of the aquifer, respectively; $\alpha = 4S_s R(R/R')^2$, the capacity ratio; and $\gamma = K'R/Ke$, the hydraulic conductivity ratio.

The sphere of radius $R$ is in contact with the aquifer through a layer with low permeability of thickness $e$ and hydraulic conductivity $K'$. The initial variation of the head $\Delta h_0$ is created in a pipe of $R'$ radius.

This solution depends on two parameters, $\alpha$ and $\gamma$, and on the dimensionless time $Kt/S_s R^2$. The series of corresponding type curves are published by Degallier and Marsily (1977). It can be useful for tests on tensiometers with a shape approaching that of a sphere in an unsaturated medium or on nearly spherical pockets opened at the bottom of an unscreened or a partly screened piezometer.

*Chapter 9*

# Multiphase Flow of Immiscible Fluids

Thus far, we have only considered the flow of one fluid in porous media. Many problems, however, involve several immiscible fluids flowing simultaneously in the same porous medium. This is the case of oil, water, and gas in a petroleum reservoir or simply of water and air in the unsaturated zone on top of a water table aquifer. We will set out the basic concepts and equations for multiphase flow and examine a certain number of special cases.

## 9.1. Theory

When several fluids occupy a given porous medium, their relationship with each other and with the porous medium will be governed by the proportion of each fluid in the medium. This will be measured by the *volumetric saturation* for each of the fluids:

$$s_i = \frac{\text{part of the porosity occupied by the fluid } i}{\text{total porosity}}$$

where $s_i$ varies between 0 and 1.

207

If we assume that the temperature does not vary significantly in the porous medium, as we did in the case of one fluid, the solution of the problem requires us to calculate six unknowns for each fluid $i$, the pressure $p_i$, the mass per unit volume $\rho_i$, the saturation $s_i$ and the three components of the filtration velocity $U_i$.

The equations we can write are:

(1)  Continuity equation, one per fluid.
(2)  Modified Darcy's law, three per fluid.
(3)  Equation of state, one per fluid.
(4)  Capillary pressure at the interface between two fluids, number of fluids minus one.
(5)  Relation between the saturations, one.

We thus have as many equations as unknowns. Let us examine them.

(a)  *Continuity equation.*   For each fluid, a mass balance equation is written:

$$\text{div}(\rho_i U_i) + \frac{\partial}{\partial t}(\rho_i s_i \omega) = 0$$

where $\rho_i s_i \omega$ is the quantity of fluid $i$ contained in a unit volume of porous medium. This equation was established in Section 3.2.1 for a single fluid; the only difference here is that the saturation $s_i$ has to be taken into account in the second term.

(b)  *Darcy's law for multiphase flow—Relative permeability.*   It is admitted that Darcy's law is valid for each fluid separately, as if it occupied a certain portion of the porous medium:

$$U_i = -\frac{k_i}{\mu_i}(\mathbf{grad}\, p_i + \rho_i g\, \mathbf{grad}\, z)$$

where $U_i$ is Darcy's velocity of the fluid $i$, and $\mu_i$, $\rho_i$, and $p_i$ are its dynamic viscosity, mass per unit volume, and pressure, respectively. Further, $k_i$ is the intrinsic permeability for fluid $i$. This permeability will, however, depend on the saturation $s_i$ of the medium by the fluid $i$. The larger the portion of the porous medium occupied by the fluid $i$, the larger the permeability linked to this fluid. A relative permeability is defined as

$$k_{ri} = \frac{k_i}{k}$$

where $k$ is the intrinsic permeability of the porous medium (see Section 4.1.a),

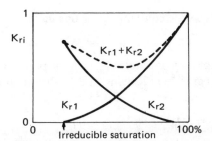

**Fig. 9.1.** Relative permeability in a two-phase system. $K_{r1}$, $K_{r2}$ are permeabilities relative to water, air respectively.

measured when the medium is saturated with a single fluid (remember that $k$ is independent of the nature of the fluid and depends only on the medium).

For example, for two fluids (e.g., air and water), Fig. 9.1 shows the shape of curves obtained. Below a certain degree of saturation of air, the air phase is no longer continuous, and the permeability to air is zero. It might be noted that the sum of the two relative permeabilities for each fluid is not constant and most of the time is less than or equal to 1: the two fluids interfere with each other.*

These curves of relative permeability are determined experimentally on samples. Unfortunately, they are subject to hysteresis, like the capillary pressure (see Section 2.2.2.c), depending on whether draining or wetting is taking place. However, this hysteresis is much less important than that of capillary pressure and is often disregarded.

The ratio $k_{ri}/\mu_i$ is sometimes called the *mobility* of the fluid $i$.

Note that the Darcy velocities $U_i$ of the fluid are not necessarily parallel; they may even be diametrically opposed, e.g., during infiltration in the unsaturated zone, where water moves downward while air is pushed upward.

(c) *State equation.* The mass per unit volume of each fluid is a function of its pressure. For a liquid a linear compressibility is generally assumed, as for water:

$$\rho_i = \rho_{i0} e^{\beta_i(p_i - p_{i0})}$$

where $\beta_i$ is the fluid compressibility coefficient (mass$^{-1}$ length time$^2$).

Hence

$$d\rho_i = \rho_i \beta_i \, dp_i$$

---

* In certain flow experiments with water, oil, and gas, sums of relative permeabilities superior to 1 (up to 2) have been measured. Attempts have been made to explain this by saying that one of the fluids works as a "lubricant" for the flow of the others. This shows that the classical theory of multiphase flow presented here is only a rough approximation.

For a perfect gas, e.g., air, the state equation is, for one mole,

$$p_i V = RT \quad \text{or} \quad \rho_i = M_i/V = (M_i/RT)p_i$$

where $V$ is the volume occupied by one mole of gas, $R$ the perfect gas constant (8.32 in SI units), $T$ the temperature in Kelvins, and $M_i$ the molar mass of the gas. Hence

$$d\rho_i = (M_i/RT)dp_i = \rho_i \beta'_i dp_i$$

with $\beta'_i = 1/p_i$ the gas compressibility coefficient (mass$^{-1}$ length time$^2$). But $\beta'_i$ is no longer a constant. For a petroleum gas, which deviates from a perfect gas, the state equation is written

$$\rho_i = (M_i/RT)p_i/Z_i$$

where $Z_i$ is the compressibility factor of the gas, which is a function of $p_i$. Hence:

$$d\rho_i = \rho_i \beta'_i dp_i \quad \text{with } \beta'_i = 1/p_i - (1/Z_i)dZ_i/dp_i$$

From these laws, the term $\partial \rho_i/\partial t$ in the continuity equation can be expressed as a function of $\partial p_i/\partial t$.

For the porous medium we have seen in Section 5.3 that the porosity $\omega$ varies with the pressure, due to a change in effective stress in the medium

$$\frac{\partial \omega}{\partial t} = (\alpha - \omega \beta_s) \frac{\partial p}{\partial t}$$

where $\alpha$ is the compressibility of the porous medium and $\beta_s$ is that of the grains of the medium (generally disregarded). The variation of $\omega$ in multiphase flow is often regarded as negligible compared to the variation in saturation. If this is not acceptable, the pressure $p$ to consider in the above equation is that of the wetting fluid, which surrounds all the grains of the medium.

(d) *Capillary pressure.* We have seen, in Section 2.2.2.c, that there is a difference in pressure across an interface separating two immiscible fluids, called *capillary pressure*:

$$p_{c_{ij}} = p_i - p_j$$

This capillary pressure is a function of the radius of the curvature $r$ of the interface between the two fluids and of the surface tension $\sigma_{ij}$ existing between them:

$$p_{c_{ij}} = \frac{2\sigma_{ij}}{r}$$

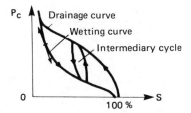

**Fig. 9.2.** Capillary pressure versus saturation.

As the radius of curvature of the menisci separating the fluids in the porous medium is a function of the saturation, the capillary pressure depends on the saturations. For air and water, $p_c(s) = p_{air} - p_{water}$.

In Section 2.2.2.c, we showed a few curves of the capillary pressure $p_c$ versus the saturation $s$. Unfortunately, these curves display hysteresis, depending on whether $s$ increases or decreases (Fig. 9.2). The same phenomena occur when there are more than two fluids. These curves are obtained experimentally.

(e) *Relationship between the saturations.* By definition, $\sum s_i = 1$. Thus, we obtain the system of equations

$$\text{div}[\rho_i k_{ri} k(\mathbf{grad}\, p_i + \rho_i g\, \mathbf{grad}\, z)] = \frac{\partial}{\partial t}(\rho_i s_i \omega)$$

$$p_i - p_j = p_{c_{ij}}$$

$$\sum s_i = 1$$

where $k_{ri}$ and $p_{c_{ij}}$ are functions of the saturations. In general, these equations are solved numerically. We shall, however, examine some simplified cases where approximate solutions are available.

Buckley and Leverett (1942) have given a classical analytical solution to the problem of the injection of a fluid in a medium initially saturated with another fluid. Both fluids and the matrix are supposed incompressible and the effect of the capillary pressure gradients on the flow field are neglected. In one dimension, it can be shown (see Subsection 9.2.2) that the sum of the fluxes (or velocities) of each fluid is a constant through space. This sum of the velocities can be explicitly calculated as well as the saturation, although the expressions are highly complex and nonlinear. A sharp front (jump in saturation) is moved inside the medium, although the saturation varies before and after the front. See Bear (1972), Morel-Seytoux (1973) and Allen (1986).

In the oil industry, the problem is even more complex, because exchanges between the phases have to be taken into account (oil and gas) as the pressure varies, and thermal problems must be considered (e.g., injection of steam into an oil reservoir). The variation in viscosity of each fluid must also be included. (See Allen, 1986.)

Fig. 9.3. Capillary entrapment.

(f)  *Capillary entrapment and fingering.*  Two-phase migration makes it possible to explain the phenomenon of "capillary entrapment," which is the cause of the formation of certain oil deposits in sedimentary porous media. Imagine a flow of oil and water. The oil is assumed to be the nonwetting fluid. A drop of oil squeezed into a pore that is too small for it looks like the illustration in Fig. 9.3. Because of the pressure gradients of the flow, it presents differences in the radii of curvature, $r_1$ and $r_2$, between its upstream and downstream sides. If the drop is to cross the narrow passage of the pore, it has to be subjected to a minimum pressure gradient producing a sufficiently small radius $r_1$. Below this gradient the drop of oil is "trapped." However, if the drop is moving, its kinetic energy may help it to cross the narrow passage. This phenomenon has been studied by Legait (1983).

Another problem posed by two-phase flow is that of instabilities or fingering. If an attempt is made to displace a fluid A by a fluid B, the result will often be neither a well-defined interface between the two fluids, nor a transition zone where the saturation varies continuously between A and B, but a penetration by, for example, the fluid A of the fluid B in the shape of a "finger" (Fig. 9.4). These fingers have a tendency to progress faster than the average front, and thus to continuously grow in size. Bubbles of the fluid B can also be left immobile behind the front, because of fingering, and remain trapped inside a medium almost saturated with fluid A.

This phenomenon is characteristic of unstable flow. The conditions of stability or instability of a two-phase flow are quite complex and depend on the viscosity, the density, the relative permeabilities, and the flow velocity.

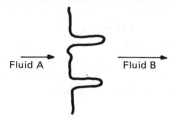

Fluid A        Fluid B

Fig. 9.4. Fingering.

## 9.2. Special Case: Flow in Unsaturated Media

The flow in the unsaturated zone can be studied either by assuming that the movement of the air can be disregarded and focusing on the movement of the water or by taking both into account. We shall briefly present the two approaches and then give some simplified solutions (see also Hillel, 1971).

### 9.2.1. Unsaturated Flow with Immobile Air Phase

Most of the time it is assumed that the air phase is immobile in unsaturated media, so only the movement in the water phase is calculated; the pressure in the air phase is equal to the atmospheric pressure, taken as zero.

First, we determine experimentally the relation between the unsaturated hydraulic conductivity $K$ and the moisture content $\theta$ (or saturation), which is taken to be a single valued function (i.e., no hysteresis), as in Fig. 9.5. In Section 2.2.2.a, we defined the moisture content $\theta$ as (volume of water)/(total volume of sample).

Next, the compressibility of the water is disregarded and Darcy's law is written using the hydraulic head:

$$\mathbf{U} = -K(\theta)\,\mathbf{grad}\,h$$

As usual, the hydraulic head is defined by

$$h = \frac{p}{\rho g} + z$$

but the pressure of the water is then negative (one talks of suction: $\psi = -p$).

Finally, the hydraulic head $h$ is used as the single unknown and the relation between hydraulic head $h$ and moisture content $\theta$ is made through an experimental relation between suction and moisture content. This relation does show phenomena of hysteresis, which may or may not be taken into account, as illustrated in Fig. 9.6. The description of the mechanism of hysteresis can be found in Topp (1971), and Mualem and Dagan (1972), its

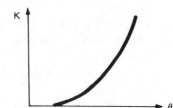

**Fig. 9.5.** Hydraulic conductivity versus moisture content.

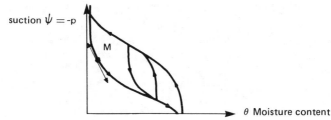

**Fig. 9.6.** Suction versus moisture content.

approximation in numerical models has been described by Mualem (1974) or Parlange (1976).

The continuity equation is then written as

$$\text{div}(\rho U) + \frac{\partial}{\partial t}(\rho\theta) = 0$$

that is,

$$\text{div } U + \frac{\partial\theta}{\partial t} = 0$$

as $\rho$ is assumed constant and the porous medium incompressible.

Darcy's law gives $U$ versus the hydraulic head. The variation of the moisture content $d\theta$ for a variation in hydraulic head $dh$ at a given fixed point remains to be expressed. If we choose a point in the plane $\psi(\theta)$ on a given cycle of wetting or draining, the variation of $\theta$ with $\psi$ is given by the slope of the tangent to this curve (Fig. 9.6):

$$d\theta = \frac{d\theta}{d\psi}d\psi = -\frac{d\theta}{d\psi}dp$$

The variation in pressure is linked to the variation in hydraulic head at a given fixed point by

$$dp = \rho g\, dh, \quad \text{thus} \quad d\theta = -\rho g\left(\frac{d\theta}{d\psi}\right)_\theta dh$$

The term $\rho g(d\theta/d\psi)_\theta$ is sometimes called the specific moisture capacity. It is, of course, a function of $\theta$. This gives

$$\text{div}(K(\theta)\,\textbf{grad}\,h) = -\rho g\left(\frac{d\theta}{d\psi}\right)_\theta \frac{\partial h}{\partial t}$$

This equation is called Richard's equation. It can also be written taking $\psi/\rho g$ as the unknown, or even $\theta$. It is very definitely nonlinear and is solved numerically [see Vauclin *et al.* (1979a,b), Neuman (1973, 1975c), Freeze (1971)].

The advantages of writing this equation in terms of hydraulic head $h$ and not in terms of moisture content $\theta$, which is equally possible, are that:

(1)  The hydraulic head is continuous when the transition is made from the saturated to the unsaturated medium. Then the medium as a whole is modeled as a continuum.

(2)  The hydraulic head is continuous even if the medium is not uniform; on the contrary, there is a discontinuity of the moisture content at the point of contact between two media of different nature.

### 9.2.2. *Unsaturated Flow with Mobile Air Phase*

In this case, the air is supposed compressible while the water and the medium are incompressible. Both fluids follow the multiphase law

$$\mathbf{U}_w = -k\frac{k_{rw}}{\mu_w}(\mathbf{grad}\,p_w + \rho_w g\,\mathbf{grad}\,z)$$

$$\mathbf{U}_a = -k\frac{k_{ra}}{\mu_a}(\mathbf{grad}\,p_a + \rho_a g\,\mathbf{grad}\,z)$$

where "w" and "a" stand for water and air. The other symbols are those defined in Section 9.1.

The continuity equations are written

$$\mathrm{div}(\mathbf{U}_w) = -\omega\,\partial s_w/\partial t$$

$$\mathrm{div}(\rho_a \mathbf{U}_a) = -\omega\,\partial(\rho_a s_a)/\partial t$$

with the state and auxiliary equations

$$\partial\rho_a/\partial t = (\rho_a/p_a)\partial p_a/\partial t \qquad (\rho_w = \mathrm{const})$$

$$s_w + s_a = 1$$

$$p_a - p_w = p_c(s_w)$$

The last equation gives the capillary pressure. This gives seven equations for seven unknowns and can be solved numerically.

In the case where the compressibility of the air is neglected ($\rho_a = \mathrm{const}$), one can add the two continuity equations and thus define a total velocity $\mathbf{U} = \mathbf{U}_w + \mathbf{U}_a$ which satisfies:

$$\mathrm{div}(\mathbf{U}) = 0$$

since $s_w + s_a = 1$

For a one-dimensional flow (e.g., vertical infiltration), this total velocity is thus a constant in space. This may simplify considerably the integration of the equations. Some analytical solutions or numerical solutions have been obtained for this case. See Brustkern and Morel-Seytoux (1970), Levan and Morel-Seytoux (1972), Noblanc and Morel-Seytoux (1972), Morel-Seytoux (1973).

Using the mobile air phase approach does not, in general, give results significantly different from the immobile approach: only in very special cases does the air pressure build up (e.g., ponding due to flooding, stratified soil profile, etc.). However, Morel-Seytoux and co-workers advocate the use of the mobile air phase approach because of the simplicity of the calculations when using the total velocity.

### 9.2.3. Solutions of the Infiltration Problem

This section summarizes the review given by Vauclin (1984).

Philip (1957) proposed an approximate analytical solution of the problem of vertical infiltration in a one-dimensional semi-infinite medium. He solves the single phase flow equation written in terms of moisture content and obtains an expression for the depth $y$ (counted positively downwards) where the water content $\theta$ is obtained at time $t$, with the initial and boundary conditions given by

$$\theta = \theta_n, \qquad y \geq 0, \qquad t < 0$$

$$\theta = \theta_0, \qquad y = 0, \qquad t \geq 0$$

$$y(\theta, t) = \sum_{i=1}^{\infty} f_i(\theta) t^{i/2}$$

The coefficients $f_i(\theta)$ are solutions of ordinary differential equations depending upon the soil characteristics. This solution becomes unreliable as $t \to \infty$.

Parlange (1971, 1972) proposed a solution for the same problem with a prescribed flux boundary condition $q_0$ at the soil surface and the same initial conditions as above. It can be written

$$y(\theta, t) = \int_{\theta}^{\theta_1(t)} D(\beta) \{ [q_0 - K(\theta_n)][(\beta - \theta_n)/(\theta_1(t) - \theta_n)]$$

$$- [K(\beta) - K(\theta_n)]\}^{-1} d\beta$$

where $\theta_1(t)$ is the water content at the soil surface, given by

$$t = \int_{\theta}^{\theta_1(t)} \frac{D(\beta)(\beta - \theta_n)}{[q_0 - K(\beta)][q_0 - K(\theta_n)]} d\beta$$

and

$$D(\theta) = -\frac{1}{\rho g} K(\theta) \left(\frac{d\psi}{d\theta}\right)_\theta$$

is the soil water diffusivity.

Green and Ampt (1911) proposed an approximate solution where they assume that the infiltrating wetting front can be defined by a water pressure $p_f$, which remains constant as the front migrates downwards. Furthermore, the soil behind the wetting front is assumed to have a uniform moisture content and thus a constant hydraulic conductivity $K_s$, which corresponds to that of a naturally saturated soil. They assume that a constant head $h_0$ is applied at the soil surface ($z = 0$) at time $t = 0$. Applying Darcy's law between the soil surface and the position $z_f$ of the front ($z$ is positive upwards) gives the infiltration rate $i$ ($i$ is negative if directed downwards):

$$h_f = \frac{p_f}{\rho_w g} + z_f \qquad h_0 = \text{constant at } z = 0$$

$$i = -K_s \frac{h_f - h_0}{z_f} = -K_s \left[\frac{p_f/\rho_w g - h_0}{z_f} + 1\right]$$

The cumulative infiltration (positive) is then $I = -z_f \Delta\theta$, where $\Delta\theta$ is the increase in moisture content in the wetted zone. Taking the derivative of $I$, one gets:

$$i = -\frac{dI}{dt} = \Delta\theta \frac{dz_f}{dt} = -K_s \left[\frac{p_f/\rho_w g - h_0}{z_f} + 1\right]$$

By integration, one obtains:

$$I = K_s t - (p_f/\rho_w g - h_0)\Delta\theta \ln\left[1 - \frac{I}{\Delta\theta(p_f/\rho_w g - h_0)}\right]$$

The pressure at the front $p_f$ can be linked to the soil characteristics in the two-phase flow theory by (Bouwer, 1964; Neuman, 1976)

$$p_f = -\int_{p_n}^{0} k_{rw}(p)\,dp$$

where $p_n$ is the initial water pressure in the soil, and $k_{rw}$ the relative permeability to water. Note that $k_{rw}$ is generally expressed as a function of the water saturation but can also be given as a function of the pressure through the suction curve.

Bouwer (1964) also suggests to use for $k_s$ one half of the saturated hydraulic conductivity to take into account the air entrapment which occurs in natural conditions.

Another expression for $p_f$ has also been suggested by Morel-Seytoux and Khanji (1974) using the total velocity defined in Subsection 9.2.2.:

$$p_f = - \int_{p_{cn}}^{0} f_w(p_c) \, dp_c$$

where $p_c$ is the capillary pressure, $p_{cn}$ is the initial capillary pressure below the wetting front, and $f_w(p_c) = 1/(1 + k_{ra}\mu_w/k_{rw}\mu_a)$ the "fractional flow".

On the other hand, empirical relationships have also been proposed to represent infiltration. These are most commonly used to analyze the results of infiltrometer tests (see Subsection 9.2.4).

Smith (1972) suggests

$$i(t) = i_c + \alpha(t - t_0)^{-\beta}, \qquad t > t_0$$

where $i$ is the infiltration rate and $i_c$, $\alpha, t_0$ and $\beta$ are constant for a given soil.

Holtan (1961) gives

$$i(t) = i_c + a(W - I)^n$$

where $i_c$, $a$, and $n$ are constant for a given soil. $W - I$ represents the available volume for storage in the unsaturated zone: $W$ is the volume of void above some impeding layer, and $I$ is the cumulative infiltration. Values for $i_c$, $a$, and $n$ for most major soils in the U.S. have been collected by the U.S. Dept. of Agriculture.

### 9.2.4. *Measurements in the Unsaturated Zone*

Predicting the flow in the unsaturated zone requires a complex series of measurements of the soil properties. These are usually made in the laboratory but can also be made in the field.

The capillary pressure versus moisture content curve is generally determined in the laboratory as follows: a sample of the soil (if possible, undisturbed) is placed in a cylinder of known volume (e.g., diameter 0.1 m, thickness 0.05 m), which is exactly filled. The lower end of the cylindrical sample lies on a porous ceramic plate, which is saturated with water and communicates with a water reservoir where the water pressure is recorded (see Subsection 2.4.2). The upper end of the sample is included in a pressure vessel, where the air pressure is controlled and measured. The high air entry pressure of the porous ceramic plate prevents the air pressure from being applied to the lower water reservoir. The difference between the air pressure and the water pressure at equilibrium is, by definition, the capillary pressure. The air pressure is increased at low moisture contents so that the water pressure always remains above atmospheric pressure to avoid boiling. To vary the moisture content of the sample for the drainage curve, one starts with a saturated sample and periodically puts the sample into an oven for partial drying. The moisture content is monitored by weighing the sample. For the imbibition curve, known amounts of water are periodically added to the sample.

The hydraulic conductivity versus moisture content curve is generally determined on an unsaturated vertical soil column where a steady-state flow has been established (e.g., a prescribed flux at the upper end). Since the flux is constant all along the column, the hydraulic conductivity is determined by measuring locally the head gradient; this is done by installing a large number of porous ceramic plugs or rings at various elevations in the column by which the water pressure in the porous medium can be measured; hence we find the head. The corresponding moisture content is measured by a continuous vertical scanning of the volum by a neutron probe (see Subsection 2.3.2.b). A transient flow situation can also be analyzed; the flux is then given by the variation in moisture content. Note that from these measurements it is also possible to determine portions of the capillary pressure curve. This does not work for very strong suctions.

The same type of measurements can be made in the field, the water pressure and the moisture content in the medium being measured by tensiometers and neutron logging, respectively (see Subsections 2.4.2 and 2.3.2.b). In general, it will be difficult to establish in the field a steady-state flow regime. Therefore the method of the "no flow boundary" is generally preferred: in summer, in general, when there is no rain, one finds by looking at the measurements of the tensiometers at several depths that there is inside the unsaturated zone a "water divide" plane (e.g., at a depth of 2 m): above this plane, the water migrates upward to compensate for evapotranspiration at the surface, and below this plane the water moves downward toward the water table. By performing two measurements of the water content at a short time interval (e.g., one week), and assuming that the no flow boundary remains exactly at the same position, it is possible to determine by continuity the flux migrating into the medium at any elevation, upward or downward, and thus to determine the hydraulic conductivity at those elevations. Since the moisture content varies with the elevation, if the soil profile is assumed to have uniform properties one can thus determine the *in situ* hydraulic conductivity versus moisture content curve, as well as the first portion of the capillary curve (for suctions below one bar). To go above one bar, indirect methods of water pressure measurements must be used, eg., the plaster blocks described in Subsection 2.4.2. (See Hillel (1971).)

Slug tests in tensiometers (Subsection 8.6.c) have also been used to measure *in situ* the hydraulic conductivity in the unsaturated zone.

Infiltration tests *in situ* are generally made using the double ring method. A first large-diameter cylinder (e.g., 0.5 m in diameter, 0.5 m in height) with its two ends opened, is pushed into the soil (e.g., 0.1 m) to insure a good seal. A second, smaller cylinder (e.g., 0.2 m in diameter and of the same height) is similarly installed at the center of the large one. A prescribed (and identical) water level is then maintained in each cylinder. However, the flux of water is only monitored in the smaller cylinder, the larger one being only there to insure that the flow beneath the smaller cylinder is directed only downwards (one-dimensional vertical flow) without any lateral migration.

Infiltration tests determine the hydraulic conductivity $K_s$ under "natural saturation," assumed to be one half of the totally saturated conditions because of air entrapment (see Subsection 9.2.3.). They can also be used to determine the empirical parameters of the infiltration formulas given in Subsection 9.2.3.

Recently, the hydraulic properties of soils have been found to have a rather large spatial variability. Geostatistical techniques (see Chapter 11) are now being used, as well as the concept of scaling. In this theory, two porous media are said to be similar if they differ only by the scale of their internal microscopic geometries. If $\alpha$ is the scaling factor (ratio of some characteristic length of the pore space, e.g., grain diameter $d_2/d_1$), it can then be shown that a soil–water property $Z$ can be scaled in the following way:

$$Z_2 = \alpha^n Z_1$$

where the exponent $n$ is $-1$ for pressure, 2 for hydraulic conductivity or flux, 1 for diffusivity, and 0.5 for sorptivity. Of course, this theory only approximately applies to real media, but it can be used to infer soil properties from a complete set of measurements at a few points and then from limited hydraulic measurements elsewhere,[*] or to introduce the scaling factors directly into the flow equations. See Vauclin (1984), Vauclin and Vachaud (1984), Nielsen et al. (1973), Warrick et al. (1977), and Russo and Bresler (1980).

## 9.3. Movement of Separating Interfaces

It may sometimes be admitted that, when two immiscible fluids move, one of them displaces the other entirely: each fluid occupies all of the porous medium in which it is found (saturation = 1), and a clear-cut interface is assumed to exist between the two fluids (Fig. 9.7). Then the continuity equation is solved (using Darcy's law) separately in each of the domains, and the dividing boundary is moved with time under the assumption that the flow is stable (no fingering).

---

[*] It is not generally possible to measure the scaling parameter $\alpha$ directly from geometrical properties: some simple hydraulic property is measured, from which the scaling parameter is inferred.

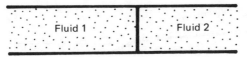

**Fig. 9.7.** Interface between two immiscible fluids.

The boundary conditions obtaining at the interface are:

(1)  Equality of pressure $p_1 = p_2$ (capillary pressure is disregarded: since the interface is assumed to be plane, it has an infinite radius of curvature).

(2)  Equality of normal Darcy velocities of fluid displacement:

$$K_i \frac{\partial h_i}{\partial n} = K_j \frac{\partial h_j}{\partial n} = -V_n$$

The displacement of the interface is given by the normal mean microscopic velocity of the two fluids at the interface, with $\omega_c$ as kinematic porosity:

$$V = \frac{V_n}{\omega_c}$$

### 9.3.1.  *Special Case: Fresh Water–Salt Water Interface in a Steady State*

This type of contact belongs in fact to the flow of miscible fluids. However, it is often dealt with by making the following two assumptions for the contact between fresh water in coastal aquifers and the sea, in a steady state: (1) the salt water is immobile, and (2) the fresh water flows over the salt water with a clear-cut interface without mixing.

This approximation is fairly valid if the flow rate is steady, i.e., with an immobile interface. In reality, there is a transition zone between the fresh and salt water, and it has a very slight thickness (of the order of 1 m), as shown in Fig. 9.8. The reason the immobile transition zone has such a small thickness is that the fresh water flows toward the coastal outlet, constantly gathering the salt diffusing into it from the immobile salt water zone.

However, if the interface moves under the influence of the tide or of variations in the outflow from the aquifer toward the sea (natural variations or withdrawals), the transition zone becomes larger and the problem must often be treated as one of miscible fluids if we want to explore what happens around the contact area (e.g., the problem of salt water intrusion into coastal aquifers).

We shall, however, choose the case of a steady state with a clear-cut interface, as in Fig. 9.9. The free surface and the salt-water wedge are flow lines. The equipotential lines are therefore at right angles to them in an isotropic

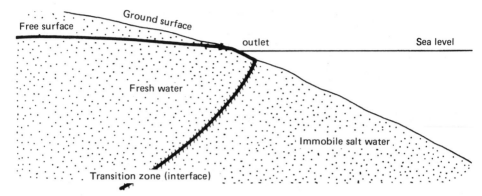

**Fig. 9.8.** Seawater intrusion

medium. At a point P (of elevation $z$) of the wedge, the equality of pressures and the immobility of the sea water allow us to write (with subscripts 1 for fresh water, 2 for salt water):

$$\left. \begin{array}{l} p_2 = -\rho_2 g z \\ p_1 = -\rho_1 g z + \rho_1 g h_1 \end{array} \right\} \quad (\rho_2 - \rho_1) g z = -\rho_1 g h_1$$

that is,

$$z = -\frac{\rho_1}{\rho_2 - \rho_1} h_1$$

The depth $z$ of the interface is related to the hydraulic head $h$ in the fresh water and to the difference in density. At a salt content of 32 g/liter (average

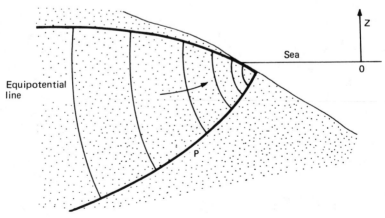

**Fig. 9.9.** Saltwater–freshwater interface.

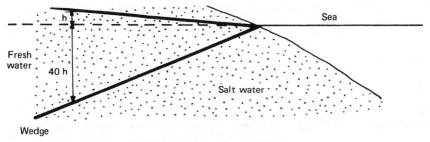

**Fig. 9.10.** Ghyben–Herzberg interface.

sea water), the mass per unit volume of sea water is close to 1025 kg/m³. This gives

$$z \simeq -40h$$

This relation is known as the Ghyben–Herzberg principle. If we further admit that the equipotential lines are vertical and that the free surface has a constant slope (both of them rather crude assumptions), the first approximation of the salt water wedge is a straight line (Fig. 9.10).

In a first approximation, it then becomes possible to estimate the probable depth of the fresh water–salt water interface in a coastal aquifer. For example, if, at 200 m from the coast, the piezometric head is 2 m above sea level, the depth of the wedge is around 80 m, unless it has already been stopped by the bedrock of the aquifer (i.e, if the aquifer is not 80 m thick) (Fig. 9.11).

However, Verruijt (1968) has calculated exactly the shapes of the free surface and the wedge in the case of an infinite homogeneous medium. For a constant seaward flow it is easy to show, by calculating the complex potential of the

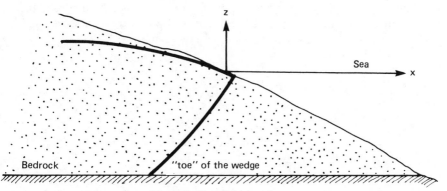

**Fig. 9.11.** Toe of the interface.

flow, that the free surface and the wedge are actually two portions of parabolas, the equations of which are, for the wedge,

$$z^2 = -\frac{2Q}{\beta K(1 + \beta)} x + \frac{Q^2}{\beta^2 K^2} \frac{1 - \beta}{1 + \beta}$$

and for the free surface,

$$z^2 = -\frac{2\beta Q}{K(1 + \beta)} x$$

which result from the following flow equation, written in terms of the complex potential $\Gamma$ (see Section 7.5)

$$Ky = (1 + \beta)\Gamma^2/2\beta Q + i\Gamma$$

with $z$ counted positively upward from the sea level, $x$ counted positively seaward from the coast, $Q$ as the flow of *fresh water* seaward in the aquifer per unit length perpendicular to the plane of the figure, $\beta = (\rho_2 - \rho_1)/\rho_1$, $y = x + iz$ (complex affix), and $K$ as the isotropic hydraulic conductivity of the medium. (For anisotropic media, the solution can be found by a transformation in the coordinate system; see Section 7.1.c.) Rumer and Shiau (1968) have given such an expression in the case where the aquifer is anisotropic in the $x$ and $z$ directions; they give, with the same notation

$$z^2 = -\frac{2Q}{\lambda K_z \beta(\beta + 1)} x + \frac{Q^2}{\beta^2 K_z^2} \frac{1 - \beta}{1 + \beta}$$

for the wedge and

$$z^2 = -\frac{2\beta Q}{\lambda K_z(1 + \beta)} x$$

for the free surface where $K_z$ and $K_x$ are the hydraulic conductivities in the $z$ and $x$ directions and $\lambda = \sqrt{K_x/K_z}$ is the anisotropy ratio. Their solution also gives the head and the streamlines in the flow domain, as

$$x = \frac{Q}{K_z}\left[\frac{\beta + 1}{\beta}\phi\psi - \phi\right]$$

and

$$z = \frac{Q\lambda}{K_z}\left[\frac{\beta + 1}{2\beta}(\phi^2 - \psi^2) - \psi\right]$$

where $\phi = -K_z h/Q$ is the potential function at location $(x, z)$ and $\psi$ is the associated stream function, defined by

$$\frac{\partial \psi}{\partial x} = -\lambda \frac{\partial \phi}{\partial z} \quad \text{and} \quad \frac{\partial \psi}{\partial z} = \frac{1}{\lambda} \frac{\partial \phi}{\partial x}$$

where $h$ is the head. Now if the aquifer is confined, i.e., the line $z = 0$ for $x < 0$ is a confining bed, the same authors give the interface, the head and the streamlines as:

$$z^2 = -\frac{2Q}{K_z \beta \lambda} x + \frac{Q^2}{K^2 \beta^2}$$

$$x = -\frac{Q\lambda}{2K_z \beta} (\phi^2 - \psi^2)$$

$$z = -\frac{Q}{K_z \beta} \phi \psi$$

Numerical solutions of the saltwater–freshwater interface problem can be found in Sa da Costa (1981), Huyakorn and Pinder (1983), and Allen (1986).

## 9.4. Multiphase Pollution Problems

This section is mainly concerned with the pollution of aquifers by petroleum products, which are the most common fluids that do not mix with water. They are difficult to treat, because pollution of the aquifers occurs through the surface and it is therefore necessary to begin by dealing with the transfer of the petroleum products through the unsaturated medium. It then becomes apparent that a rather large quantity of petroleum products is needed for the pollution to arrive at the aquifer, since a significant part is retained by capillarity first in the unsaturated zone and then in the saturated zone. A minimum oil saturation in the soil is indeed necessary, below which the oil phase cannot migrate (zero relative permeability and capillary entrapment).

Once it has arrived at the water table, the oil phase accumulates and migrates downstream. Figure 9.12, taken from Freeze and Cherry (1979) and from Schwille (1967), illustrates the process.

However, we must remember that certain components of the petroleum products (especially the aromatic ones, e.g., certain phenols) may be dissolved in water and spread as miscible fluids, rapidly polluting the aquifers since they can drastically alter the taste of water, even in very small amounts.

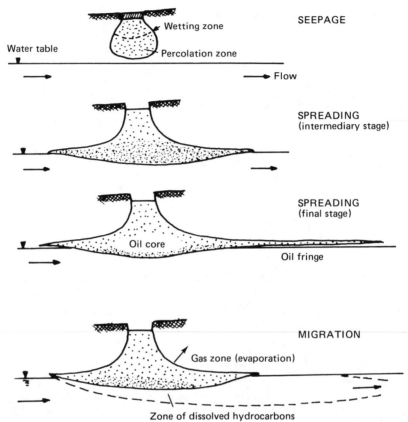

**Fig. 9.12.** Movement of spilled oil above an aquifer. [From Freeze and Cherry (1979).]

The organic phase retained by capillarity in the unsaturated zone is gradually consumed through bacterial oxidation of the hydrocarbons, but this may take several tens of years. Crude oil or fuel oil is generally more rapidly consumed than gasoline. However, some of the degradation products of the hydrocarbons, resulting from this bacterial oxidation, may themselves be soluble contaminants, further polluting the aquifer.

In order to protect an aquifer on top of which an oil spill has occurred, one first tries to dig out as much as possible of the contaminated earth and recover most of the oil. Wells are then installed to create a local drawdown of the piezometric surface of the aquifer; the oil reaching the water table will then flow in the direction of the wells and end up on top of the water in the borehole. It can be recovered by a supplementary skimming pump.

However, it is difficult to recover more than 50% of the spill because of the capillary entrapment. One then generally keeps pumping the wells for several years, even if oil is no longer found on top of the water in the borehole, in order to extract the dissolved hydrocarbons and prevent them from polluting the rest of the aquifer. The flow rate in the wells has to be adjusted so that no polluted water can leave the area due to the natural gradient in the aquifer [see Fried *et al.* (1979), Schwille (1984).]

*Chapter 10*

# Flow of Miscible Fluids: Dispersion, Retention, and Heat Transfer

We shall discuss problems where a single fluid phase is present in the medium, but where its composition or its properties vary. It may be the case of two miscible liquids (e.g., fresh water and salt water) or of a substance dissolved in variable concentration in a liquid or even of variable temperature in a fluid. We shall study three cases separately: substances that do not interact with the medium, substances that do, and finally, heat transport.

For miscible fluids we shall consider a single fluid phase and define the concentration of one substance in the other, for example, concentration of salt in water.

There are several ways of defining the concentration:

(1) The volumetric concentration, as mass of solute per unit volume of solution ($kg/m^3$, or g/liter.)

(2) The mass concentration, as mass of solute per unit mass of solution (kg/kg); the ppm or ppb (part per million or billion) is equal to $10^6$ or $10^9$ times the dimensionless unit, respectively.

(3) The molarity, number of moles of solute per unit volume of solution ($mol/m^3$). This is the standard definition of concentration in SI units.

(4)   The molality, number of moles of solute dissolved in a unit mass of solvent (mol/kg).

(5)   The equivalent per liter (epl), number of moles of solute multiplied by the valence of the species, per liter of solution. The common unit is the milliequivalent per liter (meq/liter), $10^3$ times greater than the epl.

(6)   The nuclear activity (for radionuclides) per unit volume or per unit mass of solution ($Bq/m^3$ or $Bq/kg$). The Becquerel corresponds to a quantity giving one desintegration per second of a radionuclide.

(7)   The ratio of the concentration to the maximum permissible concentration in drinking water (MPCW), often used for radionuclides as well.

In the following, we will mostly use the volumetric concentration $C$ and call it simply concentration. This concentration varies continuously in the medium; there is no longer a sharp interface between two fluids as in the case of immiscible fluids. When the fluid moves, the concentration varies in time and space. This type of displacement is called mass transport or solute transport in porous media.

## 10.1. Solute Transport of Nonreactive Substances

In order to clearly distinguish between the laws of transport and the laws of interaction between the transported substances and the medium, we shall discuss in this section the transport of substances that are not subject to any changes, exchanges, or reactions while crossing the porous medium. These are the nonreactive (or conservative) substances. This therefore excludes radioactive decay as well as adsorption.

In Section 10.2, we shall deal with the problem of reactive substances and see how special laws governing their behavior must be added to the transport equations as such.

It is important to define, at the outset, what is meant by solute transport. First of all, it concerns constituents included in the chemical combinations of elements that are soluble in water. These elements may themselves be more or less ionized* according to their ionic charge. However, these dissolved substances may also be present in the shape of electrically neutral chemicals or complexes created by aggregates of different molecules or ions.

Furthermore, salts considered to be "insoluble" may, nevertheless, be transported in a dissolved state as trace concentrations since, in reality, this "insolubility" is never total. Because certain radionuclides, for example, are

---

* Recent terminology calls any dissolved salt an ion, irrespective of whether it is electrically charged or not. Thus, for example, $CaCO_3$ in solution, not disassociated into $Ca^{2-}$, $CO_3^{2+}$, is called an electrically neutral ion complex.

toxic even in weak concentrations, these traces may be significant in calculations of radiological safety studies.

Finally, we must also consider constituents transported in the form of larger molecular aggregates, such as colloids, which may, in the end, be caught by mechanical filtration through the porous medium network (see Section 10.2.).

All these transported substances are known as "solutes," as long as they do not constitute a mobile phase distinct from the transporting fluid but integrate themselves into the single fluid phase (the water of the natural medium), possibly modifying its physical (e.g., mass per unit volume, viscosity) and chemical properties.

Solute transport is thus contrasted with the flow of immiscible fluids such as that of oil and water, which obeys completely different laws of migration.

We shall now define the laws of transport in porous and fractured media and in the unsaturated zone.

### 10.1.1. Porous Media

Traditionally, three main mechanisms of migration are recognized: convection, diffusion, and kinematic dispersion.

(a) *Convection (or advection).* This is the phenomenon where dissolved substances are carried along by the movement of fluid displacement. It is the most easily understood of the displacements. It must, however, be defined with precision:

(1)   What portion of the fluid in the porous medium is effectively mobile?
(2)   What is the real velocity of this fluid?

Indeed, in a saturated porous medium, a distinction must be made between two fluid fractions: the one that is bound to the solid by molecular forces of attraction, called adhesive water, and that which is free to circulate under the influence of the gradients of hydraulic head, called free water. In reality, especially in media with low permeability, the magnitude of the free fraction depends on the degree of the hydraulic gradient: for clays, the deviations from Darcy's law, mentioned in Section 6.4.2, are accompanied by an increase in the fraction of free water at the expense of that of adhesive water, when the hydraulic gradients increase.

It is therefore necessary to define a kinematic porosity $\omega_c$, which corresponds to the voids in the porous medium occupied by the moving water. This kinematic porosity may thus be dependent on the gradient, but such measurements have never been made, and $\omega_c$ will be assumed constant.

If it is assumed that the transport is governed only by the phenomenon of convection in the moving fluid fraction, the resulting transport equation is easily found on the macroscopic scale of the representative elementary volume by using the principle of mass balance.

Take an elementary volume $D$ of a porous medium with an outside boundary $\Sigma$. The mass balance of the transported substance in the volume $D$ is given by writing that the integral over $\Sigma$ of the mass flux of the transported substance into $D$, is equal to the change of mass of the substance in the volume $D$ per unit time.

The volumetric flux of fluid crossing the area $\Sigma$ is given by the Darcy velocity $\mathbf{U}$. This volumetric flux is transformed into mass flux of the transported substance through scalar multiplication of the Darcy velocity $\mathbf{U}$ by the volumetric concentration $C$; the left-hand side of the balance equation (mass flow entering $D$) becomes

$$\int_\Sigma C\mathbf{U}\cdot\mathbf{n}\,d\sigma$$

where $\mathbf{n}$ is the normal vector on $\Sigma$ directed toward the outside of $D$.

The mass of the transported substance contained in the element $D$ at time $t$ is the integral of the elementary volumes of fluid $\omega_c\,dv$ contained in the porous medium multiplied by the volumetric concentration $C$ in the fluid of the substance in question:

$$\int_D \omega_c C\,dv$$

The porosity $\omega_c$, which must be used here, is the kinematic porosity (i.e., the fraction of fluid that circulates) because, for the moment, we assume that it is the only one capable of containing the transported substance; elsewhere, the concentration $C$ is assumed to be zero. Thus, the assumption is made that it is possible to define, in the volume $D$, a mean concentration $C$, which is the result of the mixture of all the substances in the mobile fluid fraction of $D$.

The variation of this mass per unit time is obtained simply by taking the derivative of this expression with respect to time:

$$\frac{\partial}{\partial t}\int_D \omega_c C\,dv = \int_D \omega_c\frac{\partial C}{\partial t}\,dv$$

The passage from the first form to the second is made through Leibnitz's rule, since $D$ is fixed and $\omega_c$ is assumed constant.

The mass balance of the solute equation becomes

$$\int_\Sigma C\mathbf{U}\cdot\mathbf{n}\,d\sigma = \int_D \omega_c\frac{\partial C}{\partial t}\,dv$$

We transform the integral of the area of the left-hand side into a volume integral using Ostrogradsky's formula:

$$\int_\Sigma C\mathbf{U}\cdot\mathbf{n}\,d\sigma = -\int_D \operatorname{div}(C\mathbf{U})\,dv$$

i.e.,

$$-\int_D \operatorname{div}(C\mathbf{U})\, dv = \int_D \omega_c \frac{\partial C}{\partial t}\, dv$$

By taking out the integral signs on both sides, since $D$ is arbitrary we find:

$$-\operatorname{div}(C\mathbf{U}) = \omega_c \frac{\partial C}{\partial t} \tag{10.1.1}$$

(b) *Molecular diffusion.* This is a physical phenomenon linked to the molecular agitation. In a fluid at rest the Brownian motion projects particles in all directions of space. If the concentration of the fluid is uniform in space, each one of two neighboring points sends, on average, the same number of particles* toward the other, and the molecular agitation does not change the concentration of the solution. However, if the concentration of the solution is not uniform in space—in others words, if there is a gradient of concentration between two neighboring points—the point with the highest concentration sends out, on the average, more particles in all directions than the point with a lower concentration. The result of this molecular agitation is then that particles are transferred from zones of high concentration to those of low concentration.

Fick has found that the mass flux of particles in a fluid at rest is proportionate to the concentration gradient:

$$\phi = -d_0 \operatorname{\mathbf{grad}} C \qquad \text{Fick's law}$$

The coefficient $d_0$, known as the molecular diffusion coefficient, is isotropic and can be expressed by:

$$d_0 = \frac{RT}{N} \frac{1}{6\pi\mu r} \qquad (\text{length}^2 \ \text{time}^{-1})$$

where $R$ is the constant of perfect gases, 8.32 SI units (mass length$^2$ time$^{-2}$ kelvin$^{-1}$); $N$ is Avogadro's number, $6.023 \times 10^{23}$; $T$ is absolute temperature (kelvin = temperature $°C + 273.15$); $\mu$ is fluid viscosity; and $r$ is the mean radius of the diffusing molecular aggregates. This expression is only valid for an infinite dilution; otherwise the activity of the elements and the ionic strength of the solution has to be taken into account. However, this effect is rather small. As far as variation with temperature is concerned, since $\mu$ is also a function of temperature, it is found that $d_0$ varies in general exponentially with $T$:

$$d_0(T_2) = d_0(T_1)\exp[E(T_2 - T_1)/RT_1T_2]$$

---

* Here we are talking about particles of solutes, not about water, in a fluid without any porous medium.

where $E$ is the activation energy of the ion in solution, which is on the order of $21 \times 10^3$ J/mole for most ions. Most common ions have a diffusion coefficient on the order of $10^{-9}$ to $2 \times 10^{-9}$ m$^2$/s at 20°C; for instance, for NaCl in water at 20°C, $d_0 = 1.3 \times 10^{-9}$ m$^2$/s.

If the transport of substances in a fluid at rest is only due to Fickian diffusion, the principle of mass balance is used to establish the law of movement, exactly as above:

$$\int_\Sigma \phi \cdot \mathbf{n} \, d\sigma = -\int_D \text{div}(\phi) \, dv = \frac{\partial}{\partial t} \int_D C \, dv$$

If $\phi$ is replaced by its expression and the integrals are taken out,

$$\text{div}(d_0 \, \mathbf{grad} \, C) = \frac{\partial C}{\partial t}$$

In porous media the molecular diffusion continues all through the fluid phase (the mobile as well as the immobile one). Only the solid stops the Brownian movement of the particles, since diffusion in solids is negligible. For an immobile fluid in a porous medium this gives a diffusion coefficient in porous media that is lower than $d_0$. It is usually admitted that the ratio $d/d_0$, called the tortuosity of the medium, is equal to:

$$\frac{d}{d_0} = \frac{1}{F\omega}$$

where $F$ is the formation factor of the geophysicists, defined by the ratio of the electric resistivity of the rock over the resistivity of the contained water, and $\omega$ is the total porosity. In practice, $d/d_0$ varies from 0.1 (clays*) to 0.7 (sands).

For a fluid circulating in a porous medium it is easy to combine the phenomena of convection and diffusion, giving for the left-hand side:

$$\int_\Sigma \phi \cdot \mathbf{n} \, d\sigma + \int_\Sigma C\mathbf{U} \cdot \mathbf{n} \, d\sigma = -\int_D \text{div}(\omega\phi + C\mathbf{U}) \, dv$$

The total porosity comes into play because the integral of the diffusive flux $\phi$ over $\Sigma$ is zero over the solid [area $(1 - \omega)\Sigma$] and nonzero over the pores (area $\omega\Sigma$), whereas the Darcy velocity is defined as if the entire area $\Sigma$ were open to the flow. Now the right-hand side becomes

$$\frac{\partial}{\partial t} \int_D \omega_c C \, dv + \frac{\partial}{\partial t} \int_D (\omega - \omega_c)C' \, dv$$

---

* Neretnieks (1979) quotes measurements of $d/d_0$ up to 0.01 in highly compacted bentonite for gases, cesium, and strontium.

It is necessary to take two porosities into account: the kinematic porosity $\omega_c$, which corresponds to the mobile fraction of the fluid phase with the concentration $C$, and the porosity which corresponds to the immobile fraction $\omega - \omega_c$ ($\omega$ is the total porosity) with a concentration $C'$, which may be different from $C$.

In the case of pure convection, only the mobile fraction of the fluid could contain the transported substances, whereas here the immobile fraction necessarily contains these substances, as the molecular diffusion causes them to penetrate into the immobile fraction.

By substituting and simplifying as above, we write this equation:

$$\text{div}(\omega d \, \textbf{grad} \, C - C\textbf{U}) = \omega_c \frac{\partial C}{\partial t} + (\omega - \omega_c)\frac{\partial C'}{\partial t} \qquad (10.1.2)$$

Here, it has to be decided whether or not $C'$ is to be included in the incoming flux on the left-hand side. Where convection is concerned, it is clear that only the concentration $C$ of the mobile fraction brings solutes into the elementary volume. As for the diffusion, the immobile fractions on each side of the area $\Sigma$ of the elementary volume exchange substances according to the gradient of $C'$. Rigorously, the diffusive term should be written:

$$\omega_c \, \text{div}(d_1 \, \textbf{grad} \, C) + (\omega - \omega_c)\text{div}(d_2 \, \textbf{grad} \, C')$$

where $\omega_c$ is the fraction of the area $\Sigma$ occupied by the mobile fluid, which is then diffusing with a coefficient $d_1$, and $(\omega - \omega_c)$ is the rest of the fluid fraction of $\Sigma$ (immobile fluid) through which the diffusion of the concentration $C'$ takes place with a diffusion coefficient of $d_2$. Probably $d_1$ and $d_2$ would be, respectively, stronger and weaker than the global coefficient $d$.

We shall disregard this effect, particularly in view of the existence of the kinematic dispersion, which already makes the diffusion almost negligible.

Section 10.2.a will show how the existence of the concentration $C'$ in the immobile fraction may combine with mechanisms of adsorption over the solid phase of the porous medium.

(c)  *Kinematic dispersion.*  This is a mixing phenomenon linked mainly to the heterogeneity of the microscopic velocities inside the porous medium on whatever scale they are observed.

(1)  Inside a pore the velocities in the mobile fraction are not uniformly distributed; in laminary flow, as Poiseuille's formula suggests for a cylindrical pipe, the velocity profile is of the type given by Fig. 10.1. This causes a faster propagation of the transported substances along the axis of the pores, which, through mixing and molecular diffusion, produces a progressive spreading of the transported substances compared to the mean movement of convection.

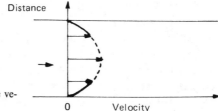

**Fig. 10.1.** Parabolic distribution of the velocities in a cylindrical pipe.

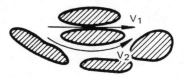

**Fig. 10.2.** Velocity differences between pores.

(2)   The differences of aperture and travel distance from one pore to another create a difference in mean velocities, as in Fig. 10.2. The fluids traveling by each of the paths mix with each other and cause a dilution of the concentration. It should be noted that this process also causes a spreading of the substances at right angles to the main direction of flow, as in Fig. 10.3.

(3)   A stratification or any features of large-scale heterogeneity such as lenses, interlayered deposits, broken or fractured zones, etc. also introduce a heterogeneity into the velocity field, which, through the same mechanisms as above, causes the substances transported by the fluid to mix and spread in all directions of space.

The kinematic dispersion is therefore in fact the product of an existing real velocity field of very complex and unknown nature, which is entirely disregarded in the convection, when the fictitious mean Darcy velocity is used (which assumes that the whole of the continuous medium is open to flow).

The division of the transport into a convection term, representing the mean displacement, and a dispersive term, integrating the effects of the heterogeneities, is quite arbitrary; the respective role of each of the terms is chiefly determined by the degree of precision with which the porous medium and the velocity field can be described.

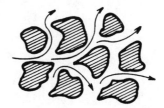

**Fig. 10.3.** Variation of the velocity direction between pores.

What mathematical form can be given to the kinematic dispersion? The answer may be either theoretical or experimental.

The classical dispersion theory has been developed primarily by Taylor (1953), De Josselin De Jong (1958), Saffman (1959, 1960), Scheidegger (1960), Bear and Bachmat (1967), Fried and Combarnous (1971) and established by considering a random distribution in space of the small channels forming the pathways through the pores of the porous medium.

The suggested mathematical formula adopts a law of transport through dispersion similar to Fick's law which accounts for the phenomena of mixing:

$$\cdot \quad \text{dispersive flux } \phi = -\mathbf{D} \text{ grad } C$$

which is applied to the whole section of the medium, like the Darcy velocity, but with a dispersion coefficient $\mathbf{D}$ which:

(1)   Is a tensor assumed to be symmetrical and of the second order.

(2)   Has as its principal directions: (a) the direction of the velocity vector of the flow (i.e., linked to the fluid and not to the medium), and (b) two other directions, generally arbitrary and at right angles to the first one.

(3)   Has coefficients that are themselves dependent on the module of the flow velocity.

If the dispersion tensor is expressed in its principal directions of anisotropy, it is limited to three components:

$$\mathbf{D} = \begin{vmatrix} D_L & 0 & 0 \\ 0 & D_T & 0 \\ 0 & 0 & D_T \end{vmatrix}$$

where $D_L$ is the longitudinal dispersion coefficient (in the direction of the flow) and $D_T$ the transverse dispersion coefficient (in the two directions at right angles to the velocity). Note that $\mathbf{D}$ is anisotropic, even if the medium has isotropic permeability: the anisotropy of the dispersion tensor stems from the fact that the spreading of the concentration is larger in the direction of the velocity than in transverse directions. For instance, if a brief injection of tracer is made through a piezometer in an aquifer, the shape of the traced water would appear as in Fig. 10.4 at different times. Inside the spotted area the concentration also decreases with time because of the spreading. Note that if only convection occurred, according to Eq. (10.1.1), the bubble of tracer would progress in the medium without any spreading. If only convection and isotropic molecular diffusion occurred as in Eq. (10.1.2), the bubble would remain spherical but would spread slightly with time. The large anisotropic spreading outlined here is due to kinematic dispersion.

This dispersion flux $\mathbf{D}$ grad $C$ is added to the diffusive flux $\omega d$ **grad** $C$ on the left-hand side of Eq. (10.1.2).

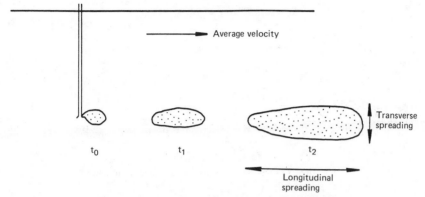

**Fig. 10.4.** Spreading of a tracer slug in an aquifer.

The value of these dispersion coefficients varies with the absolute value of the Darcy velocity or with that of the mean microscopic flow velocity:

$$|u^*| = \frac{|U|}{\omega_c}$$

For this purpose, a dimensionless Peclet number is defined:

$$P_e = \frac{|u^*|\sqrt{k}}{d_0} = \frac{|U|\sqrt{k}}{\omega_c d_0} \quad \text{or} \quad P_e = \frac{|u^*|l}{d_0}$$

where $|u^*|$ is the module of the mean microscopic velocity, $k$ is the intrinsic permeability, $d_0$ is the coefficient of molecular diffusion, and $l$ is a characteristic length of the porous medium (mean diameter of the grain or the pores, for example.)

In the laboratory of the French Petroleum Institute, O. Pfankuch (1963) has experimentally verified on small samples the validity of this dispersion law suggested by the theory, and has established an empirical relation linking the dispersion coefficient with the Peclet number. Depending on the size of the Peclet number, five flow regimes are defined, and for each of them an empirical relation between $D_L$, $D_T$, and $P_e$ is found. These five regimes are shown in Fig. 10.5. These five dispersion regimes correspond to various distributions of the roles played by molecular diffusion and kinematic dispersion:

(I)   Pure molecular diffusion.
(II)  Combination of I and III.
(III) Predominant kinematic dispersion.
(IV)  Pure kinematic dispersion.
(V)   Kinematic dispersion outside the domain where Darcy's law is valid.

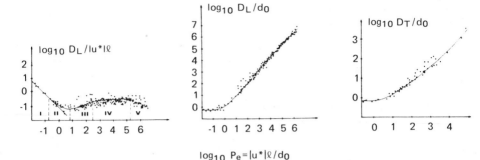

$\log_{10} P_e = |u^*|\varrho/d_0$

**Fig. 10.5.** Dispersion coefficients versus Peclet number.

In the domain of the usual velocities (domains III and IV, $P_e > 10$), the following relations are generally admitted:

$$D_L = \alpha_L|U|$$

$$D_T = \alpha_T|U|$$

where $\alpha_L$ and $\alpha_T$, which have the dimension of a length, are known as intrinsic dispersion coefficients or dispersivities. When measured in the laboratory on a column of sand $\alpha_L$ is on the order of a few centimeters. In the field, it is on the order of a meter to a hundred meters depending on the degree of heterogeneity of the formation [see Lallemand-Barrès et al. (1978) and Section 10.3]. However, $\alpha_T$ is much smaller, between $\frac{1}{5}$ and $\frac{1}{100}$ of $\alpha_L$.

A more general type of dispersion coefficient may also be adopted, explicitly taking into account the molecular diffusion $d$ so as to extend the validity of the model towards the low Peclet numbers, i.e., the states I and II, where the Darcy velocity is weak:

$$D_L = \omega d + \alpha_L|U| \qquad D_T = \omega d + \alpha_T|U| \qquad (10.1.3)$$

where $d$ is the molecular diffusion coefficient in porous media, and $\omega$ the total porosity. This term only comes into play when $|U|$ is very small.

The transport equation, now including the kinematic dispersion which takes the place of the diffusion term, becomes

$$\text{div}(\mathbf{D}\,\text{grad}\,C - C\mathbf{U}) = \omega_c\frac{\partial C}{\partial t} + (\omega - \omega_c)\frac{\partial C'}{\partial t} \qquad (10.1.4)$$

In this case the kinematic dispersion transport indeed concerns the mobile fraction with concentration $C$ and not the immobile fraction at concentration $C'$.

For the purpose of simplification, we now make the assumption that the concentration $C$ in the mobile fraction instantaneously reaches an equilibrium with the concentration $C'$ in the immobile fraction, due to the action of molecular diffusion. Because of the extraordinary interpenetration of the two fractions, it may then be admitted that

$$C' = C$$

and

$$\text{div}(\mathbf{D}\ \mathbf{grad}\ C - C\mathbf{U}) = \omega\frac{\partial C}{\partial t} \tag{10.1.5}$$

which is the usual form for the dispersion equation. If we divide by $\omega$, the total porosity, a mean fictitious velocity appears:

$$\mathbf{u}' = \frac{\mathbf{U}}{\omega}$$

Moreover, it is possible to divide the dispersion coefficient by $\omega$:

$$\mathbf{D}' = \frac{\mathbf{D}}{\omega}$$

that is,

$$D'_L = d + \frac{\alpha_L}{\omega}|U| \qquad D'_T = d + \frac{\alpha_T}{\omega}|U|$$

which can also be written

$$D'_L = d + \alpha_L|u'| \qquad D'_T = d + \alpha_T|u'|$$

This shows that the dispersivities $\alpha_L$ and $\alpha_T$ stay the same whatever form is given to the dispersion equation (with or without $\omega$ on the right-hand side).

Then the transport equation becomes

$$\text{div}(\mathbf{D}'\ \mathbf{grad}\ C - C\mathbf{u}') = \frac{\partial C}{\partial t} \tag{10.1.6}$$

which is the classical form more commonly used in the literature. Fried (1975) has shown that if the mass per unit volume $\rho$ of the solution cannot be considered as constant when the concentration varies, the dispersion equation should be written

$$\text{div}(\mathbf{D}'\ \rho\ \mathbf{grad}\ C/\rho - C\mathbf{u}') = \partial C/\partial t \tag{10.1.7}$$

In practice, this expression need only be used for studying the movement of dense brines (see also Subsection 10.1.1.d) and will not be used in the remainder of this text.

Conversely, if the immobile fluid fraction is assumed not to be invaded by the transported substances, it may be admitted that $C' = 0$, and the transport equation is reduced to

$$\text{div}(\mathbf{D} \, \mathbf{grad} \, C - C\mathbf{U}) = \omega_c \frac{\partial C}{\partial t} \tag{10.1.8}$$

It is also possible to divide both sides by $\omega_c$ in order to make a fictitious mean convection velocity $\mathbf{U}/\omega_c = \mathbf{u}^*$ appear.

This discussion shows that, contrary to general practice, it is preferable to retain the Darcy velocity, which has a precise definition and meaning for the convection term, and make the porosity (or porosities) appear explicitly on the right-hand side of the equation. We shall come back to the right-hand side in the discussion of adsorption (Section 10.2.a, fifth case).

The classical theory of dispersion was first established for homogeneous isotropic media and later extended to and used, without modification, for heterogeneous and anisotropic ones. In Section 10.1.3, we shall see how new concepts are currently being developed for these media. In particular, it appears that the dispersion tensor is no longer oriented in the direction of the velocity but at an angle to it.

(d)  *Coupling of the transport equation with that of fluid movement.*  To the transport equation must be added another needed for the calculation of the Darcy velocity $\mathbf{U}$:

$$\mathbf{U} = -\frac{\mathbf{k}}{\mu}(\mathbf{grad} \, p + \rho g \, \mathbf{grad} \, z)$$

which is the generalized Darcy equation written in terms of pressure, since $\rho$ varies with $C$. Finally, we have the continuity equation of the fluid with its state equations:

$$\text{div}(\rho\mathbf{U}) + \frac{\partial}{\partial t}(\rho\omega) = 0 \tag{3.23}$$

$$\rho = \rho(C, p) \qquad \mu = \mu(C, p)$$

$\mathbf{D}$ = a function of $\mathbf{U}$ and of the molecular diffusion coefficient in porous media $d$.

These equations are coupled and should thus be solved simultaneously. (The velocity $\mathbf{U}$ depends on the concentration, and vice versa.) Note that Eq. (10.1.7) should be used instead of (10.1.5) or (10.1.6).

(e)  *Simplification of the dispersion equation: Tracer hypothesis.*  The tracer hypothesis consists in separating the equation of the variation in concentration from that of the velocity: the concentration $C$ is assumed to be so low that the mass per unit volume $\rho$ of the fluid is almost constant. Then, the velocity $U$ does not depend on the concentration.

The flow problem is therefore solved separately, and only the dispersion equation (10.1.5) or (10.1.6) remains to be solved. There are a few analytical solutions for the latter [see Section 10.3 and Bear (1972, 1979)], but the solution must often be numerical, which causes quite a number of difficulties in numerical analysis due to the discretization, in particular the one related to the appearance of "numerical dispersion" (see Chapter 12).

(f) *Boundary conditions of the dispersion equation.* When boundary conditions are imposed on the dispersion equation, it must be remembered that this equation contains two separate terms: a diffusive and a convective term.

The characteristics of a boundary are, first of all, related to the direction of the flow crossing it:

(1)   Boundary with incoming flow. The concentration on this type of boundary is fixed by the concentration of the entering flux; $C = C_0$.

(2)   Boundary with outgoing flow (e.g., the outlet of the geologic formation toward the surface, such as a superficial aquifer, a body of fresh water or salt water, etc.). The concentration of the outside medium assumed to be well mixed does not play a dominant role on the concentration inside the medium. The concentration in the flux going out by convection is said not to vary when it crosses the boundary:

$$\mathbf{U} \cdot \frac{\partial C}{\partial n} \cdot \mathbf{n} = 0 \qquad \text{or} \qquad \frac{\partial C}{\partial n} = 0$$

where $\mathbf{n}$ is the normal to the boundary. The dispersive flux is then disregarded.

(3)   Boundary with outgoing flow, taking into account the dispersive flux. If one assumes, as in (2), that the outside domain is well mixed, and has a concentration $C_0$ independent of the flux coming from the inside domain, then there is by definition a discontinuity in concentration at the boundary, and the dispersive flux would become infinite. One must therefore consider a small buffer zone of thickness $\varepsilon$ between the two media, and assume that the concentration varies, e.g., linearly between concentration $C$ (inside the medium) and $C_0$ (outside the medium), in the buffer zone. This zone is further assumed to have no storage capacity, so that at all times the total flux coming from the inside medium is equal to the total flux leaving for the outside medium. The total flux from the inside medium is written

$$(\mathbf{D} \text{ grad } C - C\mathbf{U}) \cdot \mathbf{n}$$

and total flux to outside medium is written

$$d_0(C - C_0)/\varepsilon - C_0 U_n$$

where $\mathbf{D}$ is the total dispersion coefficient in the medium (molecular diffusion plus hydrodynamic dispersion), $\mathbf{U}$ is Darcy's velocity, $\mathbf{n}$ is the outer normal to the boundary, $U_n$ is the component of $U$ along $\mathbf{n}$, $d_0$ is the molecular diffusion coefficient in water, and $\varepsilon$ the thickness of the buffer zone. Equating these two fluxes gives a Fourier-type boundary condition in $C$ and $\partial C/\partial n$; however, the choice of $\varepsilon$ is very arbitrary, and therefore the boundary condition given in (2) is generally preferred. Note that we have written the flux to the outside medium considering molecular diffusion only and not hydrodynamic dispersion, since the buffer zone is supposed here to be made of pure water and not of a porous medium. This is of no importance as $\varepsilon$ is arbitrary. The convective flux leaving the buffer zone is indeed $-C_0 U_n$ since, by definition, the concentration of the water leaving the buffer zone becomes $C_0$.

(4) No-flow hydraulic boundary. The velocity $\mathbf{U}$ is parallel to the boundary, and the convection flux $\mathbf{U} \cdot (\partial C/\partial n)\mathbf{n}$ will always be zero even if $\partial C/\partial n$ is not.

If there is no solute flow coming in or going out by *pure diffusion* across the boundary, we write

$$\partial C/\partial n = 0$$

On the other hand, if there is a known diffusion phenomenon across this boundary, we write

$$\partial C/\partial n = f$$

(g) *Choice of dispersion coefficients.* The dispersion coefficients (or dispersivity) can be measured on a column in the laboratory. However, such coefficients are of little use in forecasting a real migration in the field, where the scales of heterogeneities are different and the coefficients much larger. Consequently, they have to be measured by tracer experiments, which are interpreted by analytical or numerical methods.

It is found, however, that if the space and time scales of the tracer experiments are changed, different values are obtained for the coefficients. This means that the problem of choosing coefficients capable of forecasting long-distance migrations is not completely solved. See Subsection 10.1.3.

(h) *Remark: Upstream Migration* For the high longitudinal dispersion coefficients, it may also be doubtful whether a theory that does not distinguish between the direction of the convective circulation and that of the concentration gradient is valid for determining the dispersion flux. If it is only a question of molecular diffusion, i.e., a phenomenon that is isotropic in all directions, it is obviously not necessary to define the flow direction, but in the case of kinematic dispersion the case is different.

Consider, for instance, an axis $l$ (longitudinal), which is parallel to the direction of the flow velocity and assumed to be oriented in the same direction.

$$U \to \mathbf{M}$$
$$\xrightarrow{\quad\;\;\vdash\quad\;\;} l$$

At a point M, the dispersion and convection fluxes are

$$\phi_l = -\alpha_L |U| \frac{\partial C}{\partial l} + CU$$

The absolute value sign may be taken out and the velocity $U$ factored out:

$$\phi_l = -U\left(\alpha_L \frac{\partial C}{\partial l} - C\right)$$

If $\partial C/\partial l < 0$, i.e., the concentration decreases downstream (e.g., the case of the substance spreading in a clean medium, propagation of the migration front), the resulting flux $\phi_l$ is always positive: that is to say, convection and dispersion cause the substance to spread downward. This shows that the dispersivity $\alpha_L$ does indeed accelerate the propagation through the influence of the velocity heterogeneities.

However, if $\partial C/\partial l > 0$, which is the case of a clean fluid sweeping through a contaminated medium, the magnitude of $\alpha_L \partial C/\partial l$ in relation to $C$ determines the sign of $\phi_l$: for strong gradients and dispersivities, $\phi_l$ may become negative, which means that the transported substances start to travel upstream against the flow. It is difficult to understand physically how the kinematic dispersion, which, after all, is a heterogeneity of the real velocities as compared to the average convective velocity, could spread the substances upstream: the real velocities in a porous medium are probably always oriented downstream rather that upstream. The only physical mechanism that could possibly explain an upstream migration of the transported substances is molecular diffusion, which would then be written

$$\phi_l = -\omega d \frac{\partial C}{\partial l} + CU$$

which would make the value of the dispersion coefficient depend on the direction of the gradient compared to that of the velocity.

This effect is diminished, if the dispersion coefficient is made to depend on the traveled distance (see Dieulin, 1980), but it appears to be one of the inconsistencies of the classical dispersion theory (Simpson, 1978).

### 10.1.2. *Fractured Media*

There are no very elaborate special theories for transport in fractured media and few experiments to support a theory. The three phenomena (convection, diffusion, dispersion) already cited exist in fractured media as well, and, if there is a porosity in the blocks between the fractures, the porosity may also play an important role.

(a) *Convection.* In the fracture network, convection works in exactly the same way as in the porous medium. Using Darcy's velocity, we write

$$- \operatorname{div}(CU) = \omega_c \frac{\partial C}{\partial t}$$

in order to identify the convection.

A set of fractures having a spacing of 10 m and an aperture of 0.2 mm has an equivalent hydraulic conductivity of $10^{-3}$ m/s (see Section 4.1.e) and a porosity of $2 \times 10^{-5}$. Compared to a porous medium of the same hydraulic conductivity with a porosity of, say 20%, the fractured medium has a kinematic porosity 10,000 times smaller: the average microscopic velocity in such a medium is thus 10,000 times larger than in the porous medium for the same hydraulic gradient.

The convection transport is therefore much faster in fractured media than in porous ones, if the rock matrix is impermeable, impervious, and compact.

(b) *Diffusion and dispersion.* These two phenomena occur in fractured media as well, the first one through molecular agitation, the second through the heterogeneities of the velocities inside a fracture (parabolic profile of the velocities as in a pore) as well as through that of the velocities from one fracture to another (different degrees of aperture) and finally through transverse mixing and dispersion, when fractures with different directions intersect.

As fractured media are quite often anisotropic, the validity of the classical assumption that the principal directions of the dispersion tensor is in the direction of the velocity is very questionable.

However, very few values for dispersion coefficients are known in fractured media. One of the rare cases where these values are known is that of the Hanford (Washington) basalts, where an accidental release of tritium polluted the aquifer over nearly 15 km.

A study made by Ahlström *et al.* (1977) gives

$$\alpha_L = 30 \text{ m}$$

$$\alpha_T = 20 \text{ m}$$

Another study of pollution by radioactive waste was made by Robertson in 1974 (quoted by Fried, 1975) at the experimental station of Snake River (Idaho). The aquifer is made up of fractured basalt and interlayered sedimentary deposits. The model has been fitted on the concentrations of chlorides and tritium, with the coefficients

$$\alpha_L = 91 \text{ m}$$

$$\alpha_T = 137 \text{ m} \qquad (\text{note } \alpha_T > \alpha_L)$$

as the pollution had spread over nearly 10 km in 10 years.

D. B. Grove (personal communication, 1973) quotes a tracer experiment with tritium in 150-m-thick fractured rocks. With an observation well at 600 m from the tracer point, he obtains $\alpha_L$ of 150 m and a porosity of $8 \times 10^{-4}$. For fractured limestone, he quotes $\alpha_L = 152$ m.

Recent developments on transport in fractured systems are now focusing on random generation of fracture networks and numerical simulation of transport in these networks. See Schwartz *et al.* (1983), Smith and Schwartz (1984), Robinson (1984), Endo *et al.* (1984), and Anderson (1985).

(c)  *Secondary porosity.*  This is understood to mean the case where the rock matrix itself, which is cut up by fractures, cannot be considered to be impermeable and compact: the transported substances migrate inside it.

If we take up the line of thought that we followed for the porous medium, we can write the transport equation including two concentrations $C$ and $C'$:

$$\text{div}(\mathbf{D}\,\mathbf{grad}\,C - C\mathbf{U}) = \omega_c \frac{\partial C}{\partial t} + (\omega - \omega_c)\frac{\partial C'}{\partial t}$$

where $\omega_c$ is the kinematic porosity of the fractures, $\omega - \omega_c$ is the porosity containing immobile water in the fractures and in the pores of the matrix, $C$ is then the concentration of the fluid in the fractures, and $C'$ is a "mean" concentration in the matrix.

We elminate the trivial case, where the migration from the fractures to the matrix is so fast or the medium so densely fractured that it might be assumed, at any instant, that the concentration $C'$ in each block of the matrix is equal to that of the fluid circulating in the fractures. As we have seen in the case of the porous medium, this extreme case would imply using a porosity for the transport equal to the total porosity of the rock (mobile fraction in the fractures plus immobile fracture plus total porosity in the matrix).

*First hypothesis: Porous matrix with almost no permeability.*  In this case, the only migration mechanism in the matrix is molecular diffusion. In each block isolated by fractures, the equation of molecular diffusion has to be solved:

$$\text{div}(\omega'd\,\mathbf{grad}\,C') = \omega'\frac{\partial C'}{\partial t}$$

where $\omega'$ is the total porosity of the matrix. This equation has as its boundary condition at the fracture planes the value of the concentration $C$ in the fractures, which itself varies with time. Then the flux exchanged with the fracture per unit surface area of contact between the two media is calculated on the contact area:

$$\phi = -\omega'd\,\mathbf{grad}\,C'$$

This term is then introduced as a source term in the transport equation in the fractures:

$$\text{div}(\mathbf{D}\,\text{grad}\,C - C\mathbf{U}) = \omega_c \frac{\partial C}{\partial t} + \alpha\phi$$

where $\alpha$ is the ratio (area of the fracture planes)/(volume of the medium). (A fracture counts as two surface areas because of its two bounding planes.)

In order to simplify these calculations, the diffusion equation in one block is solved schematically by reducing it to one dimension, giving the block a mean half-dimension equal to the mean half-distance between the fractures in all directions of space, but respecting the volume of the matrix (Fig. 10.6).

We can solve the one-dimensional diffusion equation subject to a condition of no flow at the distance $L$ and an imposed concentration on the fracture plane either numerically or analytically. The analytical solution gives $\phi$ in the form of a convolution integral of the concentration $C$ in the block. Because of this added complexity, such a calculation can only be undertaken in one, or maybe, two dimensions. A spherical solution in the blocks, assuming that they are equal to spheres of uniform radius, could also be conceived or a calculation where the dimension $L$ would be taken as infinite, if the spacing of the fractures is such that, in the time considered, the progression of the molecular diffusion front inside the blocks is small compared to $L$. See Barbreau *et al.* (1980), Neretnieks (1980), and Sudicky and Frind (1982).

*Second hypothesis: Porous and permeable matrix.* It must be said at the outset that, if the hydraulic conductivity of the matrix is of the order of that of the fractures, the fractured medium will have an equivalent hydraulic conductivity, which explicitly shows the permeability of the matrix, included in the expression given in Section 4.1.f. The transport then takes place simultaneously in the two media and may be represented by a higher dispersion coefficient, which takes into account the systematic heterogeneities of the velocity field (this coefficient must be determined by experiments).

The most difficult problem is the one where the permeability of the matrix is not zero, but small compared to that of the fractures. It might then be suggested (O'Neill 1977; Lefèbvre du Prey and Weill (1974)) that

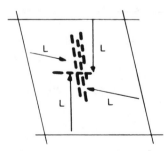

**Fig. 10.6.** Block of fractured medium ($L = 1/\alpha$).

(1)   It can be solved using one single equation for the hydraulic head by assuming that in a steady state the pressures in the two media are equal and by using the overall equivalent hydraulic conductivity.

(2)   Two velocity fields can be deduced in each medium by applying Darcy's law in each medium with its own hydraulic conductivity (the assumption is that the two media are continuous and overlapping).

(3)   Two transport equations can be written in each of the media with their own Darcy velocity and porosity.

(4)   These two transport equations are coupled by the exchange terms:

(a) a convection exchange term linked to the Darcy velocity in the matrix, where if $C$ is the concentration in the fractures, and $C'$ the one in the matrix, the term is $-\alpha U'C$ in the transport equation for $C'$ and $+\alpha U'C'$ in the transport equation for $C$, where $\alpha$ is an exchange coefficient; and

(b) a dispersion exchange term linked to the difference between the concentrations $C$ and $C'$.

### 10.1.3. New Theories of Solute Transport

The dispersion equation which we have established is known to be an approximation of reality: how good is it in practice? Working in the laboratory, on rather homogeneous columns, it is found that the measurements can be well represented by the solution of the transport equation, using dispersivities on the order of a few centimeters. But in field situations, it was soon found that things worked much less well. Interpreting a tracer test (or a real pollution case) at a given observation well, it is always possible to fit a dispersivity for which the solution of the dispersion equation will, with a good approximation, match the observations (concentration versus time). However, if a second observation well is used, at a different distance from the source, one finds in general that a different dispersivity will be needed to match the observations: the further apart the source from the observation well, the larger the dispersivity. Lallemand-Barrès and Peaudecerf (1978) synthesized all published values of dispersivities and were able to show that, on the average, the dispersivity increases with the distance between the source and the observation point. This was called the "scale effect". Their data ranged between distances of a few meters to 10 km for several rock types, and, as a rule of thumb, the dispersivity was on the order of one tenth of the traveled distance. Did this mean that dispersivity should not be regarded as a constant? It was thought initially that this might be due to different scales of heterogeneities that were encountered successively by the tracer (or the pollutant) during its migration. Dieulin (1980), Dieulin et al. (1980, 1981a,b,c) was able to show that this was not the case: in a careful field experiment, he showed that for a given scale of heterogeneity of the medium (i.e., without any change of the structure of the medium, without any additional heterogeneity), the dispersivity was increasing with time or, in other words, with the *average* traveled distance of the tracer (for a pulse injection of tracer into a parallel flow). This meant that at two observation wells, at different distances from the source, it was possible to fit the observations during a short time interval with the *same* dispersivity; for a later time interval, a larger dispersivity should be used but again identical at the two wells. Now of course, if a global fit was made for each well, using all the measurements, an average dispersivity would have to be used. At the well farther from the source, as the tracer is observed later than at the closer well, this average dispersivity would thus be larger.

Dieulin suggested that the scale effect be called a "time effect." At the same time, new theories for representing transport were being developed which also showed the dispersivity to be a function of time. These were based on a stochastic description of the velocity field in the aquifer. The reader is therefore referred to Chapter 11, particularly Section 11.10, before going any further.

As hydrodynamic dispersion is the result of the heterogeneity of the velocity in porous media, the stochastic approach seems particularly well suited for representing this variability. We will first summarize here the Lagrangian approach presented by Dieulin *et al.* (1981b,c). Transport is studied in the ordinary space $R^n (n = 1, 2,$ or 3) with the following simplifying assumptions:

(1)   The velocity variations of the fluid in the medium is the dominant mechanism, molecular diffusion is negligible.

(2)   The Eulerian microscopic velocity field **u**, which is unknown, can be regarded as a stationary random process, i.e., **u** is a vectorial stationary random function, and **u** is conservative, i.e., div **u** = 0. This means that the flow is in steady state with a constant porosity.

(3)   A slug of tracer is injected at time $t = 0$ at the origin $X = 0$ of the system (note that **u** and **X** are vectors, the components of which are denoted $u^i$ or $X^i$. A subscript $t$ will denote the time: $X_t$). The transport can be described by giving , as a function of time, the position $X_t$ of a particle* injected at $t = 0$ at the origin. Kolmogorov (1931) has shown that if the particle is transported by convection and diffusion (Brownian motion) the probability density $\rho(X, t)$ of the particle is identical to the concentration given by solving the classical transport equation for a slug injection of tracer.

Let $V(t) = u(X_t)$ be the Lagrangian velocity, i.e., the velocity of the particle following its trajectory along a flow path. Matheron (unpublished, 1981) has shown that if **u** satisfies the assumptions given above, then **V** is a stationary random function having the same probability distribution function as **u**. We

---

* This particle has no physical meaning and is just a mathematical symbol.

can now write:

$$\mathbf{X}_t = \int_0^t \mathbf{V}(\tau) \, d\tau$$

We then have:

$$\mathbf{E}(\mathbf{X}_t) = \int_0^t \mathbf{E}[\mathbf{V}(\tau)] \, d\tau = t\mathbf{E}(\mathbf{V}) = t\mathbf{E}(\mathbf{u}) = \bar{\mathbf{u}}t$$

where $\bar{\mathbf{u}} = \mathbf{E}(\mathbf{u})$. Thus the average position of the particle is just the average velocity multiplied by the time. Let us now determine the variance of this position; this variance is now a $n \times n$ matrix. Superscript T will denote the transposition of a vector:

$$\begin{aligned}
\text{Var}\,[\mathbf{X}_t] &= \mathbf{E}\{[\mathbf{X}_t - \mathbf{E}(\mathbf{X}_t)]^{\mathrm{T}}[\mathbf{X}_t - \mathbf{E}(\mathbf{X}_t)]\} \\
&= (\mathbf{X}_t^{\mathrm{T}}\mathbf{X}_t) - \mathbf{E}(\mathbf{X}_t)\mathbf{E}(^{\mathrm{T}}\mathbf{X}_t) \\
&= \mathbf{E}\left\{\int_0^t \mathbf{V}(\tau)\,d\tau \int_0^t \mathbf{V}(\tau')\,d\tau'\right\} - t^2\bar{\mathbf{u}}^{\mathrm{T}}\bar{\mathbf{u}} \\
&= \int_0^t\int_0^t \{\mathbf{E}[\mathbf{V}(\tau)^{\mathrm{T}}\mathbf{V}(\tau')] - \bar{\mathbf{u}}^{\mathrm{T}}\bar{\mathbf{u}}\}\,d\tau\,d\tau' \\
&= \int_0^t\int_0^t E\{[\mathbf{V}(\tau) - \bar{\mathbf{u}}]^{\mathrm{T}}[\mathbf{V}(\tau') - \bar{\mathbf{u}}]\,d\tau\,d\tau' \\
&= \int_0^t\int_0^t \mathbf{C}(\tau - \tau')\,d\tau\,d\tau' = 2\int_0^t (t - \tau)\mathbf{C}(\tau)\,d\tau
\end{aligned}$$

where $\mathbf{C}(t)$ is the $n \times n$ covariance matrix of the components of the *Lagrangian* velocity $\mathbf{V}$ taken with a time lag $t$.

The variance of the position of the particle is the equivalent of the "spreading" of the pulse of tracer around its mean position; it is therefore related to the dispersion coefficient. Indeed, Einstein (1905) has shown that the dispersion tensor $\mathbf{D}$ is given by:

$$\mathbf{D} = \frac{1}{2}\frac{d}{dt}\,[\text{Var}(\mathbf{X}_t)]$$

We therefore obtain here:

$$\mathbf{D} = \int_0^t \mathbf{C}(\tau)\,d\tau$$

Very important conclusions can be drawn from this simple result:

(1)   The dispersion tensor $\mathbf{D}$ is a function of time. As each component of the tensor varies with time, there is *a priori* no reason why the principal

directions of this tensor should remain constant. They will not, in general, be colinear with the average velocity.

(2)   If the covariance matrix $\mathbf{C}$ of the Lagrangian velocity is well behaved, i.e., $\mathbf{C}(t) \to 0$ sufficiently rapidly as $t \to \infty$, we can assume that the integral of $\mathbf{C}(t)$ will become constant as $t \to \infty$. Thus one can, in general, expect that after a certain time a constant dispersion tensor will be acceptable to represent the transport. This is called an asymptotic diffusive behavior.

(3)   The dispersion tensor is a direct function of the Lagrangian velocity. If the velocity is increased, e.g., by a factor of 2, but otherwise remains identical in direction, the dispersion tensor will be increased by the same factor of 2. Thus a dispersivity could be defined, but this dispersivity *is not a function of* the *properties of the porous medium only*; it is also a function of the velocity field. If a new Eulerian velocity field is created, e.g., by going from parallel flow to radial flow, or changing the vertical/horizontal velocity ratio (for instance by varying the recharge in the aquifer), then the Lagrangian velocity field (i.e., flow path) will be changed, and therefore the covariance matrix of this velocity is changed: it is no longer possible to assume that the dispersivity is an intrinsic property of the medium, independent of the flow field. This is a very important point.

(4)   These results are only valid for a slug injection of tracer. For any other source, the concept of convolution should be applied. This has not always been realized in the past and erroneous results have been obtained using a numerical model where $\mathbf{D}$ was simply made a function of time for a step injection of tracer. This is totally incorrect.

(5)   Only in the case where the probability distribution function of the Eulerian velocity field $\mathbf{u}$ is assumed to be Gaussian is it possible to show that the transport equation equivalent to the particle position is:

$$\sum_j \sum_k \int_0^t C^{jk}(\tau)\, d\tau \frac{\partial^2 C}{\partial x^j \partial x^k} - \sum_j \bar{u}^j \frac{\partial C}{\partial x_j} = \frac{\partial C}{\partial t}$$

This is similar to a dispersion equation where the dispersion tensor is made a function of time (Dieulin *et al.*, 1981b,c). For all other distributions of velocity, there is no equivalent dispersion equation for early times until the asymptotic behavior is reached. There is very little reason why the Eulerian velocity field should have a Gaussian distribution. Therefore there is at present no correct dispersion equation representing transport for early times.

Quite similar results were obtained for stratified media by Gelhar *et al.* (1979a) and Matheron and Marsily (1980) including, however, a local diffusion in the equations. The former used a spectral approach, the latter a Lagrangian approach. More recently, Gelhar and Axness (1983), Dagan (1982a), Winter *et al.* (1984) were able to relate the time-varying dispersivity tensor to the Eulerian velocity field and thus to the hydraulic conductivity of the medium. To arrive at such results, these authors were obliged to make simplifying assumptions that may—or may not—hold. Long term experiments are planned to check these theories.

In summary, the classical dispersion equation is probably only valid after large times or large distances traveled by the tracer in the medium; such distances can be equal to several times (up to 10) the average "correlation length" of the heterogeneities of the medium (i.e., on the order of 10 times the characteristic length of the geological structures of the medium). In typical sediments, this can imply that several hundreds of meters are necessary before an asymptotic behavior is reached. Meanwhile transport is only approximately represented, for a slug injection, by a time-varying dispersion coefficient and, for any other injection, by the convolution of the response to the slug injection with the actual source term.

One must also bear in mind that since the controlling parameter of transport is the Lagrangian velocity field, any field determination of the channeling properties of the medium (e.g., buried high permeability channels or highly conductive faults) will improve tremendously the understanding of transport. In fact, improving the knowledge of the heterogeneity of the medium is a prerequisite to predicting transport at early times.

## 10.2. Laws of Interactions between the Immobile Phase and the Transported Substances and Physicochemical Changes in the Substances

The purpose here is to describe the mechanisms that can turn the migration of substances in porous or fractured media into a reactive phenomenon, i.e., which tend to invalidate the laws of mass balance during the transport. The case of the porous medium and that of the fractured medium will be discussed separately.

### 10.2.1. Porous Media

The immobile phase includes mainly the solid phase, but also the immobile liquid bound to the solid by the forces of molecular attraction. Several mechanisms of interaction, transformation, or decay can make transport nonconservative (see Jackson, 1980).

*Physical mechanisms.* The transported substances can sometimes be stopped by physical filtration through the pores of the medium. This can happen even if the transported substances are much smaller than the size of the pores.

*Geochemical mechanisms.*

(1)   Combining of ions into electrically neutral molecules.

(2)   Acid/base reaction depending on the pH of the solute and on the rocks it travels through.

(3)   Oxidation–reduction reactions which condition the state of valence of the transported ions.

(4)   Precipitation–dissolution, which may immobilize or dissolve the substances.

(5)   Adsorption–desorption limited by definition, strictly speaking, solely to the ion exchanges (mainly cations), which take place on the surface of the clayey or colloidal minerals.

*Radiological mechanisms.*   These are radioactive decay (vanishing of substances), and creation of daughter products by this decay (appearance of new substances).

*Biological mechanisms.*   Biological activity in porous media can decompose or transform some elements; very often, such processes are represented by a decay reaction, like radioactive decay, with a "biological" half-life.

This whole set of mechanisms is represented by a "net source or sink term" $Q$ in the transport equation, which expresses the lack of mass balance, when the balance of fluxes entering into and accumulated in a volume $D$ is calculated, as we have seen in Section 10.1. It is written as

$$\text{div}(\mathbf{D}\,\text{grad}\,C - C\mathbf{U}) = \omega\frac{\partial C}{\partial t} + Q \qquad (10.2.1)$$

The term $Q$ represents: the disappearance of substances, if it is positive (sink), and the addition of substances, if it is negative (source). It is expressed as a *mass* of the considered substance, added (or substracted) *per unit volume of porous medium* and *per unit time*; $Q$ is the algebraic sum of the rates of each individual mechanism.

We shall try to review the laws which allow us to estimate the source or sink term.

(a)   *Filtration.*   We will first consider the case where the transported elements are actually "sieved" by the medium, i.e., when the size of the pores is smaller than that of the particles in solution. Greenberg (1971) gives the following estimate for the pore size, in clays:

(1)   Diameter of the clay particles $\sim 20{,}000$ Å (Å = ångström = $10^{-10}$ m), and sometimes much smaller.

(2)   Spacing between the sheets of the clayey minerals, 9–15 Å.

(3)    In sands, the order of magnitude of the diameter of the pores, given by the effective grain size $d_{10}$ defined in Section 2.1.e. It is usually in the range $10^{-2}$ to $10^{-1}$ mm (100,000–1,000,000 Å).

(4)    Diameter of the smallest soluble ions, such as $Na^+$ or $Cl^-$: 1–10 Å.

(5)    Diameter of the large organic molecules with high molecular weight, e.g., the chain polyethylene glycol of molecular weight 20,000: up to 500 Å.

(6)    Diameter of bacteria: 5,000–30,000 Å.

(7)    Diameter of colloids: extremely variable, usually in the range 1,000–50,000 Å.

Thus, it can be admitted that direct sieving can be effective only for very large molecules, bacteria or colloids, in clayey soils or silts.

However, the term filtration is also used to describe mechanisms where particles that are much smaller than the size of the pores are nevertheless stopped and "sedimented" in porous media. See Subsection 10.2.4.

(b)    *Adsorption and ion exchange.*    Because solutes become attached to the mineral particles, a quantity of substances bound to the solid phase should be defined. A mass concentration $F$, dimensionless, is generally used, representing the mass of substances adsorbed per unit mass of solid. In a unit volume of porous medium, the mass of solid is $(1 - \omega)\rho_s$, where $\omega$ is the total porosity and $\rho_s$ the mass per unit volume of the solid particles (e.g., the quartz grains in sands, not the bulk mass per unit volume of the medium). The mass of substances bound to the solid is then $(1 - \omega)\rho_s F$.

The source term to be introduced into the equation is the variation of this mass per unit volume per unit time:

$$Q = (1 - \omega)\rho_s \frac{\partial F}{\partial t}$$

The problem of adsorption consists in defining the relation between the concentration $F$ and $C$.

*Mechanisms of ion exchange.*    The adsorption capacity of certain minerals or colloids is due (Jackson, 1980) to the existence of nonneutralized electric charges at the surface of and/or inside these minerals. Ions with an opposite charge attach themselves to it, thus creating a "double electrical layer", which may belong to one of the following two types:

*Type* 1.    Imperfections or ion substitutions in the crystal lattice of the mineral, causing positive or negative electrical imbalance. The surface of the mineral is then called the stable electrical layer, and the ions with an opposite charge attracted by the stable layer constitute the mobile electrical layer.

*Type 2.*    The specific adsorption of certain ions by the surface of a mineral initially uncharged creates a stable electrical layer to which other ions of opposite charge become attached, thus creating the mobile layer.

Vermiculite and montmorillonite have, for example, double layers of type 1. Other clays, metallic hydroxides, and organic and inorganic colloids (silica, for example) have double layers of type 2. The latter are very much more sensitive to the pH action of the water.

An order of magnitude for the maximum adsorption capacity of the clayey minerals is around $\frac{1}{10}$ of their weight. It should rather be given in milliequivalents per unit mass, as the valences of the adsorbed ions have to be taken into account: adsorption is an electrical balance. As clays are negatively charged, the mobile electrical layer is made of cations. Orders of magnitude of cation exchange capacity are: montmorillonite, 100 meq/100 g; illite, 30 meq/100 g; kaolinite, 1–10 meq/100 g.

*First case:    Adsorption is instantaneous.*    It is then admitted that $F$ and $C$ are always in equilibrium and linked by a relation where time does not count. Experiments made up to now with *adsorption* (not necessarily with desorption) seem to prove that for clayey bodies and minerals, the time to equilibrium is of the order of a few minutes, i.e., quite negligible for common cases. James and Rubin (1979) have shown, however, that a local chemical equilibrium for calcium is obtained "only when the ratio of the hydrodynamic dispersion coefficient to the estimated molecular diffusion coefficient is near unity." This will very seldom be the case in practice.

Generally, the *entire* set of transported substances (ions) has to be taken into consideration and the concentrations $C_i$ and $F_i$ calculated for each of them. Thus, the usual transport equation for each constituent $i$ becomes

$$\text{div}(\mathbf{D}\,\mathbf{grad}\,C_i - C_i\mathbf{U}) = \omega\frac{\partial C_i}{\partial t} + (1 - \omega)\rho_s\frac{\partial F_i}{\partial t}$$

We then state that the sum of the adsorbed concentrations is equal to the total ion exchange capacity of the solid; as this exchange capacity $f_T$ is generally expressed in equivalents per gram (epg), the adsorbed mass concentration $F_i$ and the volumetric concentration $C_i$ have to be transformed into epg or epl (equivalent per liter):

$$f_i = \frac{F_i}{M_i}v_i \qquad c_i = \frac{C_i}{M_i}v_i$$

where $F_i$ is mass concentration (dimensionless), $M_i$ is molar mass of constituent $i$, $v_i$ is valence of constituent $i$, and $C_i$ is volumetric concentration (kg/m$^3$ or g/liter).

We then write

$$\sum_{1}^{m} f_i = f_T$$

where $m$ is the number of substances present.

Finally, the selectivity of the adsorption for certain substances is expressed by an equilibrium relation (mass action equation), assumed to be reached instantaneously and to be reversible (i.e., not representing irreversible fixation):

$$\left(\frac{f_j/f_T}{c_j}\right)^{v_i}\left(\frac{c_i}{f_i/f_T}\right)^{v_j} = K_{ij}$$

where $K_{ij}$ is the ion exchange selectivity coefficient of the solid matrix with respect to elements $i$ and $j$ and the dimension of $K_{ij}$ depends on the valence $v_i$ and $v_j$. The coefficients $K_{ij}$ are not, of course, independent of each other, but are more or less independent of the concentrations in the solution.

It is then possible to solve this system of equations for all elements $i$.

A good example of the application of this method can be found in Valocchi et al. (1981); they studied the ternary system $Na_{(1)}$, $Mg_{(2)}$, $Ca_{(3)}$ in the laboratory on core samples from an alluvial aquifer (sand, gravel, silt, and clay) giving $f_T = 0.1$ meq/g, $K_{12} = 1.7$ eq/liter, $K_{13} = 3$ eq/liter. They were then able to reproduce the observed concentration during an *in situ* injection test, where Mg and Ca were exchanged with Na as in Fig. 10.7.

The selectivity of ion exchange generally follows the same order of preference: divalent ions have stronger affinity than monovalent ions, and within each of these categories, the affinity is: $Cs^+ > Rb^+ > K^+ > Na^+ > Li^+$, and $Ba^{2+} > Sr^{2+} > Ca^{2+} > Mg^{2+}$ (Freeze and Cherry, 1979).

In montmorillonite clays, when 2 $Na^+$ are substituted for $Ca^{2+}$, the clay swells, and its permeability can be drastically reduced.

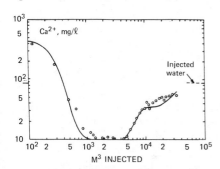
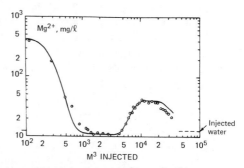

**Fig. 10.7.** Comparison between simulated and actual breakthrough of $Ca^{2+}$ and $Mg^{2+}$ at an observation well. After Valocchi *et al.* (1981). Reproduced with permission from Water Resources Research, Vol. 17, pp. 1517–1527. Copyright by the American Geophysical Union.

In cases where the transported substances are in very weak concentration, the assumption is made that the adsorption of these substances does not to any great extent change the $f_j/c_j$ ratio of other substances that are present in larger quantities. As this ratio is constant, we get

$$\frac{f_i}{c_i} = \frac{F_i}{C_i} = \left(\frac{f_T^{v_j}(f_j/f_T)^{v_i}}{c_j^{v_i} K_{ij}}\right)^{1/v_j} = \text{const} = K_{di}$$

The coefficient $K_{di}$ is called the distribution coefficient of the substance $i$ in relation to the porous medium. It assumes that the adsorption is linear, reversible and instantaneous. As $K_d$ varies with temperature, it is also known as the slope of the adsorption isotherm. Its dimension is length$^3$ mass$^{-1}$, and it is usually expressed in ml/g. We can then write

$$F_i = K_{di}C_i$$

$$\text{div}(\mathbf{D}\,\textbf{grad}\,C_i - C_i\mathbf{U}) = [\omega + (1 - \omega)\rho_s K_{di}]\frac{\partial C_i}{\partial t}$$

$$= \omega\left[1 + \frac{1 - \omega}{\omega}\rho_s K_{di}\right]\frac{\partial C_i}{\partial t} \qquad (10.2.2)$$

The term

$$R = 1 + \frac{1 - \omega}{\omega}\rho_s K_{di} \qquad \text{(dimensionless)} \qquad (10.2.3)$$

is generally called the retardation factor due to the adsorption. It is introduced as a multiplying coefficient of the porosity:

$$\text{div}(\mathbf{D}\,\textbf{grad}\,C - C\mathbf{U}) = \omega R\frac{\partial C}{\partial t} \qquad (10.2.4)$$

If both sides are divided by $\omega R$, an apparent velocity $\mathbf{U}/\omega R$ is defined, while the transport equation takes the same form as in Eq. (10.1.6): everything behaves as if the mean microscopic velocity of the convective transport were divided by $R$. Under this assumption, the displacement of each substance can be calculated independently of that of its neighbors. This approach is widely used, and is probably valid for elements in very low concentration. If there is no adsorption, $R = 1$.

*Second case: Instantaneous adsorption that is not entirely reversible.* If such a phenomenon occurs, we get adsorption–desorption isotherms with, for example, the shape shown in Fig. 10.8. The quantity, which is irreversibly fixed, may then depend on the maximum concentration $C_{\text{max}}$.

This phenomenon may be included in a numerical model, but this requires rather a large amount of calculations, because at each time step and for each mesh of the model, the new concentration $C_{t + \Delta t}$ is compared to the former

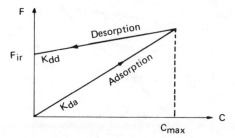

**Fig. 10.8.** Partly reversible adsorption.

concentrations $C_t$. Suppose, for example, that we start with an adsorption phase. If

$$C_{t+\Delta t} > C_t$$

then we must use the retardation factor for the mesh in question,

$$R_a = 1 + \frac{1 - \omega}{\omega} \rho_s K_{da}$$

where $K_{da}$ is the slope of the adsorption isotherm; however, if

$$C_{t+\Delta t} < C_t$$

then we must use

$$R_d = 1 + \frac{1 - \omega}{\omega} \rho_s K_{dd}$$

where $K_{dd}$ is the slope of the desorption isotherm.

It can be observed that the constant term $F_{ir}$ is eliminated in the course of the derivation [in desorption, we would write $F = F_{ir} + K_{dd}C$, but $\partial F/\partial t = K_{dd}(\partial C/\partial t)$], which shows that it is only important to determine the slope of the desorption isotherm in so far as the various desorption isotherms are, at least in the first approximation, parallel to each other (Fig. 10.9). (In reality, the isotherms are generally curved.)

It must also be noted that there is a risk involved in using this irreversibility, because it may be due to desorption kinetics. It is possible that, if the equilibrium lasts for a very long time, the situation might slowly revert to that of the single adsorption isotherm. Therefore, if irreversible desorption is used in the calculations, there is a risk of making mistakes, which may jeopardize the safety of the environment if the adsorbed substance is harmful. Then, the problem is posed of the long-term validity of measurements made in the course of laboratory experiments, which include kinetic reactions and which are necessarily of short duration.

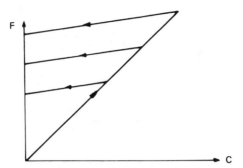

**Fig. 10.9.** Partly reversible adsorption.

*Third case: Nonlinear adsorption isotherm.* In the case where each solute moves independently of its neighbors, other instantaneous relations between $F$ and $C$ have been suggested instead of the linear isotherm. They are the following:

(1) Isotherm of the second degree:

$$F = K_1 C - K_2 C^2 \qquad K_1, K_2 > 0$$

(2) Langmuir's isotherm:

$$F = \frac{K_1 C}{1 + K_2 C} \qquad K_1, K_2 > 0$$

(3) Freundlich's isotherm:

$$F = K_1 C^{1/n} \qquad K_1 > 0 \qquad n \geq 1$$

(4) Exponential isotherm:

$$C = K_1 F e^{k_2 F} \qquad K_1, k_2 \geq 0$$

Moreover, these constants depend on the direction of the exchange (adsorption or desorption) if the phenomenon is not strictly reversible.

*Fourth case: Kinetics of noninstantaneous adsorption–desorption.* Here we must know the law of variation in time of $F$ versus $C$. Because of the complexity of the problem, the phenomenon is generally treated numerically in two stages, although some analytical solutions have been proposed. First, if $C_t$ and $F_t$ are known at the beginning of the time step, $F_{t+\Delta t}$ is calculated at the end of the time step according to the law of reaction kinetics, assuming $C_t$ to be constant during the time step:

$$Q = (1 - \omega)\rho_s \frac{F_{t+\Delta t} - F_t}{\Delta t}$$

Then this source term, assumed constant over the time step, is introduced into the transport equation.

The concentration $F$ is, as it were, an explicit term, one time step behind the concentration $C$.

If more precision is needed, $C$ and $F$ must be calculated several times during the same time step, iterating the calculations of kinetics and transport. For example, it may be assumed that the kinetics of adsorption are linear. Then we write:

$$\text{div}(\mathbf{D}\,\text{grad}\,C - C\mathbf{U}) = \omega\frac{\partial C}{\partial t} + (1 - \omega)\rho_s\frac{\partial F}{\partial t} \qquad \text{in the liquid phase,}$$

$$\frac{\partial F}{\partial t} = K_1(K_d C - F) \qquad \text{on the solid phase,}$$

where $K_1$ is the kinetic constant of linear chemical adsorption. These two equations are solved successively or simultaneously. See Harada *et al* (1980) and Pigford *et al.* (1980).

*Fifth case: Relation between the adsorption and the concentration $C'$ in the immobile fluid fraction.* In Section 10.1.1, we have a transport equation, Eq. (10.1.4), in which a concentration $C'$ appears in the immobile fraction. It is possible to add an adsorption term to this equation, which gives

$$\text{div}(\mathbf{D}\,\text{grad}\,C - C\mathbf{U}) = \omega_c\frac{\partial C}{\partial t} + (\omega - \omega_c)\frac{\partial C'}{\partial t} + (1 - \omega)\rho_s\frac{\partial F}{\partial t}$$

If we admit that there is a linear adsorption isotherm $F = K_d C$ and that the relation between $C$ and $C'$ is linear as well, $C' = K'C$, we get

$$\text{div}(\mathbf{D}\,\text{grad}\,C - C\mathbf{U}) = \omega_c\left(1 + \frac{\omega - \omega_c}{\omega_c}K' + \frac{1 - \omega}{\omega_c}\rho_s K_d\right)\frac{\partial C}{\partial t}$$

This makes a new retardation factor appear, in which the adsorption and the retention in the immobile fluid phase are merged. The same would happen if a first-order kinetic reaction was used for both $F$ and $C'$.

In practice, the coefficient $K_d$ is measured in the laboratory in a batch experiment by difference. We start with a known concentration $C_1$ in the fluid phase, into which a certain weight of rock* is introduced. After equilibrium is reached (constant concentration), the concentration $C_2$ is measured in the remaining liquid phase: the mass of the adsorbed quantity is deduced by difference. However, in fact, the quantity which has disappeared from the mobile phase (which is the only measurable) also includes the quantity

---

\* This rock must on no account be crushed, so as to avoid increasing the area of fluid–solid contact.

retained in the immobile liquid bound to the solid: the obtained coefficient $K_d$ therefore explicitly accounts for the total quantity retained in the immobile fluid, and when linear and instantaneous adsorption is introduced, it is unnecessary to consider the concentration $C'$ in the immobile phase. Similarly, in a kinetic experiment, an average kinetic constant would be obtained.

Recent work on the interaction between the solute and the rock matrix is now directed toward the coupling of geochemical codes with the transport codes. Indeed, the semi-empirical partition coefficient approach is quite insufficient to represent all the complex chemical reactions that can take place between several solutes and the medium. In general, geochemical equilibrium is assumed to take place at each time step, although including reaction kinetics is also considered in some cases. See Morel (1983), Nordstrom *et al* (1979), Graven and Freeze (1984).

(c)  *Adsorption of organics: the hydrophobic theory.*   Organics present in trace quantities in groundwater are also found to be adsorbed by the medium, i.e., to have a retardation coefficient just as ions do. However, the mechanism of sorption is different; the organics are mostly sorbed on existing solid organic compounds present in the porous medium.

Just as for ions, an equilibrium partition coefficient $K_p$, equivalent to the distribution coefficient $K_d$ (Section 10.2.1.b) is defined by:

$$F = K_p C'$$

where $C'$ is the *massic* concentration of the organic compound in water (mass per unit mass of water) and $F$ is the concentration of the organic sorbed on the solid (mass per unit mass). $K_p$ is dimensionless. As the mass per unit volume of water is assumed constant, the volumetric concentration of the organic compound in water (mass per unit volume of water) would be $C = \rho C'$. To write the transport equation in terms of $C'$, we need to divide it, e.g., Eq. (10.2.1), by $\rho$:

$$\text{div}(\mathbf{D} \, \text{grad} \, C' - C'\mathbf{U}) = \omega \frac{\partial C'}{\partial t} + \frac{Q}{\rho} = \omega \frac{\partial C'}{\partial t} + (1 - \omega)\frac{\rho_s}{\rho}\frac{\partial F}{\partial t}$$

where $Q$ is the source or sink term and $\rho_s$ the mass per unit volume of the grains of the porous medium.

Assuming instantaneous equilibrium, we can define a dimensionless retardation factor $R$ as in Eq. (10.2.3) by

$$R = 1 + \frac{1 - \omega}{\omega}\frac{\rho_s}{\rho} K_p$$

Then the transport equation is again:

$$\text{div}(\mathbf{D} \, \text{grad} \, C' - C'U) = \omega R \frac{\partial C'}{\partial t}$$

Note that $\rho_s/\rho$ is used instead of $\rho_s$ in Eq. (10.2.3) because $C'$ is the massic concentration.

Schwartzenbach and Westfall (1981) report that kinetic effects are unimportant for such sorption if the groundwater pore velocity is on the order of 1 m/day but cannot be neglected if the pore water velocity goes to 10 m/day (for chlorinated benzenes). However, to this day, no satisfactory kinetic model is available; the first-order one described in Subsection 10.2.1.b does not seem to give good results; see also Miller and Weber (1984).

The partition coefficient $K_p$ can be measured in static batch experiments or in dynamic column tests; however the so-called "hydrophobic theory" (Karickhoff et al., 1979; Schwartzenbach and Westfall, 1981, 1984) provides a method to estimate this coefficient indirectly, within a factor of 2.

(1)   For a neutral hydrophobic compound* one first measures (or finds in the literature) the dimensionless partition coefficient $K_{ow}$ between water and a reference organic solvent, namely the n-octanol:

$$K_{ow} = C_o/C_w$$

where $C_o$ and $C_w$ are the massic concentration of the compound in octanol and in water, respectively, when the water, the octanol and the compound are in contact at equilibrium. Some values of $K_{ow}$ are given in Table 10.1 (from Karickhoff, 1981).

(2)   Then a provisional dimensionless partition coefficient $K_{oc}$ is defined for a hypothetical soil made of 100% of solid organic material as found in small quantities in aquifer material. It is found that this coefficient $K_{oc}$ is very strongly correlated with the octanol–water partition coefficient $K_{ow}$ for a given compound but depends very little on the actual nature of the solid organics in the soil. Table 10.1 (from Karickhoff, 1981) gives some values of $K_{ow}$ and $K_{oc}$ for various organics. This author suggests

$$K_{oc} = 0.411 K_{ow}$$

with a correlation coefficient of 0.994. Schwartzenbach and Westfall (1984) propose a linear regression of the form:

$$\log K_{oc} = a \log K_{ow} + b$$

Values of $a$ and $b$, as well as the correlation coefficient of the regression are given in Table 10.2 for a series of major organics.

(3)   Finally the actual distribution coefficient $K_p$ is given by

$$K_p = K_{oc} f_{oc}$$

---

* An organic compound is said to be hydrophobic if it is soluble in water, but also more soluble in an organic solvent. It is neutral if it is not electrically charged (not ionized).

**Table 10.1**

Partition Coefficients for Octanol-to water ($K_{ow}$) and Sediment Solid
Organic Carbon to Water ($K_{oc}$) for Selected Organic Compounds[a,b]

| Compound | $\log K_{ow}$ | $\log K_{oc}$ |
|---|---|---|
| **Hydrocarbons and Chlorinated Hydrocarbons** | | |
| 3-methyl cholanthrene | 6.42 | 6.09 |
| dibenz[a,h]anthracene | 6.50 | 6.22 |
| 7,12-dimethylbenz[a]anthracene | 5.98 | 5.35 |
| tetracene | 5.90 | 5.81 |
| 9-methylanthracene | 5.07 | 4.71 |
| pyrene | 5.18 | 4.83 |
| phenanthrene | 4.57 | 4.08 |
| anthracene | 4.54 | 4.20 |
| naphthalene | 3.36 | 2.94 |
| benzene[b] | 2.11 | 1.78 |
| 1,2-dichloroethane[b] | 1.45 | 1.51 |
| 1,1,2,2-tetrachloroethane[b] | 2.39 | 1.90 |
| 1,1,1-trichloroethane[b] | 2.47 | 2.25 |
| tetrachloroethylene[b] | 2.53 | 2.56 |
| $\gamma$ BHC (lindane) | 3.72 | 3.30 |
| $\alpha$ BHC | 3.81 | 3.30 |
| $\beta$ BHC | 3.80 | 3.30 |
| 1,2-dichlorobenzene[b] | 3.39 | 2.54 |
| pp' DDT | 6.19 | 5.38 |
| methoxychlor | 5.08 | 4.90 |
| 22',44',66' PCB | 6.34 | 6.08 |
| 22',44',55' PCB | 6.72 | 5.62 |
| **Chloro-s-triazines** | | |
| atrazine | 2.33 | 2.17 |
| | 2.71 | 2.21 |
| | | 2.33 |
| propazine | 2.94 | 2.20 |
| | | 2.19 |
| | | 2.56 |
| simazine | 2.16 | 2.13 |
| | | 2.14 |
| | | 2.33 |
| trietazine | 3.35 | 2.74 |
| ipazine | 3.94 | 3.22 |
| | | 2.91 |
| cyanazine | 2.24 | 2.30 |
| | | 2.26 |

(*Continued*)

**Table 10.1** (*Continued*)

| Compound | log $K_{ow}$ | log $K_{oc}$ |
|---|---|---|
| Carbamates | | |
| carbaryl | 2.81 | 2.36 |
| carboturan | 2.07 | 1.46 |
| chlorpropham | 3.06 | 2.77 |
| Organophosphates | | |
| malathion[b] | 2.89 | 3.25 |
| parathion[b] | 3.81 | 3.68 |
| | | 4.03 |
| methylparathion | 3.32 | 3.71 |
| | | 3.99 |
| chlorpyrifos | 3.31 | 4.13 |
| | 4.82 | |
| | 5.11 | |
| Phenyl ureas | | |
| diuron | 1.97 | 2.60 |
| | 2.81 | 2.58 |
| fenuron | 1.00 | 1.43 |
| | | 1.63 |
| linuron | 2.19 | 2.91 |
| | | 2.94 |
| monolinuron | 1.60 | 2.30 |
| | | 2.45 |
| monuron | 1.46 | 2.00 |
| | 2.13 | 2.26 |
| fluometuron | 1.34 | 2.24 |
| Miscellaneous compounds | | |
| 13Hdibenzo[a,i]carbazole | 6.40 | 6.02 |
| 2,2' biquinoline | 4.31 | 4.02 |
| dibenzothiophene | 4.38 | 4.05 |
| acetophenone[b] | 1.59 | 1.54 |
| terbacil | 1.89 | 1.71 |
| | | 1.61 |
| bromacil | 2.02 | 1.86 |

[a] Reproduced with permission from Karickhoff (1981). Copyright 1981 Pergamon Press, Ltd.

[b] Compounds are liquids at 25°C

**Table 10.2**

Estimation of $K_{oc}$ Based on $K_{ow}$ by the Expression $\log K_{oc} = a \log K_{ow} + b$[a]

| Regression coefficient | | Correlation coefficient | Number of compounds | Type of chemical |
|---|---|---|---|---|
| $a$ | $b$ | | | |
| 0.544 | 1.377 | 0.74 | 45 | Agricultural chemicals |
| 1.00 | −0.21 | 1.00 | 10 | Polycyclic aromatic hydrocarbons |
| 0.937 | −0.006 | 0.95 | 19 | Triazines, nitroanilines |
| 1.029 | −0.18 | 0.91 | 13 | Herbicides, insecticides |
| 1.00 | −0.317 | 0.98 | 13 | Heterocyclic aromatic compounds |
| 0.72 | 0.49 | 0.95 | 13 | Chlorinated hydrocarbons alkylbenzenes |
| 0.52 | 0.64 | 0.84 | 30 | Substituted phenyl ureas and alkyl-N-phenyl carbamates |

[a] From Schwartzenbach and Westfall (1984).

where $f_{oc}$ is the dimensionless fraction of dry weight of sediment which is made of solid organic carbon compound. Schwartzenbach and Westfall (1984) indicate that only the fine fraction of the aquifer material (e.g., grain size smaller than 0.125 mm) is predominant for sorption; $f_{oc}$ should then be taken as the product of the fraction of sediment smaller than 0.125 mm times the fraction of solid organic compound in these fine sediments. Since most of the solid organics are found in the fine fraction anyway, this should not make very much difference.

Note that the relationship is only valid if $f_{oc} > 10^{-3}$, otherwise sorption of the organic compound on nonorganic solids can become significant. The linear sorption isotherm $C' = K_p F$ is approximately valid only if $C'$ remains below one half of the solubility limit of the compound.

Other methods of estimating $K_{oc}$ have also been suggested based on the solubility of the organic compound in water, or directly on its chemical formula. See Karickhoff (1981).

(4)  For an ionizable hydrophobic compound, sorption is found to vary also with pH. Several mechanisms are then responsible for sorption: ion exchange, ligand exchange, formation of ion pairs, or ion complexes (that may be transferred into the organic phase) in addition to simple partitioning. See Schwartzenbach and Westfall (1981, 1984), Schellenberg et al. (1984), Westfall et al. (1984).

(d) *Radioactive decay.* If no transport occurs, radioactive decay is expressed by the differential equation

$$\frac{\partial C}{\partial t} = -\lambda C$$

which, integrated, gives

$$C = C_0 e^{-\lambda t} \qquad \text{(exponential decay)}$$

The half-life $T$ is defined by $C/C_0 = \frac{1}{2}$, which yields

$$e^{-\lambda T} = \tfrac{1}{2} \qquad \text{or} \qquad \lambda = \frac{\ln 2}{T} = \frac{0.693}{T} \tag{10.2.5}$$

Hence, radioactive decay causes a mass $\lambda C$ per unit volume of the liquid phase to disappear per unit time. In order to restore it to a unit volume of the porous medium, it must therefore be multiplied by $\omega$. The transport equation then becomes

$$\text{div}(\mathbf{D}\,\text{grad}\,C - C\mathbf{U}) = \omega\left(\frac{\partial C}{\partial t} + \lambda C\right) \tag{10.2.6}$$

If there is a concentration $F$ in the adsorbed phase, this will also decrease according to the same law:

$$\frac{\partial F}{\partial t} = -\lambda F$$

This disappearance is expressed here in mass per unit time and per unit mass of solid. To restore it to the unit volume of the porous medium, we must multiply by $(1 - \omega)\rho_s$. Thus it becomes

$$\text{div}(\mathbf{D}\,\text{grad}\,C - C\mathbf{U}) = \omega\left(\frac{\partial C}{\partial t} + \lambda C\right) + (1 - \omega)\rho_s\left(\frac{\partial F}{\partial t} + \lambda F\right) \tag{10.2.7}$$

In the case of linear and reversible adsorption ($F = K_d C$), this becomes

$$\text{div}(\mathbf{D}\,\text{grad}\,C - C\mathbf{U}) = \omega R\frac{\partial C}{\partial t} + \omega R\lambda C \tag{10.2.8}$$

(e) *Daughter products.* If a substance $C_i$ disappears through radioactive decay, it means that it gives birth to a daughter product, i.e., a different substance $C_j$.[*] In the transport equation of the substance $C_j$, the source term

---

[*] In the following, $C_j$ can possibly be the granddaughter of $C_i$ through the action of several nuclear reactions, if the half-lives of the intermediate substances are very short compared to that of $C_j$ and, if only $C_i$ and $C_j$ are present in significant amounts.

will then represent a creation of matter. It is written as

$$\text{div}(\mathbf{D}\,\mathbf{grad}\,C_j - C_j\mathbf{U}) = \omega\frac{\partial C_j}{\partial t} - \omega\lambda_i\frac{M_j}{M_i}C_i.$$

where $M_j/M_i$ is the ratio of the atomic weights of the substances $i$ and $j$, if they are different. We can easily generalize to the case of adsorption.

Thus, the transport equation for $j$ must be solved after the one for $i$, as the distribution in time and space of the source term for element $j$ is given by the solution of the transport of $i$. For three-member decay chains, analytical solutions have been developed. The difficulty with numerical solutions lies in the agreement between the calculation time steps for the two elements. See Harada *et al.* (1980) and Pigford *et al.* (1980).

### 10.2.2. Fractured Media

All of the phenomena mentioned in connection with porous media may occur in fractured media. The only case we need to point to here is the one of adsorption in the fracture planes, when the rock matrix is assumed to be impermeable and nonporous, i.e., where, in practice, the transported substances cannot penetrate. It is then necessary to determine experimentally a distribution coefficient per unit surface area of the fracture. It is, in fact, possible to define conceptually a "concentration" $W$ adsorbed by the fracture, where $W$ is expressed in mass of substance retained per unit surface area of the fracture (Burkholder, 1975).

In order to equate this quantity $W$ of adsorbed substances with the unit of the source term, $W$ must be multiplied by the ratio $\alpha$ which we have already defined:

$$\alpha = \frac{\text{area of the fracture planes}}{\text{volume of the medium}}$$

The source term is indeed the variation of the mass of substances per unit volume of an equivalent fractured medium, per unit time.

Note that we have defined $\alpha$ by counting two planes for the walls of each fracture. Hence, the source term for adsorption is

$$Q = \alpha\frac{\partial W}{\partial t}$$

If we admit that there is still a linear relation between this concentration $W$ and the concentration $C$ in the solution, then

$$W = K_a C$$

where $K_a$, the fracture distribution coefficient, has the dimension of length (volume/surface area).

The transport equation becomes

$$\text{div}(\mathbf{D} \text{ grad } C - C\mathbf{U}) = \omega\left(1 + \frac{\alpha K_a}{\omega}\right)\frac{\partial C}{\partial t}$$

with the "retardation factor"

$$R = 1 + \frac{\alpha K_a}{\omega} \quad \text{(dimensionless)}$$

To measure $K_a$ experimentally on a core sample containing a fracture, we proceed as for the porous medium, by difference, but the mass quantity $W$ bound to the planes of the fracture is attributed to the surface area of the fracture (i.e., twice its dimension, if the two walls of the fracture are in contact with the solution).

### 10.2.3. Analytical Solutions of the Dispersion Equation

(1) If we choose a one-dimensional case (Fig. 10.10) and study the displacement of a contaminant in a semiinfinite medium, we know an analytical solution to the dispersion equation with the following initial and boundary conditions:

$$C(x) = 0, \quad \forall x > 0, \quad t = 0$$

$$C(0) = C_0, \quad t > 0$$

with the tracer hypothesis, velocity $U$ constant for one-directional flow, and dispersion coefficient $D = \alpha|U|$ a constant (only longitudinal dispersion in this one-dimensional problem). The governing equation is

$$D\frac{\partial^2 C}{\partial x^2} - U\frac{\partial C}{\partial x} = \omega R\frac{\partial C}{\partial t} \qquad (10.3.1)$$

This equation is identical in one dimension to Eq. (10.1.5) or (10.1.7) if $R = 1$ or to (10.2.4) if there is adsorption; $R$ is then the "retardation factor" of Eq. (10.2.3).

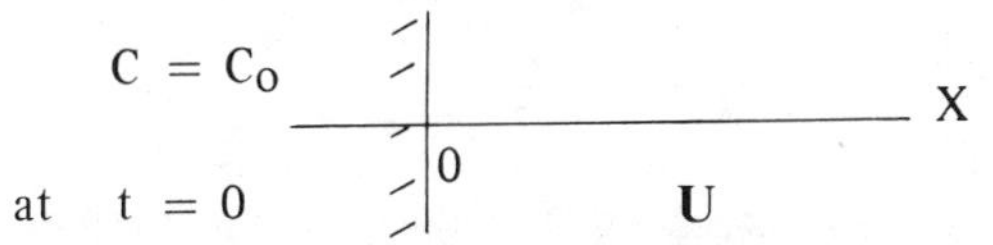

$$C = C_0$$

$$\text{at} \quad t = 0 \qquad \qquad U$$

X

0

Fig 10.10. One-dimensional transport system.

*Solution.*

$$C(x,t) = \frac{C_0}{2}\left[\text{erfc}\left(\frac{x - (U/\omega R)t}{2\sqrt{Dt/\omega R}}\right) + \exp\left(\frac{Ux}{D}\right)\text{erfc}\left(\frac{x + (U/\omega R)t}{2\sqrt{Dt/\omega R}}\right)\right]$$

(10.3.2)

where erfc is the complementary error function (see Section 8.5).

For given $x$ and after a certain time, the second term becomes negligible before the first and the expression may be simplified to

$$C(x,t) = \frac{C_0}{2}\text{erfc}\left(\frac{x - (U/\omega R)t}{2\sqrt{Dt/\omega R}}\right) = \frac{C_0}{\sqrt{\pi}}\int_{v}^{+\infty} e^{-r^2}\, dr,$$

where
$$v = \frac{x - (U/\omega R)t}{2\sqrt{Dt/\omega R}}$$

The solution given in Eq. (10.3.2) is shown in Fig. 10.11 versus the three dimensionless parameters,

$$\xi = Ut/\omega Rx \qquad \text{on the horizontal axis}$$

$$C/C_0 \qquad \text{on the vertical axis}$$

$$\eta = D/Ux \qquad \text{curve parameter}$$

(2)  If the radioactive decay is added [Eq. (10.2.6) or (10.2.8)], we get (Bear, 1979)

$$D\frac{\partial^2 C}{\partial x^2} - U\frac{\partial C}{\partial x} = \omega R\left(\frac{\partial C}{\partial t} + \lambda C\right)$$

(10.3.3)

where $\lambda$ is the coefficient of exponential decay from Eq. (10.2.5) and $R$ the retardation factor due to the adsorption as in Eq. (10.2.3).

*Solution.*

$$C(x,t) = \frac{C_0}{2}\exp\left(\frac{Ux}{2D}\right)\left\{\exp(-\beta x)\,\text{erfc}\left[\frac{x - t\sqrt{(U/\omega R)^2 + 4\lambda D/\omega R}}{2\sqrt{Dt/\omega R}}\right]\right.$$

$$\left. + \exp(\beta x)\,\text{erfc}\left[\frac{x + t\sqrt{(U/\omega R)^2 + 4\lambda D/\omega R}}{2\sqrt{Dt/\omega R}}\right]\right\}$$

(10.3.4)

where
$$\beta = \sqrt{\left(\frac{U}{2D}\right)^2 + \frac{\lambda\omega R}{D}}$$

(3)  Consider an impulse point injection of tracer of mass $dM$ into an aquifer with parallel flow in two dimensions. If $x$ is the flow direction, with the

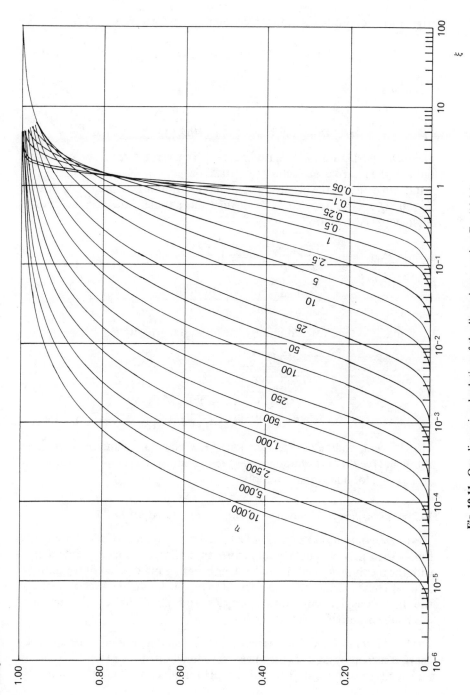

**Fig. 10.11.** One-dimensional solution of the dispersion equation, Eq. (10.3.2).

origin of the coordinates at the injection point, we get (Bear, 1979):

$$D_L \frac{\partial^2 C}{\partial x^2} + D_T \frac{\partial^2 C}{\partial y^2} - U \frac{\partial C}{\partial x} = \omega R \frac{\partial C}{\partial t} \tag{10.3.5}$$

*Solution.*

$$dC(x, y, t) = \frac{dM}{4\pi t \sqrt{D_L D_T / \omega^2 R^2}} \exp\left[ -\frac{(x - Ut/\omega R)^2}{4 D_L t/\omega R} - \frac{y^2}{4 D_T t/\omega R} \right] \tag{10.3.6}$$

where $D_L$ and $D_T$ are the longitudinal and transverse dispersion coefficients, and $R$ is the retardation factor [Eq. (10.2.3)].

If the injection at the origin is a continuous flow rate $Q$ with a concentration $C_0$ (but $Q$ small enough not to disturb the parallel flow), the solution is obtained by convolution:

$$C(x, y, t) = -\frac{C_0 Q}{4\pi \sqrt{D_L D_T / \omega^2 R^2}} \int_0^t \frac{1}{t - \tau}$$
$$\times \exp\left\{ -\frac{[x - U(t - \tau)/\omega R]^2}{4 D_L (t - \tau)/\omega R} - \frac{y^2}{4 D_T (t - \tau)/\omega R} \right\} d\tau$$

If $t$ is made to tend toward infinity (steady state), we get

$$C(x, y, \infty) = \frac{C_0 Q}{2\pi \sqrt{D_L D_T / \omega^2 R^2}}$$
$$\times \exp\left(\frac{xU}{2D_L}\right) K_0 \left[ \sqrt{\frac{U^2}{4 D_L \omega R} \left( \frac{x^2}{D_L/\omega R} + \frac{y^2}{D_T/\omega R} \right)} \right]$$

where $K_0$ is a modified Bessel function of the second kind and zero order. See also Bear (1972, 1979), Harada *et al.* (1980), Pigford *et al.* (1980), and Javandel *et al.* (1984) for additional solutions.

### 10.2.4. *Transport of Colloids in Porous or Fractured Media*

(a) *Description of the physical mechanisms involved.* Colloidal particles transported in a porous medium experience a large number of interactions with the medium, which can make their behavior quite different from that of solutes. These mechanisms are referred to as "filtration." To present them, we will first consider unretained particles and then analyze the retention mechanisms on the surfaces.

(1) *Unretained particles moving faster than the water.* The electrical charges carried by colloids are in general a function of the pH of the solution: for each type of colloid, there exists a pH for which they are uncharged, the

point of zero charge (PZC). When such very small uncharged particles are injected into a porous medium, a capillary tube, or a thin fracture, they will generally be transported without retention with the flowing liquid by convection, as well as by diffusion through Brownian motion. The diffusion coefficient $d_0$ in water is inversely proportional to the radius r of the colloid, as shown in Subsection 10.1.1.b.

In pores, capillary tubes or fractures, the velocity distribution of the water is generally more or less parabolic, the maximum velocity at the center being on the order of 1.5 times the average velocity of the water. Particles transported in the water will, by diffusion, randomly sample the velocities in the pores, but, because of their size, they will never reach the walls: their average velocity will therefore be larger than that of the water, larger colloids being faster than smaller ones. This effect was discovered by Small (1974, 1977) and is called hydrodynamic chromatography. It is used for measuring particle sizes. The ratio between particle velocity and water velocity is, however, small, in general between 1 and 1.1, in extreme cases 1.4 (Dodds, 1982). For such movements to be observed, the particles have to be much smaller than the size of the pores (or fracture aperture). Dodds (1982) gives an upper limit of particles size of 0.25 $\mu$m, for a porous medium constituted by spheres of 20 $\mu$m, i.e., roughly a factor of 100 between grain size and particle size (see Nagy *et al.*, 1981).

If the particles are charged with the same charges as the solids of the porous medium, repulsion effects will tend to increase the velocity of the particles, as they are kept further away from the walls. Small (1974, 1977) has evaluated the thickness of the repulsive layer to 0.39$\mu$m for particles of 0.357 $\mu$m in diameter, in dilute solutions. However, if the ionic strength of the solution is increased, this repulsive force decreases, and the attractive Van der Waals forces can play a role in slowing down or retaining the particles. Hydrodynamic chromatography is therefore a function of the ionic strength of the solution (Dodds, 1982).

If the charges of the particles are of opposite sign to that of the solid grains of the medium, retention mechanisms will start to play a role in the slowing down (or "filtration") of the particles. But whatever the electric charges, larger particles will always interact with the medium and be stopped or move more slowly than the water, even if their diameter is still much smaller than the average size of the pores.

(2) *Filtration of particles by the medium.*    Herzig *et al.* (1970), Wnek *et al.* (1975), Tien *et al.* (1979), Corapcioglu *et al.* (1986), and Willis (1986) among others, have described the most important interaction mechanisms, and proposed equations to represent them. Filtration includes three mechanisms: (i) particles coming into contact with the walls; (ii) particles becoming fixed to the walls; (iii) previously retained particles breaking away. We will study them successively.

(i) *Contacting*: several processes can bring the particles into contact with the solid.

*Sedimentation* occurs if the particles have a different density from that of the liquid: their velocity will be different.

*Inertia*: due to their weight, the trajectories of the particles deviate from the liquid streamlines and can come into contact with the grain.

*Hydrodynamic effects*: because the velocity field of the liquid is not uniform, the particles are subjected to a rotation couple, which modifies their velocity and trajectory. See Brenner and Gaydos (1977).

*Direct interception*: because of their size, the particles collide with the walls of the convergent pores.

*Diffusion*: Brownian motion will send particles toward the walls, or even into dead-end pores, where the velocity is nil. Diffusion is said to be negligible for particles larger than 1 $\mu$m.

(ii) *Fixing*:

*Retention sites* can be located on the surface of the solid, on edges between two convex surfaces, in constriction sites, where the particle cannot penetrate or in dead-end pores or caverns, where the velocity of the fluid is nil.

*Retentions forces* include axial pressure of the fluid (in a constriction site), friction forces (on an edge), surfaces forces (Van der Waals forces, which are always attractive, and, if the particles are charged, electrical forces which can be attractive or repulsive), and, finally, chemical forces, if chemical bonding can occur.

(iii) *Breaking away*: a moving particle may collide with a retained particle; a local variation in pressure or flow rate (due to clogging) may modify the flow field sufficiently to bring a retained particle into motion; an external change in the flow conditions will do the same in the complete medium (e.g., declogging of a filter by reversed circulation).

We will now examine how these mechanisms have been combined to form the filtration equation.

(b) *The Transport Equation For Colloids.*

(1) *Differential equation.* A general equation for the mechanisms of filtration has been proposed by Herzig *et al.* (1970), Wnek *et al.* (1975), Tien *et al.* (1979), Dodds (1982), and Dieulin (1982), among others. It is

$$\text{div}(\mathbf{D} \, \text{grad} \, C) - \text{div}(\mathbf{U}C) = \frac{\partial}{\partial t}(\omega C + \sigma)$$

where $\mathbf{D}$ is the hydrodynamic dispersion tensor in the medium, C is the concentration of colloids in the liquid phase, expressed as volume of colloids per unit volume of liquid, U is Darcy's velocity in the medium, $\omega$ is the kinematic porosity, $\sigma$ is the concentration of colloids retained by filtration in the medium, expressed as volume of colloids per unit volume of porous medium.

It is then assumed that the porosity of the medium can be reduced by filtration, with a first-order dependence on $\sigma$:

$$\omega = \omega_0 - \beta\sigma$$

Here, $\beta$ is the inverse of the compaction factor of the retained colloids, i.e., a volume $\sigma$ of retained colloids occupies a volume $\beta\sigma$ of the pores, imprisoning "dead liquid."

More complex clogging factors, including the changes in specific surface of the medium, have been proposed (Herzig et al., 1970) in order to also predict the permeability variation of the medium and, thus, the pressure gradient increase with time necessary to maintain a constant Darcy velocity U in the medium.

The variation in concentration of the retained colloids is

$$\partial\sigma/\partial t = \lambda U C F(\sigma)$$

In the hypothesis called "deep filtration," $F(\sigma) = 1$, i.e., where there is no interaction between the particles, the mechanisms stay the same as $\sigma$ increases. Otherwise, $F(\sigma)$ takes into account the variation with $\sigma$ of porosity and specific surface. $F(\sigma)$ can be taken, as a first approximation, as $1 - \beta\sigma$. $\lambda$ is the filter coefficient (see Subsection 3 below).

(2) *Analytical solutions.* Dieulin (1982) developed an analytical solution to this equation in 1 dimension assuming that the porosity $\omega$ is constant and that $F(\sigma) = 1$. It is

$$C(x,t) = \frac{C_0}{2}\left\{\exp\left[\frac{Ux}{2\omega D}(1 - \gamma)\right]\mathrm{erfc}\left[\frac{x - Ut\gamma/\omega}{2\sqrt{Dt}}\right]\right.$$
$$\left. + \exp\left[\frac{Ux}{2\omega D}(1 + \gamma)\right]\mathrm{erfc}\left[\frac{x + Ut\gamma/\omega}{2\sqrt{Dt}}\right]\right\}$$

and

$$\sigma(x,t) = \frac{C_0}{2}\lambda U\left\{\exp\left[\frac{Ux}{2\omega D}(1 - \gamma)\right]\int_0^t\mathrm{erfc}\left[\frac{x - U\gamma z/\omega}{2\sqrt{Dz}}\right]dz\right.$$
$$\left. + \exp\left[\frac{Ux}{2\omega D}(1 + \gamma)\right]\int_0^t\mathrm{erfc}\left[\frac{x + U\gamma z/\omega}{2\sqrt{Dz}}\right]dz\right\}$$

where

$$\gamma = \sqrt{1 + \frac{4\lambda\omega D}{U}}$$

$C_0$ is the constant concentration of the solution in colloids at the entrance of the medium ($x = 0$) prescribed from $t = 0$.

If the hydrodynamic dispersion can be disregarded, these equations further simplify into:

$$C(x, t) = C_0 \exp(-\lambda x), \qquad t > \frac{\omega x}{U}$$

$$\sigma(x, t) = \lambda U C_0 \exp(-\lambda x)\left(t - \frac{\omega x}{U}\right), \qquad t > \frac{\omega x}{U}$$

$$C = \sigma = 0, \qquad t < \frac{\omega x}{U}$$

Finally, for large times, the asymptotic solution is:

$$C(x, t) = C_0 \exp\left(\frac{-2\lambda x}{1 + \gamma}\right)$$

$$\sigma(x, t) = \lambda U t C_0 \exp\left(-\frac{2\lambda x}{1 + \gamma}\right)$$

the $C$ and $\sigma$ profiles become straight lines in a semi-logarithmic plot versus $x$. Such behavior was indeed observed for americium colloids filtration experiments on sand columns (Saltelli $et$ $al.$, 1984)

Herzig $et$ $al.$ (1970) also give an analytical solution for the case where the porosity and the retention vary* according to:

$$\omega = \omega_0 - \beta\sigma \qquad \text{and} \qquad F(\sigma) = 1 - \beta\sigma$$

but neglecting hydrodynamic dispersion. They found that:

$$C(x, t) = \frac{C_0 \exp(\beta\lambda U C_0 \tau)}{-1 + \exp(\lambda x) + \exp(\beta\lambda U C_0 \tau)}$$

$$\sigma(x, t) = \left(\frac{1}{\beta}\right) \frac{1 - \exp(\beta\lambda U C_0 \tau)}{1 - \exp(\lambda x) - \exp(\beta\lambda U C_0 \tau)}$$

where $\tau = t - \omega_0 x/U \geq 0$; $C = \sigma = 0$ if $\tau < 0$.

They also give similar expressions for other simple forms of $F(\sigma)$ (e.g., $1 - \beta^2\sigma^2$, $\sqrt{1 - \beta\sigma}$, $(1 - \beta\sigma)^{3/2}$). In the experiments of americium colloids filtration, it was not necessary to take into account any clogging of the

---

* See Note Added in Proof at the end of this chapter.

medium, $\omega$ was constant and $F(\sigma) = 1$ (Saltelli et al, 1984; Avogadro and Marsily, 1984).

(3) *Filter coefficients.* The filter coefficient $\lambda$ takes into account all the mechanisms described earlier. Tien et al. (1979) propose the following theoretical expression for small $\eta$

$$\lambda = \frac{3(1 - \omega)}{2d_g} \cdot \eta$$

where $\omega$ is the porosity, $d_g$: the grain diameter and $\eta$ the collection efficiency of the filter,

$$\eta = (1 - \omega)^{2/3} A_S N_{Lo}^{1/8} N_R^{15/8} + 3.375 \times 10^{-3}(1 - \omega)^{2/3} A_S N_G^{1.2} N_R^{-0.4}$$
$$+ 4A_S^{1/3} N_{Pe}^{-2/3}$$

where,

$$A_S = \frac{2(1 - p^5)}{2 - 3p + 3p^5 - 2p^6}, \qquad N_{Lo} = \frac{4H}{9\pi\mu d_p^2 U}, \qquad N_R = \frac{d_p}{d_g}$$

$$N_G = \frac{(\rho_p - \rho)d_p^2 g}{18\pi\mu U}, \qquad N_{Pe} = \frac{3\pi\mu d_p d_g U}{kT}$$

$p = (1 - \omega)^{1/3}$, $H$ is the Hamaker constant ($\sim 10^{-20}$ J), $d_p$ the particle diameter, $U$ is Darcy's velocity, $\mu$ the viscosity of the fluid, $\rho_p$ the mass per unit volume of particles, $\rho$ the mass per unit volume of fluid, and $k$ is the Boltzmann's constant ($1.38048 \times 10^{-23}$ J/K).

These expressions apply to spherical grains of the medium and to spherical colloids; similar expressions could also be developed for fractures. Other forms are suggested by Herzig et al. (1970) or Dodds (1982).

It is interesting to note that the dependence of $\lambda$ on the diameter of the grains and the particles is rather complex and depends strongly on the properties of the medium, especially the porosity. In the filtration experiments of americium colloids, the following values were obtained by fitting the analytical expression for $\sigma(x, t)$ given above on the measurements:

| | |
|---|---|
| glauconitic sand: | $\lambda = 31$ mm$^{-1}$ |
| clean sand, 100–200 $\mu$m: | $\lambda = 0.125$ mm$^{-1}$ |
| clean sand, 200–400 $\mu$m: | $\lambda = 0.032$ mm$^{-1}$. |

The calculated filter coefficients using Tien's expression fall within one order of magnitude of the measured values.

One must also bear in mind that colloids are not necessarily stable in the medium; they may dissolve, or on the other hand increase in size by coalescence or by serving as germs for precipitation. All this depends on their chemical nature and their geochemical equilibrium in the medium.

### 10.2.5. Unsaturated Media

So far we have only considered transport in a saturated medium. However, the transport equation that we have developed also applies to unsaturated media. The left-hand side of the transport equation [e.g., (10.1.5) or (10.2.1)] remains unchanged, although it is found that the hydrodynamic dispersion coefficient $D$ is now also a function of the saturation. The mass balance on the right-hand side is however modified to account for the saturation. When we first developed this mass balance, in Subsection 10.1.1.a, we stated that the total mass of solute in an arbitrary volume $D$ of saturated medium was

$$\int_D \omega C \, dv$$

Now for an unsaturated medium, the same total mass is

$$\int_D \omega s_w C \, dv \quad \text{or} \quad \int_D \theta C \, dv$$

where $s_w$ is the water saturation of the medium and $\theta$ the moisture content. We need to have the rate of change with time of this total mass. The transport equation is thus

$$\text{div}[\mathbf{D}(\theta)\mathbf{grad}\, C - C\mathbf{U}] = \omega \frac{\partial s_w C}{\partial t} = \frac{\partial \theta C}{\partial t}$$

If the flow in the unsaturated medium is in steady state, $\theta$ (or $s_w$) is a constant and can be taken out of the time derivative. However, in a transient flow situation, it is first necessary to determine $\theta$ and $\partial\theta/\partial t$ from the flow equation prior to solving the transport equation.

In the unsaturated zone, for certain conditions, the existence of an immobile water phase needs to be taken into account as in Eq. (10.1.4). In steady-state flow conditions, this would be written

$$\text{div}[\mathbf{D}(\theta)\mathbf{grad}\, C - C\mathbf{U}] = \theta' \frac{\partial C}{\partial t} + (\theta - \theta') \frac{\partial C'}{\partial t}$$

where $C$ is the concentration in the mobile water phase, $C'$ the concentration in the immobile water phase, $\theta'$ the moisture content corresponding to the mobile water phase, $\theta$ the total moisture content, and $\theta - \theta'$ the moisture content corresponding to the immobile water phase.

It is then found that the exchange between the mobile and immobile functions can often be represented by a first-order kinetic reaction (see

Subsection 10.2.1.b, fourth and fifth case):

$$\frac{\partial C'}{\partial t} = K_1(C - C')$$

Recent work on transport in the unsaturated zone can be found in *Gaudet et al.* (1977), Dagan and Bresler (1980), Arnold *et al.* (1982), and Oster (1982).


## 10.3. Heat Transfer in Porous Media

At first sight, heat transfer in porous media is governed by three separate mechanisms: (1) conduction in the solid matrix, (2) transport by the fluid phase, and (3) heat exchange between the two phases depending on their temperature difference.

The first phenomenon would produce a heat equation relative to the mean temperature $\langle \theta_s \rangle$ of the solid. The second would resemble the dispersion equation for the fluid with the fluid temperature $\langle \theta \rangle$ playing the part of the concentration. The third would be related to the exchange mechanisms between the solid and the liquid phase, which we have discussed above.

However, in practice, except for a very small number of cases, the assumption is made that the temperature of the solid and that of the fluid become identical almost at once, and that there is only one temperature $\theta$ in the porous medium. Houpeurt *et al.* (1965) have shown that the temperatures will become equal in less than a minute in a medium with grain-sizes of less than 1 mm or in less than 2 h for pebbles of 10 cm diameter.

All that has been said previously on the subject of solute transport can then be applied to heat transfer in porous media.

A single temperature is calculated for the porous medium. The transport is characterized by (1) a convection phenomenon similar to that of the solutes and (2) a phenomenon similar to that of dispersion in porous media: (a) pure conduction in the two phases, solid plus liquid, takes the place of molecular diffusion, while (b) the heterogeneity of the real velocity gives rise to an anisotropic "fictitious conductivity," equivalent to the kinematic dispersion, which experience shows to be a linear function of the absolute value of the velocity (Ledoux and Clouet d'Orval 1977; Sauty, 1978).

The principle of heat conservation (analogous to the mass balance) makes it possible to write directly:

$$\text{div}(\lambda \, \mathbf{grad} \, \theta - \rho c \mathbf{U} \theta) = \omega \rho c \frac{\partial \theta}{\partial t} + (1 - \omega)\rho' c' \frac{\partial \theta'}{\partial t} = \rho'' c'' \frac{\partial \theta}{\partial t}$$

with $\lambda$ the tensor of equivalent conductivity, $\theta$ the temperature, $\rho c$ the mass per unit volume and specific heat of the water, $\rho'c'$ the mass per unit volume and specific heat of the solid with a temperature $\theta' = \theta$, $\rho''c''$ the mass per unit volume and specific heat of the porous medium (water plus solid) $[\rho''c'' = \omega\rho c + (1 - \omega)\rho'c']$, and $\omega$ the total porosity.

The tensor of equivalent conductivity $\lambda$ combines the isotropic conductivity $\lambda_0$ of the porous medium (water plus solid) in the absence of flow and a term for the macrodispersivity linked to the heterogeneity of the velocity, which is a linear function of this velocity. We suggest using Darcy's velocity $U$ multiplied by the volumetric head capacity of the water $\rho c$ so that the proportionality coefficient has the dimension of length like the macrodispersivity in the case of dispersion. In the longitudinal and transverse axes linked to the velocity, we get

$$\lambda_L = \lambda_0 + \beta_L \rho c |U| \qquad \lambda_T = \lambda_0 + \beta_T \rho c |U|$$

It is possible to put this equation and that of the dispersion into comparable expressions in order to bring out the similarities of the dispersion coefficients. For this purpose, temperatures or concentrations are made dimensionless, as follows:

$$C \quad \text{or} \quad \theta = \frac{Z - Z_{min}}{Z_{max} - Z_{min}}$$

where $Z$ is the concentration or the temperature.

Either the Darcy velocity or the velocity of the advancing front (thermal or chemical) may be used as a reference, giving the same results if the dispersivities are compared:

*Equations relative to the Darcy velocity:*

Tracer:     $$\text{div}(\mathbf{D}\,\text{grad}\,C) - \text{div}(\mathbf{U}C) = \omega\frac{\partial C}{\partial t},$$

with     $$\mathbf{D} = \omega d + \alpha|\mathbf{U}|$$

Heat:     $$\text{div}\left(\frac{\lambda}{\rho C}\,\text{grad}\,\theta\right) - \text{div}(\mathbf{U}\theta) = \frac{\rho''c''}{\rho c}\frac{\partial \theta}{\partial t},$$

with     $$\frac{\lambda}{\rho c} = \frac{\lambda_0}{\rho c} + \beta|U|$$

*Equations relative to the velocity of the advancing convective front for tracers and heat respectively, are*

$$\text{div}(\mathbf{D}'\,\text{grad}\,C) - \text{div}(\mathbf{u}^*C) = \frac{\partial C}{\partial t} \qquad \text{with} \qquad \mathbf{D}' = d + \alpha|\mathbf{u}^*|$$

$$\text{div}\left(\frac{\lambda}{\rho''c''}\,\textbf{grad}\,\theta\right) - \text{div}(\textbf{u}^{*\prime}\theta) = \frac{\partial\theta}{\partial t} \quad \text{with} \quad \frac{\lambda}{\rho''c''} = \frac{\lambda_0}{\rho''c''} + \beta|\textbf{u}^{*\prime}|$$

In these equations, the velocities of the average front of tracer and heat are, respectively,

$$\textbf{u}^* = \frac{\textbf{U}}{\omega} \quad \text{and} \quad \textbf{u}^{*\prime} = -\frac{\rho c\textbf{U}}{\rho''c''}$$

It is then obvious that, in both cases, the dispersivity of the tracer $\alpha$ must be compared to the dispersivity $\beta$ of the heat, which are both expressed as lengths. We have tried to do this at a single experimental site (Bonnaud, Jura, France), where both tracer tests and heat transport tests were made. The aquifer consists of relatively homogeneous sands and gravels and is 3 m thick. It is confined, its transmissivity is of the order of $10^{-3}$ m$^2$/s, and its storage coefficient is between $10^{-3}$ and $10^{-4}$.

The values of the main dispersivities (Peaudecerf et al., 1975; Ledoux and Clouet d'Orval (1977); Sauty (1978b), Gringarten et al. (1979)), calculated by model calibration, are shown in Table 10.3. These results seem to prove that the dispersivities for the heat and for the tracer are comparable, even though the conductivity in the absence of flow is around 400 to 1000 times stronger than the molecular diffusion.

This seems to contradict the first laboratory experiments of such a comparison: Green [in 1963, quoted by Bear (1972)] suggests that the equivalence between the thermal and chemical dispersivities occurs for Peclet numbers of the order of 10,000, and that under 3000 the thermal dispersivity is negligible.

The Peclet numbers, chemical as well as thermal, in the experiments at Bonnaud are, at the most, a few tens:

$$P_e\ \text{tracer} = \frac{u^*l}{d_0}$$

where if $u^*$ (mean pore velocity) is $\sim 0.09$ m/hr, $l$ (mean diameter of the grains) is $\sim 2$ mm, and $d_0$ (molecular diffusion coefficient in water) is $\sim 1.0 \times 10^{-9}$m$^2$/s, then $P_e$ is $\sim 50$.

$$P_e\ \text{thermal} = \frac{u'l}{\lambda_0/\rho''c''} = \frac{Ul}{\lambda_0/\rho c}$$

where if $U$ (Darcy velocity at average radius of 6 m) is $\sim 0.03$ m/hr, $l$ (mean diameter of the grains) is $\sim 2$ mm, and $\lambda_0/\rho c$ is $\sim 2 \times 10^{-7}$m$^2$/s, then $P_e$ is $\sim 0.1$.

### Table 10.3

Comparison of Longitudinal Dispersivities in the Bonnaud Aquifer (Jura, France) for a Chemical or Thermal Tracer

| Authors | Tracer | Type of flow | Distance of tracing (m) | Dispersion | | | Conduction | | |
|---|---|---|---|---|---|---|---|---|---|
| | | | | Longitudinal dispersivity $\alpha_L$ (m) | Porosity (%) | Molecular diffusion in porous media $d$ (m²/s) | Longitudinal dispersivity $\beta_L$ (m) | Relative heat capacity of the porous medium $\rho''C''/\rho c$ (%) | Pure heat diffusivity $\lambda_0/\rho'c'$ (m²/s) |
| Peaudecerf et al. (1975) | Iodine 131 | Parallel, point tracing | 13.05 | 0.70 | 33 | | | | |
| | | Idem, other wells | 12.97 | 1.20 | 29 | | | | |
| | | Idem, | 26.02 | 2.23 | 37 | $\sim 1.0 \times 10^{-9}$ | | | |
| | | Idem | 35.52 | 2.19 | 48 | | | | |
| Gringarten et al. (1979) | INa | Radial converging tracing by piezometer | 13.00 | 1.6 | 9 | | | | |
| | | Well doublet (injection + pumping) | 13.00 | 4.0 | 9 | | | | |
| Ledoux and Clouet d'Orval (1977) | Heat | Radial single well (injection + pumping) | 13.00 | | | | 3.00 | 50 | $4 \times 10^{-7}$ |
| Sauty (1978[b]) | Heat[a] | Radial converging, multiple observations of wells | 4 to 13 | | | | 1.00 | 62.5 | $9.6 \times 10^{-7}$ |
| Sauty (1978[b]) | Heat[a] | Same experiment interpreted with constant equivalent conduction $\lambda_L = \lambda'_0$ | | | | | 0 | 62.5 | $192 \times 10^{-7}$ |

[a] The experiments of 1978 were made after a violent unclogging of the wells by air lift, which might have changed the terrain slightly. The porosities of 1975 are probably overestimated because of insufficient knowledge of the permeability, i.e., the Darcy velocity, the gradient being the only known parameter, whereas in radial flow the Darcy velocity is imposed.

**Table 10.4**

Thermal Properties of Common Materials

| Medium | $\lambda_0$ (W m$^{-1}$ K$^{-1}$) | $\rho'c'$ of the mineral (J m$^{-3}$ K$^{-1}$) |
|---|---|---|
| Dry sand | 0.4–0.8 | $1.9 \times 10^6$ |
| Wet sand | 2.5–3.5 | $1.9 \times 10^6$ |
| Dry clay | 0.8–2.0 | $2.1 \times 10^6$ |
| Wet clay | 1.2–1.7 | $2.3 \times 10^6$ |
| Granite | 2.5–3.8 | $2.1 \times 10^6$ |
| Sandstone | 1.5–4.3 | $2.3 \times 10^6$ |
| Water | 0.598 | $4.185 \times 10^6$ |
| Salt | 5.86 | $2.0 \times 10^6$ |

It then seems, as Sauty (1978b) has observed, that in the field, on the scale in question, the macrodispersivity clearly dominates the molecular diffusion, or even the conduction. This is very likely to be due to the heterogeneity of the velocity in the different layers of the formation, thus making the dispersivity obtained for each one of the tracers comparable.

However, on a larger scale, in a tracer experiment over a longer distance and time, the coefficients of dispersion and conduction reach an asymptotic value, and the dispersivity of the tracer should be around five times stronger than the thermal dispersivity. However, there are no experiments confirming this.

A few values of $\lambda_0$ and $\rho'c'$ for different rocks are given in Table 10.4. Remember that, in calculating the volumetric heat capacity of the rock, we have to take the contained water into consideration:

$$\rho''c'' = \omega\rho c + (1 - \omega)\rho'c'$$

With porosities from 10 to 20%, $\rho''c''$ is of the order of 2.1–2.5 $\times$ $10^6$ J m$^{-3}$ K$^{-1}$.

Remember also that this equation for heat transfer must be associated with a generalized form of Darcy's law and with the continuity equation in porous media, which will give the velocity **U**. This association is made through the mass per unit volume $\rho = \rho(\theta)$ and the viscosity $\mu = \mu(\theta)$.

This association through the mass per unit volume has an important consequence: hot water injected into an aquifer has a tendency to migrate toward the top of the aquifer owing to the density effect. This is one of the problems inherent in hot water storage in aquifers. Furthermore, even when no human intervention disturbs the state of flow, the flow of geothermal heat originating at the bottom may create a vertically ascending flow in an aquifer layer. However, through continuity, a descending flow must arise and thus "cells of natural convection" develop in the aquifer.

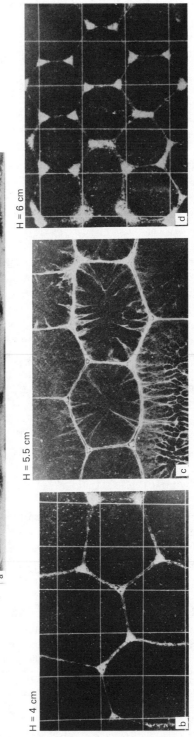

**Fig. 10.12.** Convective cells in a porous medium. From Bories (1970). Top: view of the flow lines shown in vertical section (the upper face is cold, the lower is warm). Bottom: view of the six-sided convective cells, seen from above; $H$ is the thickness of the porous layer.

H = 4 cm

H = 5.5 cm

H = 6 cm

a

b

c

d

Combarnous and Bories (1975)* have studied this phenomenon of natural convection. They have pointed out that below a minimum thermal gradient in the vertical direction, convection does not occur. When it occurs, rough six-sided cells (in two dimensions) appear and their size and migration velocity can be foreseen (see Fig. 10.12).

These phenomena of natural migration under the influence of thermal gradients may be at the origin of mineral or hydrothermal deposits.

* They define a "Rayleigh number in porous media" as

$$R_a^* = g \frac{\alpha \rho (\rho c) k}{\mu \lambda} e \Delta \theta$$

with $\alpha$ the coefficient of thermal volumetric expansion of the fluid ($10^{-3}$ to $10^{-4}°C^{-1}$), $\rho$ the mass per unit volume of the fluid, $(\rho c)$ the volumetric heat capacity of the fluid, $k$ the intrinsic permeability, $\mu$ the dynamic viscosity of the fluid, $\lambda$ the equivalent conductivity of the porous medium (immobile), $e$ the thickness of the layer, $\Delta \theta$ the difference in temperature between the top and bottom of the layer, assumed to be impermeable and at a constant temperature. Natural convection appears if $R_a^* \cos \gamma > 4\pi^2$, where $\gamma$ is the angle between the horizontal line and the layer.

### Note Added in Proof

An empirical relation between the increase in pressure drop through a filter (permeability variation due to clogging) and the concentration $\sigma$ of the retained particle is $\Delta p / \Delta p_0 = 1/(1 - j\sigma)^m$, where $j$ and $m$ are constant; to first order $\Delta p / \Delta p_0 \simeq 1 + mj\sigma$. If $l$ is the length of the filter, one can also write, to first order, $p = p_0(1 + mjUC_0T/l)$, where $U$ is Darcy's velocity, $C_0$ the concentration of particles at the inlet, $t$ the time, and $p$ the pressure applied on the filter. Values of $mj$ are in the range of 40 to 450 (Herzig et al., 1970).

*Chapter 11*

# Geostatistic and Stochastic
# Approach in Hydrogeology

The various magnitudes that are important in hydrogeology (e.g., hydraulic head, transmissivity, permeability, thickness of a layer, storage coefficient, rainfall, effective recharge, etc.) are all functions of space and are very often highly variable. However, this spatial variability is, in general, not purely random: if measurements are made at two different locations, the closer the measurement points are to each other, the closer the measured values. In other words, there is some kind of correlation in the spatial distribution of these magnitudes. Matheron (1965, 1970, 1971) has given the name of "regionalized variables" to these types of quantities: they are variables typical of a phenomenon developing in space (and/or time) and possessing a certain structure. Here, the term "structure" refers to this spatial correlation which, of course, is very different from one magnitude to the other or from one aquifer to the next.

Regionalized variables can be divided into two main categories: stationary and nonstationary. In the latter the variable has a definite trend in space: for instance, the variable decreases systematically in one direction. This is generally the case of the hydraulic head. On the contrary, there is no systematic trend in space for the stationary variables. This is in general the case of transmissivity. We shall define these terms more precisely in Sections 11.2, 11.3, and 11.7.

Here we will first address the problem of how to estimate a regionalized variable, which is the most common problem facing the hydrogeologist in the field. Having measured a variable at a set of points (e.g., heads at several piezometers, transmissivities at several wells, rainfall at several rain gauges), how do we estimate the value of the variable at all other locations in order to produce a contour map of the variable? Kriging is an optimal estimation method, and its use will be described for both the stationary and the nonstationary case.

To make this estimation we use the concept of random functions, which was introduced in Section 2.1.d. It is therefore useful to return to this section and read it first. This concept of random functions will also be used to introduce briefly, in Section 11.10, what is known as stochastic hydrogeology. This consists in regarding the parameters of the flow and transport equations as random functions and then looking at stochastic solutions to these equations, i.e., heads, velocities or concentrations, etc. which are also random functions and no longer deterministic ones, as we have assumed so far.

## 11.1. The Problem of Estimation: Definition of Kriging

Kriging is a method for optimizing the estimation of a magnitude, which is distributed in space and is measured at a network of points. Let (1) $x_1$, $x_2, \ldots, x_n$ be the locations of the $n$ points of measurement and $x_i$ denote simultaneously the one, two, or three coordinates of the point $i$ (Fig. 11.1 is a two-dimensional representation, but the theory is also applicable in one or three dimensions), and let (2) $Z_i = Z(x_i)$ be the value measured at the point $i$.

The problem of the point estimation lies in determining the value of the quantity $Z_0$ for any point $x_0$ that has not been measured. By continually modifying the position of the point $x_0$ it is thus possible to estimate the whole field of the parameter $Z$.

In hydrology, kriging has a wide variety of applications:

(1)   Calculations of rainfall, temperatures, sunshine, etc. based on measurements from climatological stations.

(2)   Interpolation of thickness or elevation of underground geological formations based on well logs.

(3)   Estimation of hydrogeological parameters such as the transmissivity of an aquifer, piezometric head, concentration of solutes based on measurements in the piezometers.

(4)   Mapping of the concentrations of polluting agents in a lake, etc.

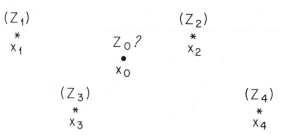

**Fig. 11.1** The point estimation problem.

However, kriging is not limited to simple point estimations of the given magnitude $Z$ but can also be used to:

(1)   Obtain the estimation variance of the magnitude $Z$, i.e., roughly, the confidence interval of this estimation,

(2)   Estimate the mean value of $Z$ on a given block, e.g., on the mesh of a model or a subdomain of any shape of a watershed basin.

(3)   Locate the best situation for a new measurement point, e.g., by minimizing the overall uncertainty in the field under consideration.

A generalization of kriging also makes it possible to create an infinite number of conditional Monte Carlo simulations of the field $Z$, i.e., different realizations of the map of $Z$, which are compatible with the measured data. These maps can be used to visualize the uncertainty of the estimation and as entries into stochastic models (Delhomme, 1979; Chiles, 1977). This will be described in Section 11.9.

For the sole purpose of making the estimation, we shall choose a probabilistic framework and assume that the magnitude $Z(\mathbf{x})$ is a random function (R.F.) $Z(\mathbf{x}, \xi)$ for which we only have one realization (see Section 2.1.d). Here, $\mathbf{x}$ denotes the point in the geometrical space and $\xi$ the state variable in the space of the realizations; $Z(\mathbf{x}, \xi_1)$ denotes a realization and $Z(\mathbf{x}_0, \xi)$ a random variable (R.V.), i.e., the whole set of realizations of the R.F. $Z$ at the point $\mathbf{x}_0$.

In order to use kriging we must try to determine, on the basis of the only sampled realization $Z(\mathbf{x}_i, \xi_1)$, both (1) the "structure" of the R.F. $Z(\mathbf{x}, \xi)$, i.e., its autocovariance function (the problem called statistical inference) and (2) the "optimal" estimation of $Z(\mathbf{x}_0, \xi_1)$ for any point $\mathbf{x}_0$.

The probabilistic method must be seen as nothing more than a language and a tool, which only leads to a system of equations, whence the desired estimation can be obtained. In most cases there is only one realization of $Z$, which is completely determined in space. Our uncertainty concerning the value of $Z$ only stems from the weakness of the available samples, and the probabilistic language only supplies a useful tool for expressing this uncertainty. When we start making hypotheses on $Z$ to make the estimation, e.g., on its stationarity, these will be working hypotheses and only required to be locally compatible with the data. The only objective proof of the validity of the procedure will come from the confirmation of the predictions it has made by measurements in the field.

In the following we give kriging equations for three cases: (1) stationary hypothesis in Section 11.2, (2) intrinsic hypothesis (stationarity of the increments) in Section 11.3, and (3) nonstationarity hypothesis in Section 11.7.

## 11.2. Kriging in the Stationary Case, Use of the Covariance

### 11.2.1. *Hypothesis of Weak Stationarity (or Second-Order Stationarity)*

An R.F. is said to be second-order stationary if (1) $E[Z(\mathbf{x}, \xi)] = m$ (constant mean) and (2) the function of autocovariance, or simply covariance, only depends on the distance and not on the points of reference:

$$\text{cov}(\mathbf{x}_1, \mathbf{x}_2) = E\{[Z(\mathbf{x}_1, \xi) - m][Z(\mathbf{x}_2, \xi) - m]\} = C(\mathbf{h}) \qquad (11.2.1)$$

where $\mathbf{h} = \mathbf{x}_1 - \mathbf{x}_2$ (distance).* If this expression is expanded, we get

$$C(\mathbf{h}) = E[Z(\mathbf{x}_1, \xi)Z(\mathbf{x}_2, \xi)] - m^2 \qquad (11.2.2)$$

The covariance function $C(\mathbf{h})$ determines the "structure" of the phenomenon (Fig. 11.2). Observe that $C(0) = \text{var}(Z) = \sigma_Z^2$ is the variance (or dispersion variance) of $Z$.

Note that the expected value is taken here over all possible realizations of $Z$, i.e., for all values of $\xi$. Saying that $E[Z(\mathbf{x}, \xi)] = m$ means that this expected value would be the same at any location $\mathbf{x}$. But it does not mean that for a particular realization $\xi_1$, $Z(\mathbf{x}, \xi_1)$ should be constant over $\mathbf{x}$: such a function would no longer be variable in space!

### 11.2.2. *Kriging with Second-Order Stationarity Hypothesis when the Mean m and the Covariance C($\mathbf{h}$) are Known*

From now on we leave out the state variable $\xi$ in order to simplify the notation. We define a process of mean zero by

$$Y(\mathbf{x}) = Z(\mathbf{x}) - m \qquad (11.2.3)$$

which gives $E(Y) = 0$. We shall estimate the value of $Y$ at the point $\mathbf{x}_0$

* See Note Added in Proof at the end of this chapter.

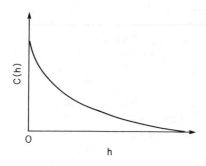

Fig. 11.2. Covariance function.

$Y_i = Y(\mathbf{x}_i)$, $i = 1, \ldots, n$, then

$$Y_0^* = Y^*(\mathbf{x}_0) = \sum_{i=1}^{n} \lambda_0^i Y_i \qquad (11.2.4)$$

*Notation.* $Y_0^*$ is the estimation of the exact unknown value $Y_0$ at the point $\mathbf{x}_0$. The asterisk shows by definition that we are dealing with an estimate. The $\lambda_0^i$ are the weights of the kriging estimator. These are the unknowns of our problem. The indices $i$ and $0$ are indices and not exponents. The index $i$ signifies that the weight $\lambda^i$ relates to the measurement made at the point $\mathbf{x}_i$; the index $0$ shows that at each point $\mathbf{x}_0$, where $Y_0$ will be estimated, there will be a different set of weights $\lambda^i$.

The estimator $Y_0^*$ is said to be "optimal" if the error of estimation $(Y_0^* - Y_0)$ is minimal. Since the real value of $Y_0$ is unknown, we will only minimize the mathematical expectation of the quadratic mean of this error of estimation:

$$E[(Y_0^* - Y_0)^2] \qquad \text{minimum} \qquad (11.2.5)$$

In other words, since $Y_0^*$ and $Y_0$ are random variables, we minimize the variance of the error of estimation $(Y_0^* - Y_0)$. Please note here that the mathematical expectation is taken for a fixed point in space $\mathbf{x}_0$ for all possible realizations of $Y_0^* - Y_0$, i.e., for all possible values of the state variable $\xi$ in the notation $Y(\mathbf{x}_0, \xi)$.

In other words, if we were able to estimate $Y_0^*$ for an infinite number of realizations, using always the same weights $\lambda_0^i$ for each realization $Y(\mathbf{x}_i, \xi)$, we would make on the average the minimum error. Of course, since in general we have only one realization, we will make at location $\mathbf{x}_0$ an estimation error that we cannot quantify. But by applying the ergodic hypothesis, we can say that on the average, over a large number of locations $\mathbf{x}_0$ where we will estimate $Y$, our error of estimation will be minimum.

We can develop the expression Eq. (11.2.5) by replacing the value of $Y_0^*$ by Eq. (11.2.4):

$$E[(Y_0^* - Y_0)^2] = E\left[\left(\sum_i \lambda_0^i Y_i - Y_0\right)^2\right] = E\left[\left(\sum_i \lambda_0^i Y_i\right)\left(\sum_j \lambda_0^j Y_j\right)\right]$$

$$- 2E\left[\sum_i \lambda_0^i Y_i Y_0\right] + E[Y_0^2]$$

$$= \sum_i \sum_j \lambda_0^i \lambda_0^j E(Y_i Y_j) - 2\sum_i \lambda_0^i E(Y_i Y_0) + E(Y_0^2)$$

By definition of the covariance function of Eq. (11.2.2), we can write

$$E(Y_i Y_j) = C(\mathbf{x}_i - \mathbf{x}_j) + m^2$$

but the mean $m$ of $Y$ is zero. Hence $E(Y_iY_j) = C(\mathbf{x}_i - \mathbf{x}_j)$ and, in particular;

$$E(Y_0^2) = C(0) = \text{dispersion variance of } Y = \text{var}(Y)$$

Then

$$E[(Y_0^* - Y_0)^2] = \sum_i \sum_j \lambda_0^i \lambda_0^j C(\mathbf{x}_i - \mathbf{x}_j) - 2\sum_i \lambda_0^i C(\mathbf{x}_i - \mathbf{x}_0) + C(0) \qquad (11.2.6)$$

Equation (11.2.6) is a quadratic function of the weights $\lambda_0^i$. To minimize this function, all the partial derivatives with respect to the $\lambda_0^i$ are equated to zero.

$$\frac{\partial}{\partial \lambda_0^i} E[(Y_0^* - Y_0)^2] = 2\sum_j \lambda_0^j C(\mathbf{x}_i - \mathbf{x}_j) - 2C(\mathbf{x}_i - \mathbf{x}_0) = 0 \qquad i = 1,\ldots,n$$

$$(11.2.7)$$

This results in a linear system of $n$ equations with $n$ unknowns.

$$\sum_j \lambda_0^j C(\mathbf{x}_i - \mathbf{x}_j) = C(\mathbf{x}_i - \mathbf{x}_0) \qquad i = 1,\ldots,n \qquad (11.2.8)$$

This system has only one solution if $C$ is a positive definite function and if the $\mathbf{x}_i$ are distinct. We assume that this is indeed the case. The solution of Eq. (11.2.8) is easily obtained by inverting the coefficients matrix or by Gaussian elimination and gives the weights $\lambda_0^j, j = 1,\ldots,n$.

Note that the left-hand side of Eq. (11.2.8) does not depend on $\mathbf{x}_0$: we only need to invert the matrix of the linear system of Eq. (11.2.8) once, when the point $\mathbf{x}_0$ is changed. Only the right-hand side of Eq. (11.2.8) is a function of $\mathbf{x}_0$. (An explicit formulation of the matrix of the kriging system for a less simple case is given in Section 11.3.2.)

### 11.2.3. Calculation of the Estimation Variance

We know now the estimated value $Y_0^*$:

$$Y_0^* = \sum_i \lambda_0^i Y_i$$

We cannot compute the error of estimation $Y_0^* - Y_0$ but only its variance:

$$\text{var}(Y_0^* - Y_0) = E[(Y_0^* - Y_0)^2] - [E(Y_0^* - Y_0)]^2$$

The second term of the right-hand side is zero, since

$$E(Y_0^* - Y_0) = E(Y_0^*) - E(Y_0) = \sum_i \lambda_0^i E(Y_i) - E(Y_0) = 0$$

as $E(Y)$ is assumed to be zero. Then

$$\text{var}(Y_0^* - Y_0) = E[(Y_0^* - Y_0)^2] \qquad (11.2.9)$$

But we have already calculated the right-hand side, which is given explicitly in

Eq. (11.2.6) as a function of the $\lambda_0^i$. When we substitute in Eq. (11.2.6) the values of the $\lambda_0^i$ obtained in Eq. (11.2.8), we get

$$\sum_j \lambda_0^j C(\mathbf{x}_i - \mathbf{x}_j) = C(\mathbf{x}_i - \mathbf{x}_0) \qquad i = 1,\ldots,n$$

$$\sum_i \sum_j \lambda_0^i \lambda_0^j C(\mathbf{x}_i - \mathbf{x}_j) = \sum_i \lambda_0^i C(\mathbf{x}_i - \mathbf{x}_0)$$

$$\mathrm{var}(Y_0^* - Y_0) = \mathrm{var}(Y) - \sum_i \lambda_0^i C(\mathbf{x}_i - \mathbf{x}_0) \qquad (11.2.10)$$

We can see that the *estimation variance* of the unknown quantity $Y_0$ is smaller than the *dispersion variance* or real variance of the R.F. $Y$. Because $Y$ was measured at the points $\mathbf{x}_i$, the uncertainty on $Y$ decreases. We can now return to our original variable $Z$ $(Z = Y + m)$:

$$Z_0^* = m + \sum_i \lambda_0^i (Z_i - m)$$

$$\mathrm{var}(Z_0^* - Z_0) = \mathrm{var}(Z) - \sum_i \lambda_0^i C(\mathbf{x}_i - \mathbf{x}_0)$$

$(11.2.11)$

Further on we shall examine how to estimate a covariance function $C$ from the data. It is also possible to establish kriging equations when the average $m$ is unknown, either in order to estimate $Z$ directly or to estimate this average $m$ [see Matheron (1970), Journel and Huitjbregts (1979)]. We proceed directly to the case called the "intrinsic case," where $m$ is unknown and where the second-order stationarity hypothesis is not satisfied.

## 11.3. Kriging in the Intrinsic Case: Definition of the Variogram

In the mining industry (estimation of ore grades), it has been shown that the hypothesis of second-order stationarity with a finite variance $C(0)$ is not satisfied by the data in certain cases. This is frequently the case in hydrology as well. The experimental variance increases with the size of the area under consideration. A less stringent hypothesis, called the "intrinsic hypothesis," has been developed to make the estimation possible.

### 11.3.1. *The Intrinsic Hypothesis*

The intrinsic hypothesis consists in assuming that even if the variance of $Z$ is not finite, the variance of the first-order increments of $Z$ is finite and these increments are themselves second-order stationary, i.e., that $[Z(\mathbf{x} + \mathbf{h}) - Z(\mathbf{x})]$ satisfies

$$\left. \begin{array}{l} E[Z(\mathbf{x} + \mathbf{h}) - Z(\mathbf{x})] = m(\mathbf{h}) \\ \mathrm{var}[Z(\mathbf{x} + \mathbf{h}) - Z(\mathbf{x})] = 2\gamma(\mathbf{h}) \end{array} \right\} \quad \text{functions of } \mathbf{h}, \text{ not } \mathbf{x}$$

where $\mathbf{h}$ is a vector in the one-, two-, or three-dimensional space and where $\gamma(\mathbf{h})$ is generally only a function of the distance $\mathbf{h}$.

Although this is not absolutely necessary, it is usually assumed that $m = 0$. If this were not the case, but $m(\mathbf{x} + \mathbf{h}) - m(\mathbf{x}) = m(\mathbf{h})$, the function $Z(\mathbf{x}) - m(\mathbf{x})$ would satisfy this condition.

The variance of the increment then defines a new function called the *variogram* $\gamma(\mathbf{h})$:

$$E[Z(\mathbf{x} + \mathbf{h}) - Z(\mathbf{x})] = 0 \tag{11.3.1}$$

$$\gamma(\mathbf{h}) = \tfrac{1}{2}\mathrm{var}[Z(\mathbf{x} + \mathbf{h}) - Z(\mathbf{x})] \tag{11.3.2}$$

Equations (11.3.1) and (11.3.2) make it possible to write

$$\gamma(\mathbf{h}) = \tfrac{1}{2}E\{[Z(\mathbf{x} + \mathbf{h}) - Z(\mathbf{x})]^2\} \tag{11.3.3}$$

where $\gamma(\mathbf{h})$ is the mean quadratic increment of $Z$ between two points separated by the distance $\mathbf{h}$.

If we compare the intrinsic hypothesis with the hypothesis of second-order stationarity, we see that Eq. (11.3.1) is equivalent to $E[Z(\mathbf{x})] = m$ (constant mathematical expectation) but that Eq. (11.3.2) is less stringent than the condition on the covariance:

$$C(\mathbf{h}) = E[Z(\mathbf{x} + \mathbf{h})Z(\mathbf{x})] - m^2$$

Is there a relation between the covariance and the variogram? In the case where both exist, i.e., in the stationary hypothesis, we can write

$$\gamma(\mathbf{h}) = \tfrac{1}{2}E[Z(\mathbf{x} + \mathbf{h})^2] - E[Z(\mathbf{x} + \mathbf{h})Z(\mathbf{x})] + \tfrac{1}{2}E[Z(\mathbf{x})^2]$$

where we can see that

$$\gamma(\mathbf{h}) = C(0) - C(\mathbf{h}) \tag{11.3.4}$$

If we know the covariance, the variogram is simply its reflection with respect to the horizontal axis and with a vertical shift (Fig. 11.3).

When var($Z$) is finite, the variogram tends towards an asymptotic value equal to this variance, which is also called the sill of the variogram (the distance at which the variogram reaches its asymptotic value is called the range). However, if the phenomenon under consideration does not have a finite variance, the variogram will never have a horizontal asymptotic value (Fig. 11.4). Not just any function $\gamma(\mathbf{h})$ can be a variogram, just as the covariance must be positive definite. It is indeed possible to show that:

(1) Minus $\gamma$ must be conditionally positive definite, i.e., for all $\mathbf{x}_1,\ldots,$ $\mathbf{x}_n \in R^m$ ($m = 1, 2,$ or 3) and for all $\lambda_1,\ldots,\lambda_n \in R$, $n$ coefficients satisfying

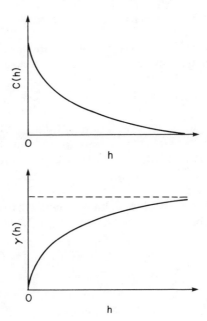

**Fig. 11.3.** Covariance and variogram.

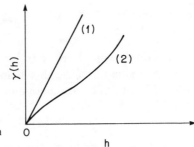

**Fig. 11.4.** Variogram of a phenomenon with an infinite variance.

$\sum \lambda_i = 0$, then

$$-\sum_i \sum_j \lambda_i \lambda_j \gamma(\mathbf{x}_i - \mathbf{x}_j) \geq 0$$

(2)   $\gamma(\mathbf{h})$ for $|\mathbf{h}| \to \infty$ must necessarily increase less rapidly than $|\mathbf{h}|^2$, i.e.,

$$\lim_{|\mathbf{h}| \to \infty} \frac{\gamma(\mathbf{h})}{|\mathbf{h}|^2} \to 0$$

In practice, only a limited class of functions is used to describe variograms. We shall present a few of them in Section 11.5.1 in connection with statistical inference (determination of the variogram from the data).

## 11.3.2. *Kriging as Used in the Intrinsic Hypothesis*

We shall also try to find the estimate $Z_0^*$ of the unknown quantity $Z_0$ by a weighted sum of all the available measurements:

$$Z_0^* = \sum_{i=1}^{n} \lambda_0^i Z_i \qquad (11.3.5)$$

Since we do not know the value of the constant mean $m$ in the intrinsic process $Z(x)$, we impose an additional condition on the estimation $Z_0^*$, namely, that its mathematical expectation be equal to that of the true R.F. $Z_0$:

$$E(Z_0^* - Z_0) = 0 \qquad \text{or} \qquad E(Z_0^*) = E(Z_0) \qquad (11.3.6)$$

Let $m$ be the unknown mathematical expectation of the process $Z$. Introducing Eq. (11.3.5) into Eq. (11.3.6), we can write

$$E\left[\sum_i \lambda_0^i Z_i\right] = E[Z_0] = m$$

or

$$\sum_i \lambda_0^i E(Z_i) = m \qquad \text{or} \qquad \sum_i \lambda_0^i m = m$$

i.e.,

$$\sum_i \lambda_0^i = 1 \qquad (11.3.7)$$

This condition is required in order to have an unbiased estimator. We now redetermine the set of weights $\lambda_0^i$ in Eq. (11.3.5), subject to the condition of Eq. (11.3.7), by imposing the condition that the error of estimation be minimal:

$$E[(Z_0^* - Z_0)^2] \quad \text{minimum} \qquad (11.3.8)$$

(or $\text{var}(Z_0^* - Z_0)$ minimum since $E(Z_0^* - Z_0) = 0$.)

Let us develop Eq. (11.3.8):

$$E[(Z_0^* - Z_0)^2] = E\left[\left(\sum_i \lambda_0^i Z_i - Z_0\right)^2\right] = E\left[\left(\sum_i \lambda_0^i Z_i - \sum_i \lambda_0^i Z_0\right)^2\right]$$

$$= E\left[\left(\sum_i \lambda_0^i (Z_i - Z_0)\right)^2\right]$$

$$= E\left[\sum_i \lambda_0^i (Z_i - Z_0) \sum_j \lambda_0^j (Z_j - Z_0)\right]$$

$$= \sum_i \sum_j \lambda_0^i \lambda_0^j E[(Z_i - Z_0)(Z_j - Z_0)] \qquad (11.3.9)$$

We then use the definition of the variogram:

$$\gamma(\mathbf{x}_i - \mathbf{x}_j) = \tfrac{1}{2}E[(Z_i - Z_j)^2]$$

$$= \tfrac{1}{2}E[((Z_i - Z_0) - (Z_j - Z_0))^2]$$

$$= \tfrac{1}{2}E[(Z_i - Z_0)^2] + \tfrac{1}{2}E[(Z_j - Z_0)^2] - E[(Z_i - Z_0)(Z_j - Z_0)]$$

$$= \gamma(\mathbf{x}_i - \mathbf{x}_0) + \gamma(\mathbf{x}_j - \mathbf{x}_0) - E[(Z_i - Z_0)(Z_j - Z_0)] \qquad (11.3.10)$$

From Eq. (11.3.10) we can calculate the mathematical expectation, which we need in Eq. (11.3.9). By substitution we find

$$E[(Z_0^* - Z_0)^2] = -\sum_i \sum_j \lambda_0^i \lambda_0^j \gamma(\mathbf{x}_i - \mathbf{x}_j) + \sum_i \sum_j \lambda_0^i \lambda_0^j \gamma(\mathbf{x}_i - \mathbf{x}_0)$$

$$+ \sum_i \sum_j \lambda_0^i \lambda_0^j \gamma(\mathbf{x}_j - \mathbf{x}_0)$$

We can factor $\sum_i \lambda_0^i$ or $\sum_j \lambda_0^j$ in the last two terms of the right-hand side of this equation, but according to Eq. (11.3.7) these two sums are equal to one. Furthermore

$$\sum_i \lambda_0^i \gamma(\mathbf{x}_i - \mathbf{x}_0) = \sum_j \lambda_0^j (\mathbf{x}_j - \mathbf{x}_0)$$

whence

$$E[(Z^* - Z_0)^2] = -\sum_i \sum_j \lambda_0^i \lambda_0^j \gamma(\mathbf{x}_i - \mathbf{x}_j) + 2\sum_i \lambda_0^i \gamma(\mathbf{x}_i - \mathbf{x}_0) \qquad (11.3.11)$$

We again find a quadratic form of the unknowns $\lambda_0^i$. The minimization of Eq. (11.3.11), subject to the linear constraint of Eq. (11.3.7), is found using the Lagrange multipliers; we simply minimize the expression

$$\tfrac{1}{2}E[(Z_0^* - Z_0)^2] - \mu\left[\sum_i \lambda_0^i - 1\right] \qquad (11.3.12)$$

where $\mu$ is a new unknown, called a Lagrange multiplier, which is added to the $n$ previous unknowns $\lambda_0^i$. It can be shown that, when the above expression is minimum, the linear condition $\sum_i \lambda_0^i = 1$ is satisfied for the value of $E[(Z_0^* - Z_0)^2]$, which is the smallest one compatible with the constraint. In Eq. (11.3.12) we have divided by 2 and put a minus sign before $\mu$ in order to simplify the following expressions, but, as $\mu$ is an unknown, this is unimportant.

The minimum of the quadratic form in $\lambda$ and $\mu$ is obtained by equating to zero its partial derivatives with respect to $\lambda_0^i$ and $\mu$. We get

$$\sum_j \lambda_0^j \gamma(\mathbf{x}_i - \mathbf{x}_j) + \mu = \gamma(\mathbf{x}_i - \mathbf{x}_0) \qquad i = 1,\dots,n$$

$$\sum_i \lambda_0^i = 1$$

$$(11.3.13)$$

Let us for once write the complete linear system of kriging in matrix form:

$$
\begin{pmatrix}
0 & \gamma_{12} & \gamma_{13} & \cdots & \gamma_{1n} & 1 \\
\gamma_{21} & 0 & \gamma_{23} & \cdots & \gamma_{2n} & 1 \\
\vdots & \vdots & \vdots & \ddots & \vdots & \\
\gamma_{n1} & \gamma_{n2} & \gamma_{n3} & \cdots & 0 & 1 \\
1 & 1 & 1 & \cdots & 1 & 0
\end{pmatrix}
\begin{pmatrix}
\lambda_0^1 \\
\lambda_0^2 \\
\vdots \\
\lambda_0^n \\
\mu
\end{pmatrix}
=
\begin{pmatrix}
\gamma_{10} \\
\gamma_{20} \\
\vdots \\
\gamma_{n0} \\
1
\end{pmatrix}
\tag{11.3.14}
$$

Note: We have denoted $\gamma(\mathbf{x}_i - \mathbf{x}_j)$ by $\gamma_{ij}$. The diagonal is zero since $\gamma_{ii} = \gamma(\mathbf{x}_i - \mathbf{x}_i) = \gamma(0) = 0$.

The matrix of the kriging equations is always regular if $-\gamma$ is conditionally positive definite. Here again the matrix only needs to be inverted once for all points $\mathbf{x}_0$.

### 11.3.3.  *Variance of the Estimation Error*

The variance of the error of estimation can also be computed:

$$
\mathrm{var}(Z_0^* - Z_0) = E[(Z_0^* - Z_0)^2]
$$

because $E(Z_0^* - Z_0) = 0$. We can calculate its value by substituting the solution of Eq. (11.3.13) in Eq. (11.3.11):

$$
\mathrm{var}(Z_0^* - Z_0) = \sum_i \lambda_0^i \gamma(\mathbf{x}_i - \mathbf{x}_0) + \mu
\tag{11.3.15}
$$

We have now, at last, obtained the usual kriging equations, which are used in the intrinsic hypothesis or even in that of second-order stationarity when the mean is unknown.

## 11.4.  A Few Remarks about Kriging

### 11.4.1.  *Kriging is Called a "BLUE"*

BLUE is an acronym for best linear unbiased estimate.

Other classes of estimators are also used in practice; for instance, nonlinear estimators can be built by prior transformation of the data. Disjunctive kriging (Matheron, 1976) and indicator kriging (see, e.g., Journal, 1984) are examples of nonlinear estimators (see also Section 11.6.2). Biased estimators can sometimes be preferred to unbiased ones, e.g., when other constraints are placed on the estimation or simply when the nonbias condition is removed. We will come back to this problem in an example in Section 11.6.2.

### 11.4.2. *Kriging is an Exact Interpolator*

If we try to compute the value of $Z$ at a point $x_k$, which has been measured, i.e., $x_k \in (x_1, \ldots, x_n)$, the kriging system gives

$$Z_K^* = Z_K$$

for a measured value, i.e., $\lambda_k^k = 1$, $\lambda_k^i = 0$, $i \neq k$, and

$$\text{var}(Z_k^* - Z_k) = 0$$

(no uncertainty at a measured point).

This contrasts with a least-squares fitting of a polynomial, which will never give the true value at the measurement points.

### 11.4.3. *Confidence Interval*

Knowing the variance of the error of estimation is, in principle, not enough to determine the confidence interval of the estimates. However, one can very often assume that the distribution of this error is normal: in that case we can say, for instance, that the 95% confidence interval is $\pm 2\sigma$, $\sigma$ being the standard deviation, i.e., the square root of the variance:

$$\sigma = \sqrt{\text{var}(Z_0^* - Z_0)}$$

Then the estimate of $Z_0$ with 95% chance is

$$Z_0^* = \sum_i \lambda_0^i Z_i \pm 2\sigma$$

We also know that many other distribution functions also satisfy a $\pm 2\sigma$ confidence interval at 95%, and consequently this expression is very often used.

### 11.4.4 *Computation of the Complete Covariance Matrix*

Instead of computing only the variance of the estimation error, $\text{var}(Z_0^* - Z_0)$, it is also possible with kriging to compute the complete covariance matrix of this error of estimation. We can show that

$$\text{cov}[(Z_1^* - Z_1), (Z_2^* - Z_2)] = -\gamma(x_1 - x_2) - \sum_i \sum_j \lambda_1^i \lambda_2^j \gamma(x_i - x_j)$$

$$+ \sum_i \lambda_1^i \gamma(x_i - x_2) + \sum_j \lambda_2^j \gamma(x_j - x_1)$$

This quantifies the relationship between the estimation error at locations $x_1$ and $x_2$.

### 11.4.5. Equations in the Kriging System do not depend on the Measured Values $Z_i$

Indeed, we only need to know the coordinates $x_i$ of the measurement points in order to calculate the weights $\lambda^i$. If the data vary in time, for instance, these weights $\lambda^i$ may be used for various situations.

### 11.4.6. Drawing Contour Maps with Kriging

By solving the kriging system we can estimate $Z_0$ at any point $x_0$. In order to draw a contour map, we generally choose a large number of points $x_0$ on a regular mesh, regardless of the position of the measurement points. The contour lines are then drawn either by hand or with a standard contouring package, which generally requires as input the value of $Z$ on just such a regular mesh. One must also plot the map of the variance of the estimation to understand the uncertainty. One can also plot the map of twice the standard deviation, which corresponds to the 95% confidence interval.

### 11.4.7. Calculating Average Values over a Mesh Instead of Point Values

Instead of estimating the value $Z_0$ at a point $x_0$, it is also possible to estimate directly any linear combination of the value of the variable $Z$, in particular its average over a given area $S_0$:

$$Z_{S_0} = \frac{1}{S_0} \int_{S_0} Z(x)\,dx$$

Here $S_0$ may be a given mesh or the entire domain (e.g., for estimates of the average rainfall on a watershed basin during a thunderstorm). The estimator of the average is built directly as a linear combination of the available data:

$$Z^*_{S_0} = \sum_i \lambda^i_0 Z_i$$

Using the same conditions of unbiased and optimal estimation, it is also possible to calculate the following new kriging system:

$$\sum_j \lambda^j_0 \gamma(x_i - x_j) + \mu = \bar\gamma(x_i, S_0), \qquad i = 1, \ldots, n$$

$$\sum_j \lambda^j_0 = 1$$

(11.4.1)

with

$$\bar\gamma(x_i, S_0) = \frac{1}{S_0} \int_{S_0} \gamma(x_i - x)\,dx$$

(average variogram between $x_i$ and the area $S_0$).

The estimation variance is given by

$$\text{var}(Z_{S_0}^* - Z_{S_0}) = \sum_i \lambda_0^i \bar{\gamma}(\mathbf{x}_i, S_0) + \mu - \bar{\bar{\gamma}}(S_0, S_0)$$

with

$$\bar{\bar{\gamma}}(S_0, S_0) = \frac{1}{S_0^2} \int_{S_0} \int_{S_0} \gamma(\mathbf{x} - \mathbf{y}) \, d\mathbf{x} \, d\mathbf{y}$$

Similarly, it is possible to use as measurements the value of $Z$ obtained as averages over a certain area; e.g., if the measurement $Z_j$ is an average of the parameter over the area $S_j$, $\gamma(\mathbf{x}_i - \mathbf{x}_j)$ will be replaced in the kriging equations by

$$\bar{\gamma}(\mathbf{x}_i, S_j) = \frac{1}{S_j} \int_{S_j} \gamma(\mathbf{x}_i - \mathbf{x}) \, d\mathbf{x}$$

and $\bar{\gamma}(\mathbf{x}_j, S_0)$ on the right-hand side by

$$\bar{\bar{\gamma}}(S_j, S_0) = \frac{1}{S_0 S_j} \int_{S_0} \int_{S_j} \gamma(\mathbf{x} - \mathbf{y}) \, d\mathbf{x} \, d\mathbf{y}$$

This makes it possible to simultaneously use data collected by different methods (measurements from core samples, slug tests, long pumping tests, etc., in the case of transmissivities, for example). Note, however, that $\gamma$ here is the variogram of point-measured data. Non-point-measured data should therefore not be used to estimate $\gamma$ (or a prior deconvolution has to be made), unless all data are measured over the same area. In such cases, $\bar{\bar{\gamma}}$ is directly determined. If measurements made as averages over different areas have to be used simultaneously both in the determination of $\gamma$ and in kriging, then co-kriging should be used (see Section 11.9).

### 11.4.8. *Kriging with Uncertain Data*

So far we have assumed that the measured values $Z_i$ are known without any uncertainty. In reality this is not always the case, but kriging can also handle uncertain data. In the case where the errors $\varepsilon_i$ linked to each measurement $Z_i$ are

(1)  nonsystematic, i.e., $E[\varepsilon_i] = 0$, $i = 1, \ldots, n$,
(2)  uncorrelated with each other, i.e., $\text{cov}(\varepsilon_i, \varepsilon_j) = 0$, $\forall i \neq j$,
(3)  uncorrelated with $Z$, i.e., $\text{cov}(\varepsilon_i, Z(\mathbf{x})) = 0$, $\forall i$, $\forall x$, and
(4)  have a known variance $\sigma_i^2$ (different for each $i$),

it is easily shown that the equations of the kriging system become, for example, for estimates of point values, as follows:

(1)   Estimator

$$Z_0^* = \sum_i \lambda_0^i Z_i$$

(2)   Kriging equations

$$\sum_j \lambda_0^j \gamma(\mathbf{x}_i - \mathbf{x}_j) - \lambda_0^i \sigma_i^2 + \mu = \gamma(\mathbf{x}_i - \mathbf{x}_0) \qquad i = 1,\ldots,n$$

$$\sum_i \lambda_0^i = 1$$

(3)   Variance

$$\mathrm{var}(Z_0^* - Z_0) = \sum_i \lambda_0^i \gamma(\mathbf{x}_i - \mathbf{x}_0) + \mu$$

The function $\gamma$ is supposed to be the true variogram of the magnitude $Z$ (without errors). This would be feasible by determining $\gamma$ on those points where the measurement error is zero only (see Section 11.5.3). The $Z_i$ are the measured values (i.e., with the measurements error, if any). The quantity $Z_0$ is the "true" (unknown) value (without measurement error).

The only change from the usual system is that the equations now have $-\sigma_i^2$ on the diagonal instead of 0. It is also possible to use both "certain" and "uncertain" data simultaneously: we simply put $\sigma_i^2$ to 0 for the certain data. If the errors $\varepsilon_i$ are correlated, the equations in the kriging system are a little more complex. The same equations can be developed for the estimation of averages instead of point values.

## 11.5.  Statistical Inference

### 11.5.1.  *Determination of the Variogram*

We have defined the variogram in the case where the mean is constant by

$$\gamma(\mathbf{h}) = \tfrac{1}{2}E\{[Z(\mathbf{x} + \mathbf{h}) - Z(\mathbf{x})]^2\}$$

To estimate the variogram we simply use the measurement points $Z_i$ and assume ergodicity on the increments (i.e., that space averages can be used to estimate the averages in the whole set of realizations).

First we define a certain number of classes of distances between the measurement points, e.g.,

$$0 < d_1 < 1\,\mathrm{km} \qquad 1 < d_2 < 2\,\mathrm{km} \qquad 2 < d_3 < 3\,\mathrm{km}$$

$$3 < d_4 < 5\,\mathrm{km} \qquad 5 < d_5$$

Then, taking all possible pairs of points $i$ and $j$ for each class of distance, we calculate:

(1)  The number of pairs present in the class.
(2)  The average distance in the class.
(3)  The average square increment $\frac{1}{2}(Z_i - Z_j)^2$.

With a set of 50 measurements we obtain, for example:

| Class | $d_1$ | $d_2$ | $d_3$ | $d_4$ | $d_5$ |
|---|---|---|---|---|---|
| Number of elements | 500 | 350 | 250 | 100 | 25 |
| Average distance | 0.7 | 1.3 | 2.4 | 3.8 | 6.2 |
| $\frac{1}{2}(Z_i - Z_j)^2$ average | 130 | 275 | 350 | 570 | 400 |

Note that the number of pairs that can be formed from a set of $n$ points is $n(n-1)/2$; for 50 points this gives 1225 pairs. However, generally they are not evenly distributed. There are more pairs at short than at long distances. The variogram becomes more and more uncertain as $|\mathbf{h}|$ increases. At large distances certain points may play a privileged role and introduce errors into the estimation. It may be necessary to eliminate a few measurements when calculating the variogram (Fig. 11.5).

However, we have seen that all functions cannot be variograms. In a class of acceptable analytical functions we choose a given form and fit the parameters of this function on the observed points. The main types of variograms commonly used are: linear; in $|\mathbf{h}|^\lambda, \lambda < 2$; spherical; exponential; Gaussian; cubic. The forms and equations of these variograms are given in Fig. 11.6,

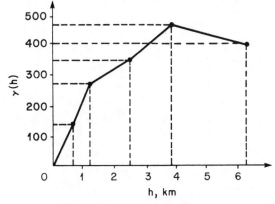

**Fig. 11.5.** Experimental variogram.

adapted from Delhomme (1976). For example, the variogram in Fig. 11.5 would be interpreted as a spherical one and the two parameters, $\omega$ and $a$, would be fitted by hand on the data (Fig. 11.7). Note that a piecewise linear variogram (i.e., made of segments of straight lines) is not acceptable; it is not in general a positive definite function. See Armstrong and Jabin (1981).

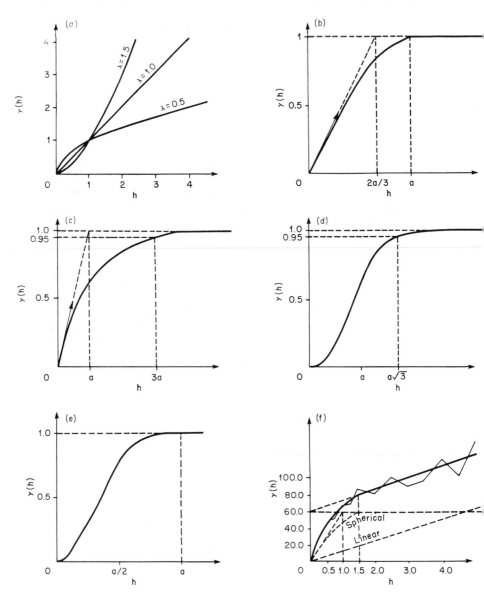

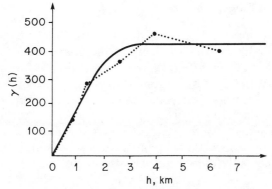

**Fig. 11.7.** Spherical variogram fitted to the experimental one, $\omega = 430$, $a = 3$ km.

### 11.5.2. *Behavior of the Variogram for Large h*

Note that an unbounded variogram, e.g., a linear one, suggests that the field has infinite variance and that there is no covariance function: the intrinsic hypothesis is the only acceptable one here. But if the variogram reaches a sill, as for example in Fig. 11.7, then the covariance function exists for the phenomenon in question.

---

**Fig. 11.6.** Common variogram models (from Delhomme (1976)). Here, $h$ denotes the length of the vector **h**. The expressions given are for $\gamma(h)$.

(a) model in $h^\lambda$     $\omega h^\lambda$        $\lambda < 2$

(b) spherical model     $\omega\left[\dfrac{3}{2}\left(\dfrac{h}{a}\right) - \dfrac{1}{2}\left(\dfrac{h}{a}\right)^3\right]$     $h < a$

                        $\omega$                        $h > a$

(c) exponential model     $\omega[1 - \exp(-h/a)]$

(d) Gaussian model     $\omega\{1 - \exp[-(h/a)^2]\}$

(e) cubic model     $\omega\left[7\left(\dfrac{h}{a}\right) - 8.75\left(\dfrac{h}{a}\right)^3 + 3.5\left(\dfrac{h}{a}\right)^5 - 0.75\left(\dfrac{h}{a}\right)^7\right]$

                                                            $h < a$

             $\omega$                        $h < a$

(f) fitting on a "linear plus spherical" model (example)

$$13.3\,h + 60\left[\dfrac{3}{2}\left(\dfrac{h}{1.5}\right) - \dfrac{1}{2}\left(\dfrac{h}{1.5}\right)^3\right] \qquad h < 1.5$$

$$13.3\,h + 60 \qquad\qquad\qquad\qquad\qquad h > 1.5$$

### 11.5.3. *Behavior Close to the Origin*

Theoretically, for $\mathbf{h} = 0$, $\gamma(\mathbf{h}) = 0$ regardless of the variogram. However, very often variograms exhibit a jump at the origin, as in Fig. 11.8. This apparent jump at the origin is called the nugget effect, as it originated in the mining industry. Indeed, if a core contains a nugget, the concentration will be very high, whereas neighboring cores even with high mineral concentration will never be as rich: there is an "erratic" component in the behavior.

Such behavior is very frequently found when data are analyzed (e.g., transmissivities). To take it into account, one just adds the quantity $C$ to the variogram fitted on the data as if $C$ were the origin:

$$\gamma(\mathbf{h}) = C[1 - \delta(\mathbf{h})] + \gamma'(\mathbf{h})$$

where $\delta(\mathbf{h})$ is the Kronecker $\delta$ ($\delta = 1$ if $\mathbf{h} = 0$, $\delta = 0$ if $\mathbf{h} \neq 0$) and $\gamma'(\mathbf{h})$ is the variogram fitted on the data with $C$ as origin.

This nugget effect can also be attributed to measurement errors or to the fact that the data have not been collected with a sufficiently small interval to show the underlying continuous behavior of the phenomenon (Fig. 11.9).

A horizontal variogram, i.e., $\gamma(\mathbf{h}) = C$, $\forall\, \mathbf{h} > 0$, is called a variogram with pure nugget effect. It expresses a purely random phenomenon without spatial structure.

When the variogram has a nugget effect, kriging is still an exact interpolator, as stated in Section 11.4.2., but the estimation is discontinuous at the measurement points, i.e., if $\mathbf{x}_k$ is a measurement point and $Z_k = Z(\mathbf{x}_k)$, then $Z^*(\mathbf{x}_k) = Z_k$, but $Z^*(\mathbf{x}_k + \mathbf{dx}) \neq Z_k$ even if $\mathbf{dx} \to 0$. However, the estimation is continuous everywhere else.

Let $C$ be the nugget effect. Let us write the variogram $\gamma(\mathbf{h}) = C(1 - \delta) + \gamma'$.

An alternative that gives strictly identical results is to subtract the nugget effect from the variogram, considering that it only represents measurement errors, and use an uncertainty $\sigma_i^2 = C$ on the diagonal of the kriging system, as explained in Section 11.4.8, whereas the kriging matrix and the right-hand side are both built using $\gamma'$ instead of $\gamma$. It is easily seen that this new kriging system is identical to the original one, by adding the last line of the system multiplied by $C$ to all the others.

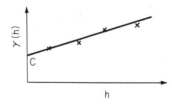

Fig. 11.8. The nugget effect.

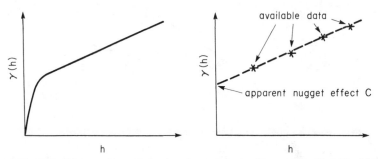

**Fig. 11.9.** Underlying continuous behavior of a variogram with a nugget effect. The left-hand side shows the "true" variogram.

However, it is also possible to write the kriging equations using $\gamma$ on the left-hand side (with zero on the diagonal) and $\gamma'$ on the right-hand side. De Smedt et al. (1985) have shown that this corresponds to the case where $\gamma'$ is said to be the true variogram of the phenomenon (without measurement errors) and $\gamma$ the variogram of the noisy data. In this case, the estimate $Z_0^*$ is continuous everywhere but is no longer exact at the measurement points. The variance of the error of estimation is also reduced with respect to the normal system. It is given by:

$$\text{var}(Z_0^* - Z_0) = \sum_i \lambda_0^i \gamma'(\mathbf{x}_i - \mathbf{x}_0) + \mu + C$$

Finally, if a horizontal variogram is used (pure nugget effect), one finds that $\lambda^i = n^{-1}, \forall i, n$ being the number of measurement points. The estimation $Z_0^*$ is then constant over the domain and equal to the average of all measurements.

Much work is presently being done to establish procedures that improve the quality and the robustness of the determination of the variogram. See for instance Armstrong (1984) and Diamond and Armstrong (1984).

### 11.5.4. Anisotropy in the Variogram

It may be useful to compute the variogram while assuming that $\gamma(\mathbf{h})$ is also a function of the direction of the vector $\mathbf{h}$. Of course, this requires more data points in order to be significant. We could, for instance, use four (or eight) classes of directions and plot each variogram separately, as in Fig. 11.10.

Generally, variograms do not show anisotropy like Fig. 11.10. If they do, then (1) this may be a sign that the assumption of stationarity (or even intrinsic behavior) does not hold (such cases are dealt with in Section 11.7); or, (2) if the intrinsic or the stationary hypothesis is valid, this anisotropy can be eliminated by an appropriate linear transformation in the coordinate system. This will permit us to krige as usual in the new system.

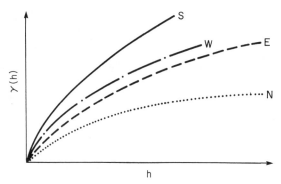

**Fig. 11.10.** Directional variogram.

## 11.6. A Few Additional Remarks about Kriging

### 11.6.1 *Kriging with a Moving Neighborhood*

For this case, instead of kriging with all the measurement points, $i = 1,\ldots,n$, we use only those that are situated inside a given neighborhood of the point we want to estimate (Fig. 11.11). This neighborhood can be defined in several different ways:

(1)  As all the data points at a distance less than $R$ from the point $x_0$.

(2)  As the set of $m$ points closest to the point $x_0$.

(3)  Even better, by selecting the points according to some criterion, e.g., the quality of the data.

There may be several reasons for using a moving neighborhood:

(1)  The variogram is best known for small values of **h** and becomes less and less certain as **h** increases; therefore it is more efficient to use data close enough to $x_0$ that $|x_i - x_j|$ remains in the range where the uncertainties on the variogram are still small.

(2)  By using only a limited number of neighboring points, the kriging system has fewer equations and therefore less effort is required to invert the

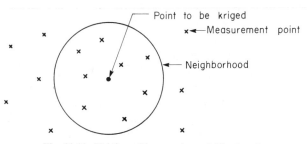

**Fig. 11.11.** Kriging with a moving neighborhood.

matrix; however, each time the set of neighboring points is modified one must compute and invert the matrix of the kriging system again.

(3)   By this procedure, one can sometimes relax the condition of stationarity (or "intrinsic" behavior) of the phenomenon. One only requires pseudo-stationarity in a limited area around the estimated point and can allow for a long-range trend in space.

(4)   If data from distant points are used in the estimation, their weights will be very small and possibly negative. In certain cases, this could lead to negative estimated values.

For example, if the object were to krige the elevation of the ground in a mountainous area, the stationarity would be a question of scale (Fig. 11.12).

### 11.6.2. *Kriging of the Logarithms*

Instead of working with the magnitude $Z$ it is sometimes possible to krige its logarithm. There are several reasons for choosing this procedure.

First, some magnitudes have log-normal probability distribution functions. In such cases the spatial structure is much better (the variogram shows a stronger correlation) if we use the logarithm of the variable instead of its natural value. This is true for the transmissivities, for instance.

Second, if we take mean values over a mesh, the arithmetic mean of the logarithm gives, in fact, the geometric mean of the natural values, and it so happens that the latter is a better estimator of the true average than the former for the transmissivities (see Section 4.4).

It might, however, be desirable to return to a nonbiased estimator. If the probability distribution function of $Z$ is log-normal and we estimate $Y = \log Z$, then $\exp[E(Y)]$ is not an estimator of the average $E(Y)$ but of the median. Conversely,

$$E(Z) = \exp[E(Y) + \tfrac{1}{2}\text{var}(Y)]$$

However, another bias is introduced by the condition of unbiased estimation,

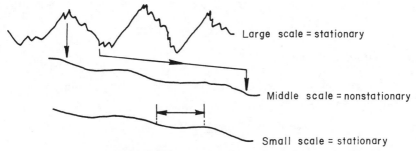

Fig. 11.12. Stationarity versus scale of observation.

$E(Y_0^* = E(Y)$. It can be shown that the optimal estimator of $Z_0^*$ is

$$Z_0^* = \exp[Y_0^* + \tfrac{1}{2}\mathrm{var}(Y_0^* - Y_0) - \mu]$$

However, this expression is not very robust (i.e., insensitive to the assumptions or to error) because it supposes that all the probability distribution functions of $Z$ (with $n$ variables) are log-normal as well. If this is not the case, it is preferable to use an iterative process (Journel and Huijbrechts, 1978; Journel, 1980).

When kriging average transmissivities over blocks, it is preferable, however, to obtain an *unbiased* estimation of $Y = \ln T$, and to use $T = \exp(Y)$ as a *biased* estimation. It is indeed impossible to have an unbiased estimation of both $Y$ and $T$, and we have seen in Section 4.4 that the geometric mean is the optimal estimation of the average transmissivity, in two dimensions, for uniform flow conditions.

### 11.6.3. Verification of the Validity of the Model

To check the validity of all the assumptions used in kriging (e.g., stationarity, good estimation of the variogram), it is preferable to test the ability of the model to predict known data.

(1)   One value is taken out of the set used in kriging, e.g., at point $i$ (say $Z_i$).

(2)   We compute the predicted value $(Z_i^*)$ at point $i$, obtained by kriging with the other data.

(3)   We can then exactly estimate the error of kriging at this point and compare it with the variance of estimation at the same point (or rather the standard deviation $\sigma_Z$).

(4)   By doing this successively for all data points, one can check that there is no systematic bias,

$$\frac{1}{n}\sum_{i=1}^{n}(Z_i - Z_i^*) \simeq 0$$

and that the kriging errors are coherent with the predicted variance,

$$\frac{1}{n}\sum_{n}\left(\frac{Z_i - Z_i^*}{\sigma_Z}\right)^2 \simeq 1$$

### 11.6.4. Network Optimization

The variance of the estimation is a powerful tool for optimizing a network, because in the expression:

$$\mathrm{var}[Z_0 - Z_0^*] = \sum_i \lambda_0^i \gamma(\mathbf{x}_i - \mathbf{x}_0) + \mu$$

the measured values $Z_i$ at each measurement point $\mathbf{x}_i$ are not included. One can therefore conceptually add a new "fictitious" measurement point, compute the map of the variance of estimation with this new point, and compare it with the previous map. One can then locate the additional measurement points in the area where the variance of estimation is high.

The quality of the measurement (variance or "uncertainty" of the measurement) can also be included as shown before.

Conversely, a network that is too costly to maintain can be reduced by maintaining only the observation points that give the most acceptable map of estimation variance in view of the objectives of the observation (general surveillance or local zone of interest).

Furthermore, if the purpose of the network is to estimate the average of $Z$ over the entire domain (e.g., computation of rainfall averages), then the variance of estimation is only a single figure, not a map for the entire domain. The fictitious point can then be moved around over the entire domain, and the "increase in precision" can be plotted. It is defined by

$$G(\mathbf{x}) = \frac{\text{var}[\bar{Z}^* - \bar{Z}] - \text{var}[\bar{Z}^* - \bar{Z}]'}{\text{var}[\bar{Z}^* - \bar{Z}]}$$

where the prime shows the variance computed with the fictitious point.

Having selected the first best additional point, one can repeat the calculation for a second point, and so on. The suppression of one or several measurement points can be decided according to the increase of the global variance.

## 11.7. Nonstationary Problems

### 11.7.1. *Definition*

In nonstationary problems the mathematical expectation of $Z$ is no longer a constant: $E[Z(\mathbf{x})] = m(\mathbf{x})$, and the variogram cannot be calculated directly from the data since $m(\mathbf{x})$ is unknown.

$$\gamma = \tfrac{1}{2}\text{var}[Z(\mathbf{x} + \mathbf{h}) - Z(\mathbf{x})] = \tfrac{1}{2}E\{[Z(\mathbf{x} + \mathbf{h}) - Z(\mathbf{x})]^2\} - \tfrac{1}{2}[m(\mathbf{x} + \mathbf{h}) - m(\mathbf{x})]^2$$

If we then try to calculate the variogram as shown in the preceding section, i.e., directly from the data, by

$$\gamma'(\mathbf{h}) = \frac{1}{2n_h}\sum(Z_i - Z_j)^2$$

where $n_h$ is the number of pairs $(Z_i - Z_j)$ separated by distance $\mathbf{h}$, we find that the variogram $\gamma'(\mathbf{h})$ is anisotropic because the mathematical expectation $m$ is anisotropic and $Z$ has a main direction of drift, e.g., the direction of flow, for hydraulic head.

In this direction, if $m$ is a linear spatial function, a parabolic function is added to the true variogram $\gamma$. Thus the calculation of the variogram in several direction (see Section 11.5.4.) makes it possible to detect the importance of the nonstationarity.

There are several procedures for solving nonstationary problems. We start with three solutions of special cases before turning to the general one in Section 11.7.3.

### 11.7.2. Special Cases

In certain cases, it is possible to:

(1)   Assume that $Z$ is "locally stationary," that is to say, that the variogram stays isotropic for a certain neighborhood, and we can krige with the intrinsic hypothesis in that area using a moving neighborhood (see Section 11.6.1 and Chiles, 1977).

(2)   Assume that the mathematical expectation $m(\mathbf{x})$ is known. It may, for instance, be deduced from other types of measurements. Its mathematical expression might also be known for physical reasons (e.g., the general shape of the drawdown in the vicinity of a borehole for the hydraulic heads), and then the constants of this expression may be fitted on the model. We then verify that the residues $Z(\mathbf{x}) - m(\mathbf{x})$ are stationary and can be kriged under the assumptions of the intrinsic hypothesis. It is, however, incorrect to fit a polynomial expression (by least squares) arbitrarily on the data, assimilate it to $m(\mathbf{x})$, and work on the residues. Indeed, the fitting by simple least squares assumes that the residues are independent and therefore that no spatial structure exists. It is usual in statistics to test the independence of these residues with the Durbin–Watson test. It is thus contrary to the hypothesis to try to find a variogram for them. It is nevertheless possible to use generalized least squares if we take this spatial correlation into account iteratively [see Neuman, (1984)].

(3)   Assume that the variogram $\gamma$ is stationary and known. This is an extension of simple kriging which is called "universal kriging" but which is usually difficult to apply because the variogram must be known. However, assume that we know it and that it is stationary and not a function of $\mathbf{x}$:

$$\gamma(h) = \tfrac{1}{2}\text{var}[Z(\mathbf{x} + \mathbf{h}) - Z(\mathbf{x})] \qquad (11.7.1)$$

We have seen that we cannot compute this variogram directly from the data because the average $m(\mathbf{x})$ is not known. As usual, the kriging estimation is

$$Z_0^* = \sum_i \lambda_0^i Z_i \qquad (11.7.2)$$

but the condition for having an unbiased estimator is different:

$$E(Z_0^*) = E(Z_0)$$

$$E\left(\sum_i \lambda_0^i Z_i\right) = E(Z_0) \qquad \text{but} \qquad E(Z) = m(\mathbf{x})$$

whence

$$\sum_i \lambda_0^i m(\mathbf{x}_i) = m(\mathbf{x}_0) \tag{11.7.3}$$

The average $m(\mathbf{x})$ is not known, but we make the assumption that it is regular and that it may be represented *locally* by a known set of basis functions. Polynomial expressions are commonly used for this purpose. For example, in two dimensions, we write

$$m(\mathbf{x}) = a_0 + a_1 X + a_2 Y + a_3 X^2 + a_4 XY + a_5 Y^2 + \cdots$$

where $X$ and $Y$ are the coordinates of point $\mathbf{x}$ in two dimensions or

$$m(\mathbf{x}) = \sum_k a_k p^k(\mathbf{x})$$

where the $p(\mathbf{x})$ are polynomials in $X$ and $Y$.

In order to ascertain that the estimator is unbiased, we impose that Eq. (11.7.3) is satisfied by any value of $a_k$:

$$\sum_i \lambda_0^i \left[\sum_k a_k p^k(\mathbf{x}_i)\right] = \sum_k a_k p^k(\mathbf{x}_0)$$

or

$$\sum_k a_k \left[\sum_i \lambda_0^i p^k(\mathbf{x}_i)\right] = \sum_k a_k p^k(\mathbf{x}_0)$$

which is satisfied if

$$\sum_i \lambda_0^i p^k(\mathbf{x}_i) = p^k(\mathbf{x}_0), \qquad k = 1, \ldots, m \tag{11.7.4}$$

These conditions are the equivalent of the single condition $\sum_i \lambda_0^i = 1$, which was imposed in the stationary case. We then minimize the estimation variance $\text{var}(Z_0^* - Z_0)$, subject to the $m$ conditions [Eq. (11.7.4)], in the same way. The estimation variance is again a function only of the variogram $\gamma$ because of Eq. (11.7.4), and the equations in the kriging system become

$$\left.\begin{aligned} \sum_j \lambda_0^j \gamma(\mathbf{x}_i - \mathbf{x}_j) + \sum_k \mu_k p^k(\mathbf{x}_i) &= \gamma(\mathbf{x}_i - \mathbf{x}_0) & i = 1, \ldots, n \\ \sum_i \lambda_0^i p^k(\mathbf{x}_i) &= p^k(\mathbf{x}_0) & k = 1, \ldots, m \end{aligned}\right\} \tag{11.7.5}$$

where the $\mu_k$ are Lagrange multipliers. The solution of Eq. (11.7.5) gives the $\lambda_0^i$ for calculating $Z_0$ with Eq. (11.7.2) and the $\mu_k$ for calculating the estimation variance with

$$\text{var}(Z_0^* - Z_0) = \sum_i \lambda_0^i \gamma(\mathbf{x}_i - \mathbf{x}_0) + \sum_k \mu_k p^k(\mathbf{x}_0) \qquad (11.7.6)$$

Note that the drift $m$ is fitted only locally and that it does not appear directly in the estimation of $Z$ but only in the calculation of the variance. Therefore it is not the same thing to fit one polynomial expression on the system as a whole and to krige the residues. Each new point $\mathbf{x}_0$ has a new fit for the drift $m(\mathbf{x}_0)$, the coefficients of which (the $a_k$) are never calculated. For this reason we generally use only low-degree polynomial expressions (linear or quadratic in $X$ and $Y$).

However, the serious problem with universal kriging is that the "true" variogram $\gamma(\mathbf{h})$ must be known and cannot be estimated directly from the data. Although efforts have been made to calculate $\gamma$ iteratively (assume that $\gamma$ is known, krige, verify $\gamma$ once $m$ is known), this is not practical. This is why universal kriging is used only if (1) there is a drift in part of the system, e.g., towards the boundaries (the variogram is fitted in the center, where the phenomenon is stationary, and is then used to krige the entire domain); or (2) if there is no drift at all in a given direction in the field as a whole. (Then the variogram is determined from the data in this direction only and we use it in all the other directions while assuming that the "true" variogram is stationary and isotropic. However, it is very difficult to verify the validity of such an assumption.)

### 11.7.3. General Solution: Intrinsic Random Functions of Order $k$

(a)  *Redefinition of the intrinsic hypothesis.*  Kriging with the intrinsic hypothesis, which we have described above for the stationary case, may be summarized as follows:

(1)  Define the weights $\lambda_0^i$ such as

$$Z_0^* = \sum_{i=1}^n \lambda_0^i Z_i \qquad (11.7.7)$$

(2)  Write the condition

$$\sum_{i=1}^n \lambda_0^i = 1 \qquad (11.7.8)$$

(3)  Take the (minimal) estimation error given by the kriging system to be (by Eq. (11.7.8))

$$Z_0^* - Z_0 = \sum_{i=1}^n \lambda_0^i Z_i - Z_0 = \sum_{i=1}^n \lambda_0^i (Z_i - Z_0) \qquad (11.7.9)$$

(4)   Assume that the difference $(Z_i - Z_0)$ or $[Z(\mathbf{x} + \mathbf{h}) - Z(\mathbf{x})]$, called first increment of $Z$, is stationary. It can then be shown that the variance of the estimation error with kriging depends only on the variogram:

$$\text{var}(Z_0^* - Z_0) = E[(Z_0^* - Z_0)^2] = -\sum_{i=1}^{n} \sum_{j=1}^{n} \lambda_0^i \lambda_0^j \gamma(\mathbf{x}_i - \mathbf{x}_j) + 2 \sum_{i=1}^{n} \gamma(\mathbf{x}_i - \mathbf{x}_0)$$

(11.7.10)

We can formulate all these equations again by arranging them slightly differently. We define $\lambda_0^0 = -1$ (i.e., the value of $\lambda_0^i$ for $i = 0$) and associate the point $\mathbf{x}_0$ with the value $i = 0$. Then Eq. (11.7.8)–(11.7.10) can be written as

$$\sum_{i=0}^{n} \lambda_0^i = 0$$

(11.7.8a)

$$Z_0^* - Z_0 = \sum_{i=0}^{n} \lambda_0^i Z_i$$

(11.7.9a)

$$E[(Z_0^* - Z_0)^2] = -\sum_{i=0}^{n} \sum_{j=0}^{n} \lambda_0^i \lambda_0^j \gamma(\mathbf{x}_i - \mathbf{x}_j)$$

(11.7.10a)

Equation (11.7.9a), subject to the condition of Eq. (11.7.8a), is called an increment of zero order of the random function $Z$. The intrinsic hypothesis assumes that this increment is stationary. The variance of the estimation error is then a linear function of the variogram. Finally, it is possible to determine the variogram directly from the data, as shown in Section 11.5.

This method can be called the procedure for the intrinsic random functions of zero order, which will be generalized below.

(b)   *Intrinsic random functions of order* 1 *(IRF–1) and of order* 2 *(IRF–2)*. We treat orders 1 and 2 simultaneously and the estimation runs as follows.

(1)   We define the weights $\lambda_0^i$ such as

$$Z_0^* = \sum_{i=1}^{n} \lambda_0^i Z_i$$

(11.7.11)

(In fact, we are looking for the optimal weights $\lambda_0^i$). We define likewise $\lambda_0^0 = -1$.

(2)   We impose three conditions (first order):

$$\left. \begin{array}{c} \displaystyle\sum_{i=0}^{n} \lambda_0^i = 0 \\[2mm] \displaystyle\sum_{i=0}^{n} \lambda_0^i X_i = 0 \\[2mm] \displaystyle\sum_{i=0}^{n} \lambda_0^i Y_i = 0 \end{array} \right\}$$

(11.7.12)

or six conditions (second order):

$$\sum_{i=0}^{n} \lambda_0^i = 0 \qquad \sum_{i=0}^{n} \lambda_0^i X_i^2 = 0$$

$$\left.\sum_{i=0}^{n} \lambda_0^i X_i = 0 \qquad \sum_{i=0}^{n} \lambda_0^i Y_i^2 = 0 \right\} \qquad (11.7.13)$$

$$\sum_{i=0}^{n} \lambda_0^i Y_i = 0 \qquad \sum_{i=0}^{n} \lambda_0^i X_i Y_i = 0$$

where $x_i = (X_i, Y_i)$ are the two coordinates (in two dimensions) of the point $x_i$.

(3)  Then the error of estimation is given by

$$Z_0^* - Z_0 = \sum_{i=0}^{n} \lambda_0^i Z_i \qquad (11.7.14)$$

The quantity $\sum_{i=0}^{n} \lambda_0^i Z_i$, subject to the condition of Eq. (11.7.12) or Eq. (11.7.13), is called a generalized increment of order 1 (or order 2) because it filters a polynomial expression of order 1 (or order 2). Assume that we define

$$Z_i' = Z_i + a_0 + a_1 X_i + b_1 Y_i \qquad i = 0, \ldots, n$$

Then

$$\sum_{i=0}^{n} \lambda_0^i Z_i' = \sum_{i=0}^{n} \lambda_0^i Z_i + a_0 \left( \sum_{i=0}^{n} \lambda_0^i \right) + a_1 \left( \sum_{i=0}^{n} \lambda_0^i X_i \right)$$

$$+ b_1 \left( \sum_{i=0}^{n} \lambda_0^i Y_i \right) = \sum_{i=0}^{n} \lambda_0^i Z_i$$

if the conditions of Eq. (11.7.12) are satisfied. The generalized increment of $Z \pm$ any polynomial expression of the first order is unchanged. The same would be true for the second order.

In one dimension, to first order, if the measurement points are equally spaced one can take a set of three $\lambda$'s as, for instance $\lambda^1 = 1, \lambda^2 = -2, \lambda^3 = 1$. They satisfy the constraints $\sum_1^3 \lambda^i = 0$ and $\sum_1^3 \lambda^i x_i = 0$ if $x_1 = a$, $x_2 = 2a$, $x_3 = 3a$. Then $\sum_1^3 \lambda^i Z_i = Z_1 - 2Z_2 + Z_3$. This is by definition a second-order difference. Generalized increments of order $k$ are therefore just a generalization, in two or more dimensions, of this simple concept.

(4)  We make the assumption that the generalized increments of the first or the second order of $Z$ are stationary (intrinsic hypothesis of order 1 or 2). It is then possible to show, exactly as in the hypothesis of zero order, that the variance of the estimation error may be expressed in the following form:

$$\text{var}(Z_0^* - Z_0) = \text{var}\left( \sum_{i=0}^{n} \lambda_0^i Z_i \right) = \sum_{i=0}^{n} \sum_{j=0}^{n} \lambda_0^i \lambda_0^j K(x_i - x_j)$$

where $K$ is a new function, called the "generalized covariance" of the first or second order of the IRF $Z$; $K$ is stationary, i.e., $K$ is only a function of $\mathbf{h} = \mathbf{x}_i - \mathbf{x}_j$.

(5)  If we assume that $K(\mathbf{h})$ is known, the equations in the kriging system become (when we replace $\gamma$ by $-K$ in the preceding expressions) for first order

$$\sum_{j=1}^{n} \lambda_0^j K(\mathbf{x}_i - \mathbf{x}_j) - \mu_1 - \mu_2 X_i - \mu_3 Y_i = K(\mathbf{x}_i - \mathbf{x}_0) \qquad i = 1,\ldots,n$$

$$\sum_{i=1}^{n} \lambda_0^i = 1 \qquad \sum_{i=1}^{n} \lambda_0^i X_i = X_0 \qquad \sum_{i=1}^{n} \lambda_0^i Y_i = Y_0$$

(11.7.15)

[where $\mathbf{x}_i = (X_i, Y_i)$] and for second order,

$$\sum_{j=1}^{n} \lambda_0^j K(\mathbf{x}_i - \mathbf{x}_j) - \sum_{k=0}^{5} \mu_k p^k(\mathbf{x}_i) = K(\mathbf{x}_i - \mathbf{x}_0) \qquad i = 1,\ldots,n$$

$$\sum_{i=1}^{n} \lambda_0^i p^k(\mathbf{x}_i) = p^k(\mathbf{x}_0), \qquad k = 0,\ldots,5$$

(11.7.16)

where $p^0(\mathbf{x}),\ldots,p^5(\mathbf{x})$ designate the 6 polynomials in $X_i$, $Y_i$ from Eq. (11.7.13). The estimation variance is given for first order by

$$\mathrm{var}(Z_0^* - Z_0) = E[(Z_0^* - Z_0)^2] = K(0) + \mu_1 + \mu_2 X_0 + \mu_3 Y_0 - \sum_{i=1}^{n} \lambda_0^i K(\mathbf{x}_i - \mathbf{x}_0)$$

(11.7.17)

and for second order by

$$\mathrm{var}(Z_0^* - Z_0) = K(0) + \sum_{k} \mu_k p^k(\mathbf{x}_0) - \sum_{i=1}^{n} \lambda_0^i K(\mathbf{x}_i - \mathbf{x}_0) \qquad (11.7.18)$$

When these new equations are compared with those of universal kriging, Eqs. (11.7.4)–(11.7.6), they prove to be identical. This is not surprising if we bear in mind the filtering properties of the generalized covariance. The IRF hypothesis assumes that the drift is locally linear (or quadratic) and we can krige as soon as we know the generalized covariance $K(\mathbf{h})$.

Observe that here $K(0)$ is usually zero unless we use integrated values, in which case we have shown that this term is given by

$$K(0) = \frac{1}{S_0^2} \int_{s_0} \int_{s_0} K(\mathbf{x} - \mathbf{y}) \, d\mathbf{x} \, d\mathbf{y}$$

IRF $k$ of a higher order can also be defined, but practical experience shows that it is enough to use IRF–1 and IRF–2 in most cases.

(c)  *Statistical inference of the generalized covariance.*  In order to identify the variogram in the case of an IRF–0, it was only necessary to calculate $\gamma(\mathbf{h}) = \frac{1}{2} E\{[Z(\mathbf{x} + \mathbf{h}) - Z(\mathbf{x})]^2\}$, since the first increment (of zero order) was stationary. To do it, we only used the measurement points 2 by 2. Then an analytical expression was fitted on the experimental variogram (cf. Section 11.5). For $K(\mathbf{h})$ we actually proceed in a similar way but the fitting becomes automatic. The statistical inference runs as follows.

(1)   Choice of a point $\mathbf{x}_0$ where, naturally, $Z_0 = Z(\mathbf{x}_0)$ is known. We then choose $n$ points $\mathbf{x}_i$ close to $\mathbf{x}_0$ with $Z_i = Z(\mathbf{x}_i)$ known, $i = 1,\ldots,n$. A moving neighborhood is generally used, just as in kriging.

(2)   A generalized increment of $k$ order is built; i.e., we compute

$$G(\mathbf{x}_0) = \sum_{i=0}^{n} \lambda_0^i Z_i \tag{11.7.19}$$

where the weights $\lambda_0^i$ satisfy the conditions for being generalized increments of order $k$, e.g., Eq. (11.7.13) at the second order. To calculate a set of weights $\lambda$ that fulfill the conditions of Eq. (11.7.13), several methods can be used. We can, for instance, calculate the $\lambda$ that minimize

$$\left( \sum_{i=0}^{n} \lambda_0^i Z_i \right)^2 \tag{11.7.20}$$

subject to the conditions of Eq. (11.7.13). Just as in kriging, these weights are obtained by equating to zero the partial derivatives of Eq. (11.7.20) with respect to $\lambda$ while taking Eq. (11.7.13) into account through the Lagrange multipliers. We can also calculate the $\lambda$ terms as solutions to a problem of universal kriging (see Section 11.7.2.c) by using any variogram $\gamma$ and taking $\lambda_0^0 = -1$. Another solution is to take a small number of points and solve Eq. (11.7.13) directly.

(3)   We assume that the increments $G(\mathbf{x}_0)$ are stationary, of zero average (for all $\mathbf{x}_0$ and for all sets $\mathbf{x}_i$ of neighboring points). Then the variance of these increments $G(\mathbf{x}_0)$ may be expressed as a function of the (still unknown) generalized covariance $K(\mathbf{h})$ by

$$\mathrm{var}[G(\mathbf{x}_0)] = [G(\mathbf{x}_0)]^2 = \sum_{i=0}^{n} \sum_{j=0}^{n} \lambda_0^i \lambda_0^j K(\mathbf{x}_i - \mathbf{x}_j) \tag{11.7.21}$$

(4)   We assume that $K(\mathbf{h})$ can be expressed as a preselected function of $\mathbf{h}$, which only depends linearly on unknown coefficients $A_i$; a usual form is

$$K(\mathbf{h}) = A_0[1 - \delta(\mathbf{h})] + A_1|\mathbf{h}| + A_s h^2 \ln|\mathbf{h}| A_3 |\mathbf{h}|^3 \tag{11.7.22}$$

where $\delta$ is the Kronecker symbol (see p. 304) and $A_0$ is the nugget effect. In order for $K(\mathbf{h})$ to be a generalized covariance, the $A_i$ must satisfy

$$A_0 \geq 0, \qquad A_1 \leq 0, \qquad A_3 \geq 0, \qquad A_s \geq -\sqrt{-24 A_1 A_3/\pi^2}$$

in one dimension and

$$A_0 \geq 0, \qquad A_1 \leq 0, \qquad A_3 \geq 0, \qquad A_s \geq -\tfrac{3}{2}\sqrt{-A_1 A_3}$$

in two dimensions. In practice, experience has shown that this limited class of generalized covariance functions $K(\mathbf{h})$ is quite sufficient for the study of most problems. Sometimes it is not even necessary to use all the terms: $A_0, A_1$ or $A_0, A_1, A_s$, or simply $A_3$, may suffice.

We can then write Eq. (11.7.21) as follows:

$$[G(\mathbf{x}_0)]^2 = \sum_{i,j} \lambda_0^i \lambda_0^j \{ A_0[1 - \delta(\mathbf{x}_i - \mathbf{x}_j)] + A_1|\mathbf{x}_i - \mathbf{x}_j|$$
$$+ A_s|\mathbf{x}_i - \mathbf{x}_j|^2 \ln|\mathbf{x}_i - \mathbf{x}_j| + A_3|\mathbf{x}_i - \mathbf{x}_j|^3 \} \quad (11.7.23)$$

The second, third, and fourth terms are called the linear, spline, and cubic terms, respectively.

(5)   The calculation of $G(\mathbf{x}_0)$ by Eq. (11.7.19) is repeated a great number of times (several hundreds or thousands) while varying the point $\mathbf{x}_0$ as well as the neighboring points $\mathbf{x}_i, i = 1, \ldots, n$ of each point $\mathbf{x}_0$. Often the $\mathbf{x}_i$ are chosen in increasingly large circles surrounding the point $\mathbf{x}_0$, as in Fig. 11.13. The linear combinations $\lambda$ must be correlated as little as possible (not have many points in common between two of them).

(6)   The coefficients $A_i$ are determined by simple regression:

$$\min_{A_i} \sum_{\mathbf{x}_0} \left\{ G(\mathbf{x}_0)^2 - \sum_{i,j=0}^{n} \lambda_0^i \lambda_0^j [A_0(1 - \delta(\mathbf{h})) + A_1\mathbf{h} + A_2\mathbf{h}^2 + A_3\mathbf{h}^3] \right\}$$

where $\mathbf{h} = |\mathbf{x}_i + \mathbf{x}_j|$. The $A_i$ are calculated by canceling the first derivative of the preceding expression with respect to the $A_i$ (linear system of four equations with four unknowns). We can also, if we so desire, weight this sum in order not to give too much weight to the large $G(\mathbf{x}_0)$ values, which would present too great a variance.

(7)   Once $K(\mathbf{h})$ is known, the whole kriging procedure is verified by recalculating, one by one, all the measurement points as we have explained in Section 11.6.3. Consequently, to krige with the I.R.F. $k$, we must (1) choose the $k$ order (0, 1, or 2); (2) choose the form of $K(\mathbf{h})$ [selected terms in Eq. (11.7.22)]; (3) calculate $K(\mathbf{h})$ and verify its validity as for the variogram. It is possible to

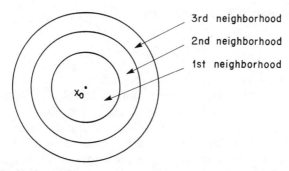

3rd neighborhood
2nd neighborhood
1st neighborhood

$x_0$

**Fig. 11.13.** Neighborhood for calculating the generalized covariance.

make several test runs in order to choose the best $k$ order or form of $K(\mathbf{h})$, but experience also allows us to select the most suitable values simply by studying the data. A computer program, BLUEPACK 3-D (Renard *et al.*, 1985) has recently been developed to apply nonstationary geostatistics to two- or three-dimensional problems using these principles (see also Delfiner, 1976; Kitanidis, 1983).

## 11.8. Examples of Kriging

We shall describe two examples where kriging is used in hydrogeology: transmissivities and heads.

### 11.8.1 *Kriging Transmissivities* (after Delhomme, 1974)

The aquifer of Fig. 11.14 consists of confined eocene sands in the Aquitaine basin (France). Data were available in 86 wells; 29 came from pumping tests and were considered exact and the 57 other data points were only the specific capacity in the wells, which was also available in the 29 wells with pumping tests. A linear regression was then made between the log of the specific capacity $(Q/s)_j$ and the log of the transmissivity $T_j$ at these 29 wells. The 57 other wells were given a transmissivity value by means of this regression as well as an "uncertainty," estimated to be the variance of the regression:

$$\sigma_j^2 = \sigma^2 \left\{ 1 + \frac{1}{n} + \frac{[\ln(Q/s)_j - \overline{\ln(Q/s)}]^2}{\sum\limits_{i=1}^{n} [\ln(Q/s)_i - \overline{\ln Q/s}]^2} \right\} \qquad j = 1,\dots,m \quad (11.8.1)$$

where

$$\sigma^2 = \frac{1}{n-2} \sum_{i=1}^{n} [\ln T_i - a\ln(Q/s)_i - b]^2$$

with $n$ as the number of pairs in the regression

$$\overline{\ln Q/s} = \frac{1}{n} \sum_{i=1}^{n} \ln(Q/s)_i$$

and $a$ and $b$ as coefficients of the linear regression.

The variogram was then estimated with the 86 values of transmissivities. This was done both on $\ln T$ and on the natural values (Fig. 11.15). The former have a much better "structure" than the latter. This is due to the fact that the transmissivities are very often log-normally distributed, as can be seen on the histogram (Fig. 11.16). A linear variogram with a small nugget effect was

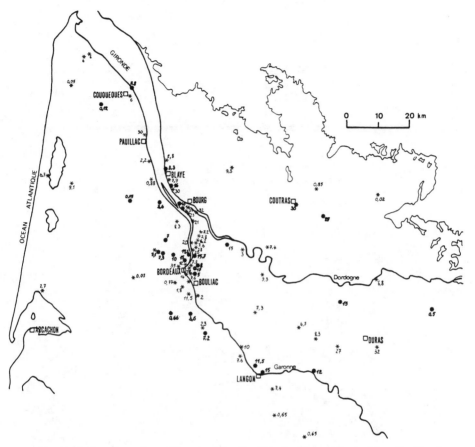

**Fig. 11.14.** Eocene aquifer and location of wells; ●, well with pumping test, *, well with only specific capacity.

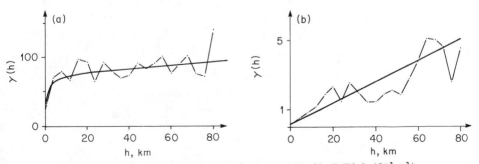

**Fig. 11.15** Variogram of transmissivity: (a) of $T$ and (b) of ln $T$. $T$ is in $10^{-3}$ m$^2$/s.

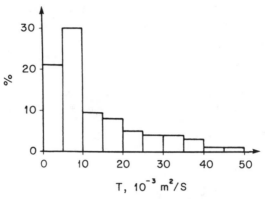

**Fig. 11.16.** Histogram of transmissivity values.

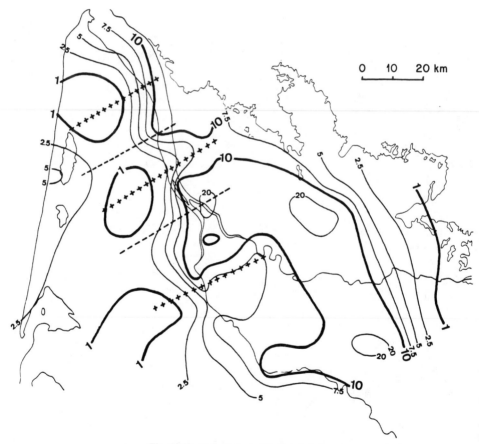

**Fig. 11.17.** Kriged map of transmissivity.

adjusted on the data, such that

$$\gamma(\mathbf{h}) = 0.15(1 - \delta) + 0.0625|\mathbf{h}| \qquad (11.8.2)$$

was the variogram for ln $T$, with $T$ in $10^{-3}$ m$^2$/s, $\mathbf{h}$ in km, and $\delta = 1$ if $\mathbf{h} = 0$.

The variogram shows that the intrinsic hypothesis holds. Intrinsic kriging was then done using the "exact" data (in 29 wells), the "uncertain" data, and the variance of each [Eq. (11.8.1)]; the variogram of Eq. (11.8.2) was also used. The estimation was done on a square grid of $2 \times 2$ km, using all the 86 data points without a moving neighborhood. Figure 11.17 gives the kriged contour map of $T$ and Fig. 11.18 gives the contour map drawn manually by a hydrogeologist. As the kriging was done on ln $T$, the 95% confidence interval

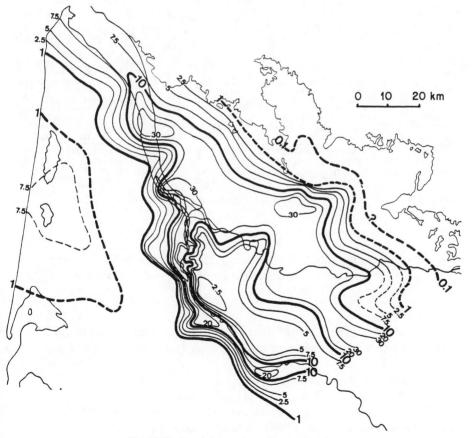

**Fig. 11.18.** Hand-drawn map of transmissivity.

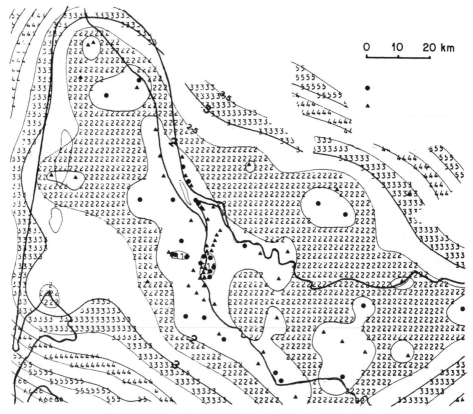

**Fig. 11.19.** Uncertainty of kriged map of transmissivity ($e^\sigma$).

is $\pm 2\sigma$, $\sigma$ being the standard deviation of the estimation error.

From $\ln T = (\ln T)^* \pm 2\sigma$, where the asterisk indicates estimation one gets

$$T^*/e^{2\sigma} \le T \le T^* e^{2\sigma} \qquad \text{with} \qquad T^* = e^{(\ln T)^*}$$

Note that $T^*$ is here the median estimator and that the correction for obtaining an unbiased estimator was purposefully not applied, as explained in Section 11.6.2.

Figure 11.19 gives the contour map of $e^\sigma$. For instance, if $e^\sigma = 3.15$, then $e^{2\sigma} = 10$, i.e., the uncertainty on $T$ is of one order of magnitude.

Kriging was also done to estimate the average of $\ln T$ directly on the meshes of a digital model, which in this case used a nested square grid.

## 11.8.2. *Kriging Heads*

The example concerns an unconfined aquifer in chalk, at Origny Sainte Benoite (Aisne, France). The aquifer is drained by three rivers to the north, west, and south. A piezometric survey was made on December 31, 1976, in 88 piezometers. An additional 64 measurement points were introduced into the kriging by taking water levels in the rivers surrounding the aquifer at regular intervals, since these rivers acted as prescribed head boundary conditions for the aquifer.

Since heads are typically nonstationary, generalized covariances were used at order 1 (locally linear drift). The generalized covariance was found to be in $\mathbf{h}^3$: $K(\mathbf{h}) = a|\mathbf{h}|^3$. It was, however, necessary to use a different coefficient $a$ for different zones of the aquifer, as the spatial variability was greater under the plateau than under the plains. The values of $a$ were adjusted by fitting the estimation error of kriging on the true estimation error when the validity of the model was verified as shown in Section 11.6.3.

The map of $a$, the kriged map of the heads and the standard deviation of the estimation error are given in Fig. 11.20.

## 11.9. Co-Kriging

Co-kriging is an estimation technique useful when two (or more) variables which are correlated are measured in the field and can be estimated together. For instance, if in an aquifer the concentration in the water of several metals is correlated, then it is possible to estimate in one location the amount of metal $Z_1^*$ based on the measurements not only of $Z_1$, but also of metals $Z_2$ or $Z_3$. Another example is that of the estimation of transmissivity in an aquifer based not only on the measurements of the transmissivity itself, obtained by pumping tests, but also simultaneously on the measurements of the specific capacity, which is correlated to the transmissivity. Co-kriging is then a more elaborate and accurate method for using both types of data than the simple regression and kriging with uncertain data, which was presented in Sections 11.8.1 and 11.4.8.

### 11.9.1. *Co-Kriging Equations*

Let $Z_1(\mathbf{x})$ and $Z_2(\mathbf{x})$ be two regionalized variables that are correlated. The estimation of $Z_1$ (and if necessary $Z_2$) by co-kriging is again given by a best linear unbiased estimate in the form

$$Z_1^*(x_0) = \sum_{j=1}^{n} \lambda_1^j Z_1(\mathbf{x}_j) + \sum_{l=1}^{m} \lambda_2^l Z_2(\mathbf{x}_l) \qquad (11.9.1)$$

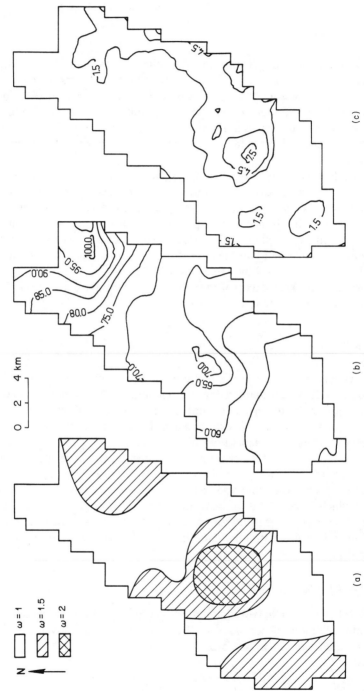

**Fig. 11.20.** Kriging of heads: (*a*) coefficient of the generalized covariance, (*b*) kriged head, meters, and (*c*) standard deviation of kriged heads, meters.

Here, $Z_1^*(\mathbf{x}_0)$ is the estimate of $Z_1$ at location $\mathbf{x}_0$, $Z_1(\mathbf{x}_j)$ are the measured values of $Z_1$, $j = 1, \ldots, n$, and $Z_2(\mathbf{x}_l)$ are the measured values of $Z_2$, $l = 1, m$. (Note that $Z_1$ and $Z_2$ need not be measured at the same location, and $m$ and $n$ may be different); the $\lambda$'s are the co-kriging weights. However, we have to slightly change the notation here compared to the previous sections: $\lambda_j^i$ is the kriging weight for variable $Z_j$ measured at location $\mathbf{x}_i$, and we omit the index 0 used earlier ($\lambda_0^i$) which meant that these weights were related to the estimation of $Z$ at location $\mathbf{x}_0$. It is clear however that these $\lambda$'s will change for each new location $\mathbf{x}_0$ to be estimated. It would also be necessary to have another index to show that they refer to the estimation of $Z_1^*$ since one could also wish to estimate $Z_2^*$ from the same measurements. To keep the notation simple, we will only consider here the estimation of one variable $Z$.

We will however generalize Eq. (11.9.1) for $N$ variables by writing

$$Z_i^*(\mathbf{x}_0) = \sum_{\alpha=1}^{N} \sum_{j=1}^{n_\alpha} \lambda_\alpha^j Z_\alpha(\mathbf{x}_j) \qquad (11.9.2)$$

As in simple kriging, the $\lambda$'s will be determined by writing the two usual conditions: nonbias and optimality.

(a)  *Unbiased estimation.*  We will assume that $E(Z_\alpha) = m_\alpha$, $\alpha = 1, \ldots, N$ (valid both for the stationary and intrinsic hypotheses). Then the condition $E(Z_i^*) = m_i$ can be written

$$m_i = \sum_\alpha m_\alpha \sum_j \lambda_\alpha^j \qquad (11.9.3)$$

If all variables have a different expected value, then the estimation of $Z_i^*$ requires:

$$\sum_{j=1}^{n_i} \lambda_i^j = 1 \qquad (11.9.4)$$

$$\sum_{j=1}^{n_\alpha} \lambda_\alpha^j = 0, \qquad \alpha \neq i$$

But if the variable $Z_k$ has the same expected value as $Z_i$, a case which is very common in practice, then Eq. (11.9.3) gives:

$$\sum_{j=1}^{n_i} \lambda_i^j + \sum_{l=1}^{n_k} \lambda_k^l = 1 \qquad (11.9.5)$$

$$\sum_{j=1}^{n_\alpha} \lambda_\alpha^j = 0, \qquad \alpha \neq i \text{ and } k$$

(b)  *Optimality.*  When the optimality condition was developed for simple kriging in Sections 11.2.2 or 11.3.2 by imposing the condition that the variance of the estimation error $E\{[Z(\mathbf{x}_0) - Z^*(\mathbf{x}_0)]^2\}$ be minimum, the covariance or the variogram of the variable of interest was introduced by the need to evaluate terms like $E[Z(\mathbf{x}_i)Z(\mathbf{x}_j)]$. In co-kriging, additional terms like $E[Z_\alpha(\mathbf{x}_i)Z_\beta(\mathbf{x}_j)]$ will also have to be evaluated. We will therefore introduce a new function to quantify the spatial correlation of the variable $Z_\alpha$ with the variable $Z_\beta$ as a function of the distance $\mathbf{h}$ between the points where $Z_\alpha$ and $Z_\beta$ are measured. In the stationary case, this will be called the cross-covariance function $C_{\alpha\beta}(\mathbf{h})$ and, in the intrinsic case, the cross-variogram $\gamma_{\alpha\beta}(\mathbf{h})$. These will be precisely defined in section 11.9.2.

We will also need in the co-kriging equations the usual covariance or variogram of each variable $Z_\alpha$ which we will denote $C_{\alpha\alpha}(\mathbf{h})$ or $\gamma_{\alpha\alpha}(\mathbf{h})$.

The co-kriging equations can very easily be developed given these new functions. We will first write them for the estimation of $Z_1^*(\mathbf{x}_0)$ (Eq. 11.9.1) from $n$ measurements of $Z_1$ and $m$ measurements of $Z_2$ (i.e., with the nonbias condition 11.9.4). If $Z_1$ and $Z_2$ are both second-order stationary, the co-kriging system can be written

$$\sum_{j=1}^{n} \lambda_1^j C_{11}(\mathbf{x}_i - \mathbf{x}_j) + \sum_{l=1}^{m} \lambda_2^l C_{12}(\mathbf{x}_i - \mathbf{x}_l) - \mu_1 = C_{11}(\mathbf{x}_0 - \mathbf{x}_i)$$
$$\text{for} \quad i = 1,\ldots,n \qquad (11.9.6a)$$

$$\sum_{j=1}^{n} \lambda_1^j C_{21}(\mathbf{x}_k - \mathbf{x}_j) + \sum_{l=1}^{m} \lambda_2^l C_{22}(\mathbf{x}_k - \mathbf{x}_l) - \mu_2 = C_{12}(\mathbf{x}_0 - \mathbf{x}_k) \qquad (11.9.6b)$$
$$\text{for} \quad k = 1,\ldots,m$$

with

$$\sum_{l=1}^{n} \lambda_1^l = 1 \quad \text{and} \quad \sum_{l=1}^{m} \lambda_2^l = 0 \qquad (11.9.6c)$$

The variance of the estimation error is:

$$\text{var}[Z_1^*(\mathbf{x}_0) - Z_1(\mathbf{x}_0)] = E\{[Z_1^*(\mathbf{x}_0) - Z_1(\mathbf{x}_0)]^2\}$$
$$= -\sum_{j=1}^{n} \lambda_1^j C_{11}(\mathbf{x}_0 - \mathbf{x}_j) - \sum_{l=1}^{m} \lambda_2^l C_{12}(\mathbf{x}_0 - \mathbf{x}_l) + \mu_1 \qquad (11.9.7)$$

If $Z_1$ and $Z_2$ are both intrinsic, then the kriging system will be obtained by replacing $C_{ij}$ by $-\lambda_{ij}$ in Eqs. (11.9.6) and (11.9.7). This can easily be generalized for more than two variables: for instance, the kriging system for the estimation in the intrinsic hypothesis of $Z_i^*(\mathbf{x}_0)$ from Eq. (11.9.2) is given by

$$\sum_{\alpha=1}^{N} \sum_{l=1}^{n_\alpha} \lambda_\alpha^l \gamma_{\beta\alpha}(\mathbf{x}_j - \mathbf{x}_l) + \mu_\beta = \gamma_{i\beta}(\mathbf{x}_0 - \mathbf{x}_j)$$
$$\text{for} \quad j = 1,\ldots,n_\beta \quad \text{and} \quad \beta = 1,\ldots,N \qquad (11.9.8a)$$

and

$$\sum_{l=1}^{n_\alpha} \lambda_\alpha^l = \begin{cases} 1 \text{ if } \alpha = i \\ 0 \text{ if } \alpha \neq i \end{cases} \quad \alpha = 1,\dots,N \qquad (11.9.8b)$$

The variance of the estimation error is:

$$\text{var}[Z_i^*(\mathbf{x}_0) - Z_i(\mathbf{x}_0)] = \sum_{\alpha=1}^{N} \sum_{l=1}^{n_\alpha} \lambda_\alpha^l \gamma_{i\alpha}(\mathbf{x}_0 - \mathbf{x}_l) + \mu_i$$

These expressions can also easily be generalized for kriging the average over a mesh or with uncertain data. A matrix formulation of co-kriging was also proposed by Myers (1982, 1983, 1984).

### 11.9.2. Cross-Covariances or Cross-Variograms

In the hypothesis where the variables $Z_1(\mathbf{x})$ and $Z_2(\mathbf{x})$ are both second-order stationary, the cross-covariance $C_{12}(\mathbf{h})$ of $Z_1$ and $Z_2$ is defined by

$$C_{12}(\mathbf{h}) = E\{[Z_1(\mathbf{x}) - m_1][Z_2(\mathbf{x} + \mathbf{h}) - m_2]\} \qquad (11.9.9)$$

where $m_1 = E(Z_1)$ and $m_2 = E(Z_2)$.

One can show that, in general, $C_{12}(\mathbf{h}) \neq C_{21}(\mathbf{h})$, but that $C_{12}(\mathbf{h}) = C_{21}(-\mathbf{h})$ (Journel and Huijbregts, 1978).

In the intrinsic hypothesis, the cross-variogram $\gamma_{12}(\mathbf{h})$ is defined by

$$\gamma_{12}(\mathbf{h}) = \tfrac{1}{2}E\{[Z_1(\mathbf{x} + \mathbf{h}) - Z_1(\mathbf{x})][Z_2(\mathbf{x} + \mathbf{h}) - Z_2(\mathbf{x})]\} \qquad (11.9.10)$$

In the stationary case, where both $C_{12}$ and $\gamma_{12}$ exist,

$$\gamma_{12}(\mathbf{h}) = C_{12}(0) - \tfrac{1}{2}[C_{12}(\mathbf{h}) + C_{21}(\mathbf{h})]$$

Thus one sees that the variogram is always symmetric:

$$\gamma_{12}(\mathbf{h}) = \gamma_{21}(\mathbf{h}) = \gamma_{12}(-\mathbf{h}) = \gamma_{21}(-\mathbf{h})$$

Using the cross-covariance has therefore more possibilities than using the cross-variogram in the stationary case. Experimental cross-covariance or cross-variogram can be determined in a fashion similar to that used for ordinary variograms as shown in Section 11.5.1.

The main difficulty in using co-kriging is that these functions have to satisfy constraints in order to be acceptable, just as we indicated that covariance functions need to be positive definite and that, minus the variograms, must be conditionally positive definite.

Let $G$ be a weighted sum of all the measurements such as the one used in co-kriging

$$G = \sum_\alpha \sum_j \lambda_\alpha^j Z_\alpha(\mathbf{x}_j)$$

then

$$\text{var}(G) = \sum_\alpha \sum_\beta \sum_j \sum_l \lambda_\alpha^j \lambda_\beta^l C_{\alpha\beta}(\mathbf{x}_j - \mathbf{x}_l)$$

or

$$\text{var}(G) = -\sum_\alpha \sum_\beta \sum_j \sum_l \lambda_\alpha^j \lambda_\beta^l \gamma_{\alpha\beta}(\mathbf{x}_j - \mathbf{x}_l)$$

with $\sum_j \lambda_\alpha^j = 0$ for the latter, as in Section 11.7.3.a.

Imposing that $\text{var}(G)$ in both cases be positive for any $\lambda$'s or points $\mathbf{x}$ gives the constraints that must be met simultaneously by the covariances, cross-covariances or variograms and cross-variograms used in a co-kriging system. It is not at all simple to select functions which satisfy these constraints.

(a) *Linear Model.* A first method (Francois-Bongarson, 1981; Waker-nagel, 1985) is to assume that the $N$ variables $Z_\alpha$ can be considered as linear combinations of $M$ hypothetical variables $Z_i'$, which would be uncorrelated:

$$Z_\alpha(\mathbf{x}) = \sum_{i=1}^M a_{\alpha i} Z_i'(\mathbf{x}), \quad \text{for} \quad \alpha = 1, \ldots, N \qquad (11.9.11)$$

Even if the variables $Z_i'$ are independent (i.e., their cross-covariance or cross-variogram is zero) the variables $Z_\alpha$ would be correlated. Let $C_i'(\mathbf{h})$ or $\gamma_i'(\mathbf{h})$ be the covariances or variograms of the variables $Z_i'$. Then one finds easily that the covariances and cross-covariances (or variograms and cross-variograms) of the variables $Z_\alpha$ are

$$C_{\alpha\beta}(\mathbf{h}) = \sum_{i=1}^M b_{\alpha\beta}^i C_i'(\mathbf{h})$$

or                                                                             (11.9.12)

$$\gamma_{\alpha\beta}(\mathbf{h}) = \sum_{i=1}^M b_{\alpha\beta}^i \gamma_i'(\mathbf{h})$$

where $b_{\alpha\beta}^i = a_{\alpha i} a_{\beta i}$ (thus $b_{\alpha\beta}^i = b_{\beta\alpha}^i$).

Then the constraint on the covariances or variograms becomes simply that the matrices $[b_{\alpha\beta}^i]$ for $i = 1, \ldots, M$ must be positive definite. This can be achieved by imposing that the second-order minors of these matrices be positive definite:

$$\begin{vmatrix} b_{\alpha\alpha}^i & b_{\alpha\beta}^i \\ b_{\beta\alpha}^i & b_{\beta\beta}^i \end{vmatrix} > 0 \quad \begin{aligned} &\text{for each} \quad i = 1, \ldots, M \\ &\text{and for} \quad \alpha = 1, \ldots, N \quad \text{and} \quad \beta = 1, \ldots, N \end{aligned} \quad (11.9.13)$$

This bring about the condition

$$b_{\alpha\alpha}^i > 0, \quad b_{\beta\beta}^i > 0, \quad |b_{\alpha\beta}^i| = |b_{\beta\alpha}^i| < \sqrt{b_{\alpha\alpha}^i b_{\beta\beta}^i} \qquad (11.9.14)$$

For example, suppose that we have two variables $Z_1$ and $Z_2$, and the variograms of $Z_1$, $Z_2$, and $Z_1 Z_2$ all have the same form, which we will call $\gamma$. This means that $M = 1, N = 2$. Thus

$$\gamma_{11}(\mathbf{h}) = b_1\gamma(\mathbf{h}) \qquad\qquad \text{with} \quad b_1 > 0$$

$$\gamma_{22}(\mathbf{h}) = b_2\gamma(\mathbf{h}) \qquad\qquad \text{with} \quad b_2 > 0 \qquad\qquad (11.9.15)$$

$$\gamma_{12}(\mathbf{h}) = \gamma_{21}(\mathbf{h}) = b_{12}\gamma(\mathbf{h})$$

Then the only condition is $b_{12} < \sqrt{b_1, b_2}$.

If the variograms $\gamma_{11}$, $\gamma_{22}$, and $\gamma_{12}$ are each the sum of two components (e.g., a nugget effect plus a variogram $\gamma$, $C[1 - \delta(\mathbf{h})] + b\gamma(\mathbf{h})$, then a similar constraint would apply on the $C$'s.

In practice, experimental variograms and cross-variograms are determined using all pairs of available measurement points, as in Section 11.5.1, using Eq. (11.9.10). Then these experimental variograms are adjusted with a linear combination of the *same* basic variograms, and one makes sure that the constraints (11.9.14) or (11.9,15) are satisfied.

(b) *Nonlinear Model.* Myers (1982) proposes an alternative to the linear model. He shows that, if we define

$$U_{\alpha\beta}(\mathbf{x}) = Z_\alpha(\mathbf{x}) + Z_\beta(\mathbf{x})$$

then the variogram of $U$ is

$$\gamma_{U_{\alpha\beta}}(\mathbf{h}) = \tfrac{1}{2}[\gamma_{\alpha\beta}(\mathbf{h}) - \gamma_\alpha(\mathbf{h}) - \gamma_\beta(\mathbf{h})] \qquad\qquad (11.9.16)$$

One can therefore calculate and fit separately the variograms of $Z_\alpha, Z_\beta$, and $U_{\alpha\beta}$, and calculate $\gamma_{\alpha\beta}$ from (11.9.16).

It is only necessary to verify that

$$|\gamma_{\alpha\beta}(\mathbf{h})| \leq \sqrt{\gamma_\alpha(\mathbf{h})\gamma_\beta(\mathbf{h})}$$

This approach can be extended to the generalized covariances (in the nonstationary case, see Section 11.7; see also Matheron (1973)). Examples of the use of co-kriging in soil science are given by Vauclin *et al.* (1983); Abourifasso and Marino (1984) give examples of co-kriging used to calculate transmissivities and specific capacities in an aquifer.

## 11.10. Stochastic Partial Differential Equations

Stochastic partial differential equations can be used to study groundwater flow in three different cases:

(1)  When the boundary conditions or initial conditions are prescribed as stochastic processes.* This could, for instance, be the case when the water level

---

* A stochastic process is a phenomenon that can be described by one or several dependent or independent random variables.

in a river or the infiltration rate at an outcrop are considered as stochastic processes.

(2) When a source/sink term is a stochastic process (e.g., recharge).

(3) when the coefficients of the equations are stochastic processes. This can be the case for transmissivity, for example, when the uncertainty associated with its estimation suggests that it should be considered as a stochastic process.

In all cases some of the inputs of the flow equation are assumed to be random, i.e., they will be defined by their probability distribution functions or its first moments. The dependent variable of the flow equation, e.g., the head, is then also a random function, and the solution of the stochastic flow equation is then, by definition, the probability distribution function of the dependent variable.

Let us give a few simple examples. Annual recharge in an unconfined aquifer depends directly on the rainfall. If a given annual recharge is prescribed and the flow conditions in the aquifer are known (i.e., boundary conditions, parameters), one can deterministically define the head at any location in the aquifer. For every different annual recharge there will be a different value of the head. If this problem is treated stochastically, given the variability of rainfall, the annual recharge will be considered as a random function and defined by its mean and variance. Can one then directly calculate the mean and variance of the head at each location? This could be very useful, for instance, for predicting the probability of the water table rising above a certain elevation (e.g., the bottom of an underground excavation).

We can take another example involving transmissivities. They are known to be rather variable in an aquifer, generally log-normally distributed and with a variance of $\ln T$ on the order of one or more. Given a set of measurement points of $T$ in an aquifer, kriging gives an optimal *estimation* of $T$ as well as the variance of the estimation error. If one only uses the kriged $T$ map to predict the flow in the aquifer, one completely disregards the residual uncertainty of $T$. It would make much more sense to consider $T$ as a stochastic process and to try to determine the expected value and the variance of the head directly at each point, thereby determining the uncertainty in the predicted flow.

Our last example is one of transport in porous media. Hydrodynamic dispersion is known to be the result of the variability of the pore water velocity in the medium at every scale. Rather than defining a single average velocity in the aquifer and then describing the variability of velocity by an empirical dispersion coefficient, is it not more efficient to characterize the variability of the velocity by its variance and covariance and to represent hydrodynamic dispersion by solving a stochastic transport equation, where this velocity is a stochastic process?

All these approaches have been used recently.

Having defined a few properties of stochastic partial differential equations, we shall give a brief outline of several methods for solving them.

### 11.10.1 *Properties of Stochastic Partial Differential Equations*

For equations where the parameters are random functions one must first define what the derivative of a stochastic process is. If $K$ is a random function, the quadratic mean derivative $K'$ is defined by

$$\lim_{\Delta x \to 0} E\left[ \frac{K(x + \Delta x) - K(x)}{\Delta x} - K'(x) \right] = 0$$

Also, $K'$ is a random function.

The complete solution of a stochastic partial differential equation consists in obtaining all the probability distribution functions, at every location and at all times, of the unknown random function, e.g., the head. This is almost always impossible to achieve. Therefore one will generally look for (1) the probability distribution function of the unknown at several particular locations or (2) the moments of the unknown function: expected value, variance, covariance. These moments can sometimes only be approximately evaluated.

To obtain even such approximate and limited solutions, one must often make some hypotheses on the stochastic processes in question.

(1) If these hypotheses concern the input parameters (e.g., boundary conditions, source terms, coefficients), the corresponding solution is said to be "honest."

(2) If these hypotheses concern the unknown solution, whose form is a priori unknown, the solution is said to be "dishonest." This does not mean that a dishonest solution is necessarily incorrect if these hypotheses (e.g. stationarity,..) are based on valid physical reasoning: dishonest solutions can, on the contrary, sometimes be more precise than honest ones. It is only when the assumptions are not physically based that they may be invalid [see Keller (1964), Lumley and Panofsky (1964), and Schweppe (1973)].

We shall now briefly examine some methods of solution.

### 11.10.2. *Spectral Method*

This method is applicable to second-order stationary stochastic processes for both inputs and outputs. If $Y(x)$ is second-order stationary, the spectrum (or spectral density) of $Y$ is the Fourier transform of its autocovariance function:

$$\varphi(k) = \frac{1}{2\pi} \int_{-\infty}^{+\infty} e^{-iks} \operatorname{cov}[Y(x+s), Y(x)] \, ds \qquad (11.10.1)$$

Using the inverse Fourier transform, one can also write

$$\operatorname{cov}[Y(x+s), Y(x)] = C(s) = \int_{-\infty}^{+\infty} e^{iks} \varphi(k) \, dk \qquad (11.10.2)$$

The following "representation theorem" will be used: if the second-order stationary stochastic process $Y(x)$ is of zero mean $[E(Y) = 0]$ and of covariance $C(s)$, then one can define a complex associated process $Z$ (i.e., $Z \in C$ if $Y \in R$) that satisfies

$$Y(x) = \int_{-\infty}^{+\infty} e^{ikx} \, dZ(k) \qquad (11.10.3)$$

$$\begin{aligned} E[dZ(k_1) \, dZ^*(k_2)] &= 0 & \text{if} & \quad k_1 \neq k_2 \\ E[dZ(k_1) \, dZ^*(k_1)] &= \varphi(k_1) & \text{i.e. if} & \quad k_1 = k_2 \end{aligned} \qquad (11.10.4)$$

Equation (11.10.3) is a Fourier–Stieltjes integral and the asterisk in Eq. (11.10.4) denotes the complex conjugate.

We shall give a simple example of the use of the spectral method, from Gelhar (1976), Bakr et al. (1978), and Gutjahr et al. (1978). Let us consider a one-dimensional steady-state flow in an infinite medium. The flow equation is written as

$$\frac{d}{dx} \left[ K(x) \frac{dH}{dx} \right] = 0 \qquad (11.10.5)$$

We assume that $K(x)$ is a second-order stationary stochastic process, i.e., we shall treat the class 3 type of stochastic partial differential equation. If we integrate Eq. (11.10.5) once, it gives

$$K(x) \frac{dH}{dx} = -q \qquad (11.10.6)$$

where $q$ is the *constant* flow rate in the flow tube. Dividing by $K$ and defining $W = 1/K$,

$$\frac{dH}{dx} = -qW \qquad (11.10.7)$$

Let us define the expected value of $H$ and $W$ and their fluctuation around the average by

$$\begin{aligned} \bar{H} &= E(H) & h &= H - \bar{H} & \text{thus} & \quad E(h) = 0 \\ \bar{W} &= E(W) & w &= W - \bar{W} & \text{thus} & \quad E(w) = 0 \end{aligned}$$

By substituting in Eq. (11.10.7) and taking its expected value, we get

$$\frac{d\bar{H}}{dx} + \frac{dh}{dx} = -q(\bar{W} + w)$$

$$E\left(\frac{dH}{dx}\right) = -qE(W) \qquad \text{thus} \qquad \frac{d\bar{H}}{dx} = -q\bar{W} \qquad (11.10.8)$$

$$\frac{dh}{dx} = -qw \qquad (11.10.9)$$

Assuming $h$ to be second-order stationary ("dishonest" hypothesis) and using the "representation theorem," we can define two complex stochastic processes such as

$$h(x) = \int_{-\infty}^{\infty} e^{ikx}\, dZ_h(k) \qquad w(x) = \int_{-\infty}^{+\infty} e^{ikx}\, dZ_w(k)$$

We then take the first derivative of $h$ and introduce it into Eq. (11.10.9):

$$\frac{dh}{dx} = \int_{-\infty}^{\infty} e^{ikx} ik\, dZ_h(k)$$

$$ik\, dZ_h(k) = -q\, dZ_w(k) \qquad \text{thus} \qquad dZ_h(k) = i\frac{q}{k}\, dZ_w(k)$$

From Eq. (11.10.4) we can calculate the spectrum of $h$:

$$\varphi_h(k) = E[dZ_h(k)\, dZ_h^*(k)]$$

$$= E\left\{\left[i\frac{q}{k} dZ_w(k)\right]\left[-i\frac{q}{k} dZ_w^*(k)\right]\right\} = \frac{q^2}{k^2} E[dZ_w(k)\, dZ_w^*(k)] \qquad (11.10.10)$$

$$\varphi_h(k) = \frac{q^2}{k^2}\, \varphi_w(k)$$

We have now solved our problem. Equation (11.10.8) gives us the first moment of $H$, $\bar{H} = -q\bar{W}x + \text{constant}$, and Eq. (11.10.10) gives us the spectrum of $H$ given the spectrum of $W$. Using Eq. (11.10.2), one can also determine the covariance and variance of $h$ from the spectrum. For instance, if the following covariance is used for $W$ as suggested by Gutjahr et al. (1978),

$$\text{cov}[w(x + s), w(x)] = \sigma_W^2(1 - |s|/l)e^{-|s|/l}$$

where $\sigma_W^2$ is the variance of $w$ and the distance $l$ is called the correlation length, one obtains

$$\varphi_w(k) = \frac{2k^2\sigma_W^2 l^3}{\pi(1 + k^2 l^2)^2}$$

$$\text{cov}[h(x + s), h(x)] = q^2\sigma_W^2 l^2(1 + |s|/l)e^{-|s|/l}$$

$$\sigma_h^2 = C(0) = q^2 l^2 \sigma_W^2$$

In this simple example the covariance and variance of $h$ are constant all over the medium.

Gelhar *et al.* (1974, 1977, 1979a,b), Gelhar and Axness (1983), and Gelhar (1986) have used this spectral method extensively, mainly for the transport equation in their later articles.

### 11.10.3. *The Method of Perturbations*

We shall use the same example as before, i.e., Eq. (11.10.5). Let $K$ be second-order stationary with $E(K) = \bar{K}$ and the "fluctuation" $k = K - \bar{K}$, $E(k) = 0$. We shall also assume that the "fluctuation" $h$ of the solution $H$ is second-order stationary ("dishonest" hypothesis) with $E(H) = \bar{H}$ and $h = H - \bar{H}$, $E(h) = 0$. We develop $K$ and $H$ to the first order, i.e., add to $\bar{K}$ and $\bar{H}$ a "small perturbation," i.e., a fraction of their fluctuation:

$$K = \bar{K} + \beta k \qquad H = \bar{H} + \beta h \qquad (11.10.11)$$

Given $k$, we now look for $h$. We can introduce Eq. (11.10.11) into Eq. (11.10.5) and develop in $\beta$, disregarding the terms in $\beta^2$ (assuming $\beta$ to be small):

$$\bar{K}\frac{d^2\bar{H}}{dx^2} + \beta\left(\bar{K}\frac{d^2h}{dx^2} + \frac{dk}{dx}\frac{d\bar{H}}{dx} + k\frac{d^2\bar{H}}{dx^2}\right) = 0$$

If this is to hold for any small $\beta$, each of these two terms must be equal to zero. Thus:

$$\bar{K}\frac{d^2\bar{H}}{dx^2} = 0 \quad \text{or} \quad \frac{d\bar{H}}{dx} = -q/\bar{K} \quad \text{and} \quad \bar{H} = -qx/\bar{K} + \text{const}$$

and substituting this result in the second term,

$$\frac{d^2h}{dx^2} = \frac{q}{\bar{K}^2}\frac{dk}{dx}$$

or

$$\frac{dh}{dx} = \frac{q}{\bar{K}^2}k + a \qquad (11.10.12)$$

where $a$ is a constant. We take the expected value to be

$$E\left(\frac{dh}{dx}\right) = \frac{d}{dx}E(h) = \frac{q}{\bar{K}^2}E(k) + E(a)$$

As $E(h) = E(k) = 0$, we can see that $E(a) = 0$. Then Eq. (11.10.12) gives directly

$$\text{cov}\left(\frac{dh}{dx}\right) = \frac{q^2}{\bar{K}^4}\text{cov}(k)$$

However, for a stationary random function with a differentiable covariance one can write

$$\text{cov}\left(\frac{dh}{dx}\right) = -\frac{d^2}{ds^2}\text{cov}(h)$$

Thus, if we can assume that

$$\frac{d}{dx}\text{cov}[h(x), h(x + s)]\bigg|_{s \to -\infty} = 0 \quad \text{and} \quad \text{cov}[h(x), h(x + s)]\bigg|_{s \to -\infty} = 0$$

then with two integrations we find

$$\text{cov}[h(x + s), h(x)] = -\frac{q^2}{\bar{K}^4}\int_{-\infty}^{s}\int_{-\infty}^{y}\text{cov}[k(x), k(x + u)]\,du\,dy$$

We have again found the expected value and covariance of the head, but this time we must assume that $\sigma_K^2$ is small, otherwise the first-order development in $\beta$ is not valid. To overcome this difficulty in the case of a permeability where $\sigma_K^2$ is generally rather large (see Section 4.4), Gelhar has suggested the use of the logarithm of $K$. Equation (11.10.5) is written as

$$K\frac{d^2H}{dx^2} + \frac{dK}{dx}\frac{dH}{dx} = 0 \quad \text{or} \quad \frac{d^2H}{dx^2} + \frac{d}{dx}(\ln K)\frac{dH}{dx} = 0$$

If $F = \ln K$ is second-order stationary, one again writes $F = \bar{F} + \beta f$ and $h$ is expressed as a function of the covariance of $f$; Gelhar has shown that in one dimension the error involved in the method of perturbations is less than 10% if $\sigma_F^2 \leq 1$ (by comparing it to the exact spectral method).

Tang and Pinder (1977) have used the method of perturbations for the transport equation. Sagar (1978) applied it to the flow equation. Gelhar and Axness (1983) have used it for the same equations together with the spectral method. Winter et al. (1984) applied it in the second order to the transport equation.

### 11.10.4. Simulation Method (Monte Carlo)

This is probably the most powerful method, where fewer assumptions are required. However, it is a numerical method, which may require much central processing unit (CPU) time and a careful examination of the results. The principle of the method is very simple. Let $Z(\mathbf{x}, \xi)$ be a stochastic process, $\mathbf{x}$ being the coordinates in space and $\xi$ the state variable. Remember that $Z(\mathbf{x}, \xi_1)$ is called a realization of $Z$. One first generates "simulations" of $Z$ in the probabilistic sense, i.e., a large number of realizations of $Z$. To do so, we must know the probability distribution function of $Z$ and its covariance (or variogram) if $Z$ is spacially correlated.

Note that the knowledge of the probability distribution function of $Z$ was not necessary in the two previous methods.

Then, for each of these realizations the parameter represented by $Z(\mathbf{x}, \xi_i)$ is completely determined and known (e.g., the permeability or the source term or the boundary conditions). Thus, the flow equation can be solved numerically for each realization, giving the value of the dependent variable, e.g., $h(\mathbf{x}, \xi_i)$. It is then possible to statistically analyze the ensemble of calculated solutions $h(\mathbf{x}, \xi_i)$ for $i = 1, \ldots, N$: expected value, variance, histogram, and distribution function for each location $\mathbf{x}$. It is no longer necessary to assume that $h$ is stationary; these statistics can be calculated at each point. The covariance or variogram can also be determined if $h$ is found to be stationary or intrinsic.

There are some difficulties associated with the simulation method. First, a large number $N$ of realizations is necessary in order to get meaningful statistics: from 50 to several hundreds or thousands. Second, as $N$ is necessarily finite, one can always calculate an experimental variance or covariance, even for a phenomenon where they do not exist. It is preferable to check that when $N$ increases, these statistics indeed become constant. Third, the solution can be a function of the mesh size: because the numerical solution requires us to estimate an average of $Z(\mathbf{x}, \xi_i)$ over a mesh, this estimate becomes less variable as the mesh becomes larger simply because of the integration. Thus, the variability of the solution $\mathbf{h}(\mathbf{x}, \xi)$ will also be affected. Furthermore, one must realize that if $C$ (or $\gamma$) is the correlation structure of $Z$ in space, then the correlation structure of the average of $Z$ over a mesh will be the integrated covariance or variogram (see Section 11.4.7). This has not always been recognized in the past.

The main difficulty with the simulation method is how to generate the realizations $Z(\mathbf{x}, \xi_i)$. Freeze (1975) assumed that $Z$ (in this case, the hydraulic conductivity on a one-dimensional flow problem) was not spatially correlated. When the probability distribution of $Z$ was known, independent values were drawn randomly in each mesh. In two dimensions, Smith and Freeze (1979) and Smith and Schwartz (1980, 1981a,b) imposed a correlation structure on $Z$ (the hydraulic conductivity) using the method of the "nearest neighbor." The correlation is imposed by a kind of "moving average" of the value of $Z$, taken in adjacent meshes. Binsariti (1980) generated the complete covariance matrix of $Z$ and took a vector of independent random numbers and solved for the correlated $Z(\mathbf{x}, \xi_i)$ by triangulation of the covariance matrix using Cholesky's method (Section 12.4.1.b). See also Neuman (1984). Meija and Rodriguez-Iturbe (1974) used spectral methods.

Delhomme (1979) used the method of the turning bands, developed by Matheron (1973), which is a very powerful tool in two dimensions (see also Chiles (1977) and Mantoglou and Wilson (1982)). Delhomme also used conditional simulations of $Z$ instead of simple simulations. This is a great improvement on the Monte Carlo method for practical problems. Indeed, the stochastic process $Z$ is then said to be conditioned by the measurements $Z(x_j)$ in space: all the realizations $Z(x, \xi_i)$ must have the measured values $Z(x_j)$ at each point $x_j$, where a measurement has been made. The method used to generate these conditional simulations is based on kriging.

Nonconditional simulations are suitable for studying the theoretical variability of a process: the statistics of $Z$ are assumed to be known, but no measured values are available. On the contrary, conditional simulations take the measured values into account and the considered variability is only that stemming from the uncertainty in the estimation of $Z$ between measurement points. Conditional simulations are thus a logical follow-up to kriging. Delhomme (1979) used them for transmissivities and mentioned that the transmissivity could be further conditioned by the inverse problem (see Section 12.6). Such conditioning is also discussed by Neuman and Yakowitz (1979), Neuman (1984) and Dagan (1982,a,b).

### Note added in Proof

Here, $h$ is by definition a vector in space. In general, the covariance $C$ or variogram $\gamma$ are only functions of the length of this vector. In some cases, however, they also depend on the direction of the vector $h$. To simplify, we shall keep the notations $C(h)$ or $\gamma(h)$ even if $h$ only refers to a length. (See also Section 11.5.4.)

# Chapter 12

# Numerical Solutions of the Flow and Transport Equations

## 12.1. Selection of a Numerical Technique and a Code

It may become necessary to use numerical solutions to the flow and/or transport equations, rather than analytical solutions, for one or more of the following reasons.

(1)   The flow domain is bounded, with boundaries of complex shape that play a role during the time for which a solution is sought. The available analytical solutions deal with infinite or semiinfinite media; the method of images cannot be used or is too complex to use to represent the role of the boundaries.

(2)   The problem is nonlinear (e.g., the transmissivity varies with the head in an unconfined aquifer) and no analytical solutions are available.

(3)   The properties of the medium vary in space, whereas analytical solutions assume the medium to be homogeneous or the geometry of the heterogeneities to be very simple.

(4)   The geometry and magnitude of the source term are too intricate to be represented by a point source, a line source, or an integral of these along a simple path.

(5)   An analytical solution can be found, but its expression is so complex (e.g., sum of infinite series, integral of complex functions) that the numerical calculation of its values requires far more effort (programming and CPU time) than the direct use of a numerical solution of the original problem.

In some of these instances, it may be advantageous to use a semianalytical method that consists in solving the problem first analytically in the Laplace transform domain and then computing the inverse Laplace transform numerically; this may be of interest for a transport equation involving first-order kinetic reactions, for instance, for which the Laplace transform is very suitable (see Talbot, 1979). Another semianalytical method involves the use of Green's functions (see for instance Roach (1982), and Herrera (1985)).

When numerical solutions are required, one must first decide (1) what numerical method to use (essentially, finite differences, finite elements, boundary elements) and (2) how to obtain a code (program it, or get access to an existing one).

There is no universally agreed answer to the first question; for each of the three methods quoted above, one can say the following.

(1)   *Finite differences* are easy to understand and to program; they are very well suited to solving regional aquifer flow problems, in one or two dimensions, in multilayered systems, or in three dimensions. Although they can in principle handle meshes of any shape and size, they are restricted in practice to simple meshes: regular squares, nested squares, rectangles, or

rectangular parallelepipeds in three dimensions. They can handle heterogeneities of the properties of the medium very well, provided that the shape of these heterogeneities can be adequately described by the shape of the meshes; anisotropy must be restricted, for all practical purposes, to directions parallel to the sides of the meshes. They are not very well suited to solving the transport equation, unless the methods of characteristic and particle tracking are used (see Section 12.5.2).

(2) *Finite elements* are less easy to explain and far less easy to program. As this method is more flexible than that of finite differences, a finite-element program can be more complex to use (more input data, e.g., on the geometry of the meshes, thus more possibilities for error) and may require more computer time. However, the shape of the meshes is much less restricted: in practice, triangles and quadrilaterals are used in two dimensions, and tetrahedra or parallelepipeds of any angle in three dimensions. This makes it possible to describe much more satisfactorily the shape of the boundaries of the medium and that of the heterogeneities or the source functions. It also makes finite elements ideally suited to solving problems with moving boundaries, e.g., with a free surface, an abrupt interface between fresh water and sea water, or between two immiscible fluids. Finite elements can handle any directions of anisotropy, and these directions may even change from one element to the next or with time. In practice, for flow problems, finite elements can be used for regional studies but are best suited to local civil engineering problems like dewatering of an excavation, mine drainage, and flow around a dam, where the shape of the boundaries and heterogeneities must be represented with precision. Note that when seepage forces must be calculated as input for structural analysis, it is often necessary to compute them on the same network that will be used for the structural calculations: virtually all of these use finite elements. For solving the transport equation, finite elements are far superior to finite differences, as they can handle the anisotropy of the dispersion tensor and the mesh size can be adapted to the magnitude of the velocity; it is thus possible to seek a compromise between stability and numerical dispersion.

(3) *Boundary elements* or boundary integral methods have been proposed recently for solving the flow equation. Their main advantage is that the precision of the calculations is not a function of the size of the elements used, contrary to what happens with finite differences of finite elements. Thus, a few very large (even infinite) elements can be used, so that the method is very efficient in terms of computer time. In a first step the numerical solution is only calculated along the boundaries of the elements; if the solution is also explicitly required inside an element, its value is calculated in a second step by numerical integration inside the element. The main restriction is that the properties of the medium in a given element are assumed constant: if the heterogeneities of the medium are such that a large number of elements are

required to describe them adequately, then the boundary integral method loses its superiority, and finite differences or finite elements can just as well be used. This method is therefore much less flexible and general than the previous ones [see Brebbia (1978), Liu and Liggett (1979), Liggett and Liu (1979), Tal and Dagan (1983), and Herrera (1984)].

The second question, how to obtain a code, is a matter of personal judgement, and only a few hints can be given here. Programming a simple finite difference code, for a one- or two-dimensional problem with simple meshes (squares or rectangles) and simple boundaries can be done from scratch in a few days. However, the code will not be easy to use; to make it "user-friendly" (i.e., give it, e.g., simple inputs, error checking, and messages, graphical outputs) may require a couple of months. A more complex multilayer or three-dimensional finite-difference code with several options (e.g., nonlinearities) may require 6 months to a year, as does a user-friendly two-dimensional finite-element code. A very complex three-dimensional finite-element transport code may require 1 or 2 years, and a multicomponent, multiphase, and three-dimensional oil-reservoir model may represent an effort of 5–10 (or more) man-years. One must remember that any new code must be carefully tested and validated against known analytical (or other numerical) solutions before it can be used for any serious purpose. This testing can be quite lengthy.

However, a very large number of codes are now available, either free of charge or at the cost of the duplication of a card deck, a tape, or a floppy disk, or even at a cost covering part of the expenditure for their development. In order to make such codes easily available, a clearing-house has been put together (Bachmat *et al.*, 1980).[†] A computer file of available codes in groundwater modeling has been set up for flow, transport, management, data processing, etc., where more than 500 codes are described. A search through this file will reveal the available codes best suited to solving a given problem and adapted to a hand-held calculator, a microcomputer, or a main-frame computer.

In the rest of this chapter we shall briefly describe the methods of finite differences and finite elements, the resolution of large linear systems, and finally, the use of numerical models for regional groundwater flow studies. Even if one decides to use an existing code to solve a groundwater flow or transport problem, it is essential to clearly understand the principle and limitations of numerical models in order to be able to use them efficiently.

---

[†] The International Groundwater Modeling Center at Holcomb Research Institute, Butler University, Indianapolis, Indiana 46208 or at TNO–DGV, Institute of Applied Sciences, P.O. Box 285, 2600 Delft, The Netherlands. Inquiries can be directed to either of these institutes.

## 12.2. Finite Differences

Three different methods, at least, can be used to present finite differences. We will illustrate the first two, the methods of differentials and mass balance, by a simple example and proceed more rigorously with the third one, the method of integrated finite differences. The results will be, for the same equation, identical.

We shall first consider the simple flow equation in a confined aquifer, in two dimensions [Eq. (5.3.10)], which is written as

$$\frac{\partial}{\partial x}\left(T_x \frac{\partial h}{\partial x}\right) + \frac{\partial}{\partial y}\left(T_y \frac{\partial h}{\partial y}\right) = S\frac{\partial h}{\partial t} + q$$

where T is the transmissivity, which can vary in space, (length$^2$ time$^{-1}$), $h$ is the unknown, the head (length); $S$ is the storage coefficient, which can vary in space (dimensionless), and $q$ is the source/sink term, representing at each point the algebraic sum of the density of recharge to or discharge from the aquifer, which also varies in space. It is expressed in flow rate per unit area (length time$^{-1}$) and is positive for a sink and negative for a source.

We look for the solution of this equation on a finite bounded domain with prescribed boundary conditions (see Section 6.3). We shall consider prescribed head boundaries (i.e., Dirichlet) or prescribed fluxes (i.e., Neumann). Any one of these two boundary conditions can be prescribed on different segments of the boundaries. The corresponding values of the prescribed head or fluxes are assumed to be known, as well as the values of T, $S$, and $Q$ in the entire domain. For the first simple examples we shall assume that the equation is solved in steady state (i.e., $\partial h/\partial t = 0$) and that T is isotropic. On the domain where the equation is to be integrated, a square grid is superimposed (Fig. 12.1) with its size governed by the desired precision of the numerical approximation of the true solution (the smaller the size of the squares, the better the approximation). The coordinates $x$ and $y$ in the domain are taken along the sides of the grid.

The principle of finite differences is to look for the numerical value of the heads in each of the centers of the squares, assumed to represent an "average" value of the true head in each square. The squares are numbered from 1 to $r$; $H_1$ to $H_r$ will be the heads at the nodes (the centers of the squares) and $T_1$ to $T_r$, $S_1$ to $S_r$, and $Q_1$ to $Q_r$ the transmissivity, storage coefficient, and source terms in each square, assumed to be the average of $T$ and $S$ and the integral of $q$ over the square, respectively. If a node $i$ falls on a prescribed head boundary, $H_i$ will be known at this node; if the side of a square represents a prescribed flux boundary, the flow entering into the square through that side will be known. Let us look (Fig. 12.2.) at five adjacent nodes inside the grid, which we shall number for the sake of convenience C, N, E, S, W (for center, north, east, south,

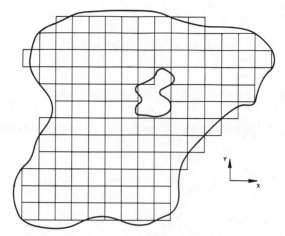

**Fig. 12.1.** Finite-difference square grid on a bounded domain.

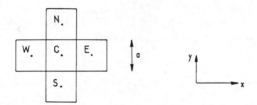

**Fig. 12.2.** Five adjacent nodes of a finte-difference grid.

west), respectively, although in reality they will have numbers falling between 1 and $r$ and depending on the numbering system used, which is generally from west to east and north to south. We shall use three different methods to establish the finite differences approximation of the continuous partial differential equation for flow in a steady state, which is

$$\frac{\partial}{\partial x}\left(T\frac{\partial h}{\partial x}\right) + \frac{\partial}{\partial y}\left(T\frac{\partial h}{\partial y}\right) = q \qquad (12.2.1)$$

We shall not address the problem of consistency óf the finite difference approximation, which consists in showing that when the mesh size tends towards zero the approximate solution $H$ tends towards the true solution $h$. For finite differences and finite elements it can be shown that this is indeed so.

The method is well described by Varga (1962), Remson *et al.* (1971), Thomas (1973), Prickett (1975), Narashiman and Witherspoon (1976), Trescott *et al.* (1976), Mercer and Faust (1981), and Wang and Anderson (1982).

### 12.2.1. Approximation of the Derivatives by Differences

The name "finite differences" has its origin in the fact that derivatives are approximated by differences. If $a$ is the size of a square, we can write

$$\text{approximation of } \frac{\partial h}{\partial x} \text{ between nodes W and C} = \frac{H_c - H_w}{a}$$

$$\text{approximation of } \frac{\partial h}{\partial x} \text{ between nodes C and E} = \frac{H_e - H_c}{a}$$

$$\text{approximation of } \frac{\partial h}{\partial y} \text{ between nodes S and C} = \frac{H_c - H_s}{a}$$

$$\text{approximation of } \frac{\partial h}{\partial y} \text{ between nodes C and N} = \frac{H_n - H_c}{a}$$

We now need to approximate the second-order derivatives or, more precisely, the derivatives like $(\partial/\partial x)(T\,\partial h/\partial x)$. Let $T_{nc}, T_{ec}, \dots$ be the value of the transmissivity evaluated between N and C, E and C, etc. (see Section 12.2.5 for their evaluation from the transmissivity $T_n$, $T_c, \dots$ in each square). We can write, for node C,

$$\text{approximation of } \frac{\partial}{\partial x}\left(T\frac{\partial h}{\partial x}\right) \text{ in C}$$

$$= \left[\left(T\frac{\partial h}{\partial x}\right) \text{ between C and E} - \left(T\frac{\partial h}{\partial x}\right) \text{ between W and C}\right]\Big/a$$

$$= \left[T_{ec}\frac{H_e - H_c}{a} - T_{wc}\frac{H_c - H_w}{a}\right]\Big/a$$

$$= T_{ec}(H_e - H_c)/a^2 + T_{wc}(H_w - H_c)/a^2$$

Similarly, we get

$$\text{approximation of } \frac{\partial}{\partial y}\left(T\frac{\partial h}{\partial y}\right) \text{ in C} = T_{nc}\frac{H_n - H_c}{a^2} + T_{sc}\frac{H_s - H_c}{a^2}$$

Adding these two terms, multiplied by $a^2$, and given Eq. (12.2.1), we obtain

$$T_{nc}(H_n - H_c) + T_{ec}(H_e - H_c) + T_{sc}(H_s - H_c) + T_{wc}(H_w - H_c) = a^2\bar{q} = Q_c$$

$$(12.2.2)$$

where $\bar{q}$ would be the average of the source term $q$ over the square. However, $a^2\bar{q}$ is then equal to $Q_c$, the integral of $q$ over the square.

This is the finite-difference equation for node C of the original partial differential equation. Note that it is a linear equation in $H_i$; if there are $p$ nodes in the grid where the head is not prescribed (i.e., that in $r - p$ nodes lying on the boundaries the head is prescribed), then our problem has $p$ unknowns, and we can write $p$ linear equations similar to Eq. (12.2.2) for these $p$ nodes. The solution in each node is thus obtained by solving a linear system of $p$ equations with $p$ unknowns, which is mathematically trivial. (Section 12.4.). Note that the ordering number of the $p$ unknowns will not, in general, be from 1 to $p$: depending on the numbering system used and the position of the prescribed head boundaries, these numbers will fall between 1 and $r$, the total number of squares.

The finite difference equations for the nodes adjacent to a prescribed flux boundary are slightly different, but we shall examine them later.

### 12.2.2. Mass-Balance Equation

Instead of starting from the partial differential equation (12.2.1), we can establish the finite-difference equation (12.2.2) directly using only Darcy's law and the principle of mass balance. Consider square C in Fig. 12.2. In a steady state the principle of mass balance imposes that the algebraic sum of the mass fluxes crossing each of the four sides of square C be equal to the mass entering or leaving C by recharge or discharge, i.e., the integral of the source/sink term $\rho q$ over square C, where $\rho$ is the mass per unit volume of the fluid. Using Darcy's law, we can evaluate these fluxes directly. We will keep the same notation as in Eq. (12.2.1) and assume that these fluxes are positive when they leave C, while $q$ is positive when it is a sink (discharge), so that the mass balance equation becomes

$$\text{sum of the fluxes of mass} + \text{integral of } \rho q = 0$$

and

$$\text{flux of mass leaving C through one side}$$

$$= (\text{area of side}) \times (\text{velocity}) \times (\text{mass per unit volume of fluid})$$

$$= ae\left(-K\frac{\partial h}{\partial n}\right)\rho$$

where $e$ is the thickness of the aquifer, $K$ the hydraulic conductivity, $n$ the normal to the side directed outwards, $a$ the size of a square, and $T = Ke$. The mass flux leaving C through one side is then given by $a\rho T\, \partial h/\partial n$.

For each side, we then get

$$\text{flux leaving the side between W and C} = -a\rho T_{wc}\frac{H_w - H_c}{a}$$

$$\text{flux leaving the side between N and C} = -a\rho T_{nc}\frac{H_n - H_c}{a}$$

$$\text{flux leaving the side between E and C} = -a\rho T_{ec}\frac{H_e - H_c}{a}$$

$$\text{flux leaving the side between S and C} = -a\rho T_{sc}\frac{H_s - H_c}{a}$$

and the integral of $\rho q$ over square C is $\rho Q_c$ if $\rho$ is constant.

Then, if we write the mass-balance equation and simplify by $\rho$, assumed constant, we finally obtain exactly the same equation as Eq. (12.2.2). This will help us to establish the form of the finite-difference equation for the nodes adjacent to a prescribed flux boundary. Indeed, each of the terms like $+ T_{nc}(H_n - H_c)$ represents a volumetric flux entering the side of square C between N and C. Therefore, if a side of a square is a prescribed flux boundary, it is sufficient to substitute the prescribed value of this flux, calculated along the given side, for the difference expression that one would normally have. For instance (Fig. 12.3), if the side north of C is a prescribed flux boundary, i.e., there is no node north of C, then the finite-difference equation becomes

$$T_{ec}(H_e - H_c) + T_{sc}(H_s - H_c) + T_{wc}(H_w - H_c) - F_n = Q_c$$

or

$$T_{ec}(H_e - H_c) + T_{sc}(H_s - H_c) + T_{wc}(H_w - H_c) = Q_c + F_n$$

where $F_n$ is the prescribed flowrate (length$^3$ time$^{-1}$) crossing the side north of C, i.e., the integral of the prescribed flux over the side, and $F$ is counted as positive when leaving the domain. Such an equation is still a linear equation in $H_i$, but with one unknown less than the usual equation. Similarly, if one of the heads, e.g., $H_e$, in a finite-difference equation is a prescribed value (representing a prescribed head boundary condition), then the term $T_{ec}H_e$ is known and is transferred to the right-hand side of the equation, leaving only the unknowns

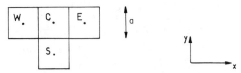

**Fig. 12.3.** Prescribed flux boundary.

on the left-hand side. This is, in general, of no consequence for solving the complete linear system of $p$ equations with $p$ unknowns. It can be shown, indeed, that the matrix of this linear system is always regular (i.e., can be inverted and has a unique solution), provided that for a steady state, there is at least one mesh in the domain where a prescribed head boundary condition is imposed.

### 12.2.3. *Integrated Finite Differences*

This is a more rigorous method of establishing finite, difference equations. To be more general, we shall now assume the transmissivity of the medium to be anisotropic with $x$ and $y$ as the principal directions of anisotropy and the mesh to be formed of polygons of any shape or number of sides. Let $D_i$ be one of these polygons with I its center (or node) and J and K the nodes of two adjacent polygons (Fig. 12.4). The exact definition of the "center" of a polygon (e.g., its center of gravity) is of no importance at this time; we shall give examples later.

In the entire domain, and therefore also in $D_i$, the partial differential flow equation

$$\frac{\partial}{\partial x}\left(T_x \frac{\partial h}{\partial x}\right) + \frac{\partial}{\partial y}\left(T_y \frac{\partial h}{\partial y}\right) = q$$

should be satisfied at every point $(x, y)$. The principle of integrated finite differences is that only the integral of this equation over each of the polygons $D_i$ must be satisfied. We write

$$\iint_{D_i}\left[\frac{\partial}{\partial x}\left(T_x \frac{\partial h}{\partial x}\right) + \frac{\partial}{\partial y}\left(T_y \frac{\partial h}{\partial y}\right)\right] dx\, dy = \iint_{D_i} q\, dx\, dy \qquad i = 1,\ldots,p$$

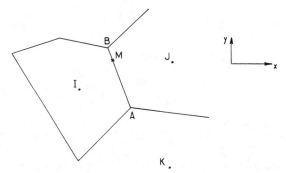

**Fig. 12.4.** A polygon for an integrated finite-difference approximation.

where $p$ is the number of polygons where the head is unknown (i.e., not prescribed). In other words, the partial differential equation no longer has to be satisfied at every location but only in the average over each polygon of the grid. Using very simple mathematics, like Taylor's series expansions, we shall now establish rigorously the general form of a finite difference equation. With Green's first identity, the spatial integral over $D_i$ is first transformed into a contour integral over the perimeter $\Gamma_i$ of $D_i$; if $A$ is any vector, we can write

$$\iint_{D_i} \text{div } A \, dx \, dy = \int_{\Gamma_i} (A, n) \, ds$$

where $n$ is the outer normal to $\Gamma_i$ and $(A, n)$ the scalar product. Here, we obtain

$$\int_{\Gamma_i} \left[ T_x \frac{\partial h}{\partial x} n_x + T_y \frac{\partial h}{\partial y} n_y \right] ds = \iint_{D_i} q \, dx \, dy \qquad (12.2.3)$$

where $n_x$ and $n_y$ are the direction cosines of $n$, and $ds$ is an element of $\Gamma_i$.

Let us evaluate this integral over one side, AB, of $D_i$ (Fig. 12.4). Note that it represents, by definition, the flowrate exchanged between polygons $D_i$ and $D_j$ across AB. To evaluate the derivatives $\partial h/\partial x$, $\partial h/\partial y$ along AB, we shall need three adjacent nodes, I, J, and K, in general. Only in simple cases will the two nodes I and J be sufficient, as we shall see later. We shall also show how to select node K, when necessary. It is only important now to realize that the same node K must be selected when the flowrate along AB is evaluated for the benefit of the equation of polygons $D_i$ or $D_j$. Otherwise, mass balance would not be conserved in the entire domain.

Let $h_i$, $h_j$, and $h_k$ be the actual heads at nodes I, J, and K, and M be any point of AB. Using Taylor's first-order series expansion, we can write

$$h_i = h_m + (x_i - x_m) \left( \frac{\partial h}{\partial x} \right)_m + (y_i - y_m) \left( \frac{\partial h}{\partial y} \right)_m$$

$$h_j = h_m + (x_j - x_m) \left( \frac{\partial h}{\partial x} \right)_m + (y_j - y_m) \left( \frac{\partial h}{\partial y} \right)_m$$

$$h_k = h_m + (x_k - x_m) \left( \frac{\partial h}{\partial x} \right)_m + (y_k - y_m) \left( \frac{\partial h}{\partial y} \right)_m$$

Assuming $h_i$, $h_j$, $h_k$ to be known, this is a linear system with three unknowns, $h_m$, $(\partial h/\partial x)_m$ and $(\partial h/\partial y)_m$, which can easily be solved.

If we further assume that the actual heads $h_i$, $h_j$, and $h_k$ can be approximated by the finite difference values $H_i$, $H_j$, and $H_k$ which will be evaluated at the nodes I, J, and K, we get

$$\left(\frac{\partial h}{\partial x}\right)_m = \frac{(H_j - H_i)(y_k - y_i) - (H_k - H_i)(y_j - y_i)}{(x_j - x_i)(y_k - y_i) - (x_k - x_i)(y_j - y_i)}$$ (12.2.4)

$$\left(\frac{\partial h}{\partial y}\right)_m = -\frac{(H_j - H_i)(x_k - x_i) - (H_k - H_i)(x_j - x_i)}{(x_j - x_i)(y_k - y_i) - (x_k - x_i)(y_j - y_i)}$$

Note that these expressions are not functions of the coordinates of M and thus are constant along AB. As the direction cosines are also constant along AB, we can write

$$\int_{AB}\left(T_x\frac{\partial h}{\partial x}n_x + T_y\frac{\partial h}{\partial y}n_y\right)ds = \left(\frac{\partial h}{\partial x}\right)_m n_x \int_{AB} T_x\,ds + \left(\frac{\partial h}{\partial y}\right)_m n_y \int_{AB} T_y\,ds$$

Let $T_{xab}$ and $T_{yab}$ be the integrals of the directional transmissivities $T_x$ and $T_y$ along AB. Then we obtain

$$\int_{AB}\left(T_x\frac{\partial h}{\partial x}n_x + T_y\frac{\partial h}{\partial y}n_y\right)ds = C_{ij}(H_j - H_i) + C_{ik}(H_k - H_i)$$

where $C_{ij}$ and $C_{ik}$ are functions only of the geometry and transmissivities:

$$C_{ij} = \frac{T_{xab}n_x(y_k - y_i) - T_{yab}n_y(x_k - x_i)}{(x_j - x_i)(y_k - y_i) - (x_k - x_i)(y_j - y_i)}$$ (12.2.5)

$$C_{ik} = \frac{-T_{xab}n_x(y_j - y_i) + T_{yab}n_y(x_j - x_i)}{(x_j - x_i)(y_k - y_i) - (x_k - x_i)(y_j - y_i)}$$

Similar expressions would be obtained for the other sides of polygon $D_i$. If we finally define the integral source/sink term over the polygon by

$$Q_i = \iint_{D_i} q\,dx\,dy$$

then the finite difference equation, for each polygon, would have the form

$$C_{ij}(H_j - H_i) + C_{ik}(H_k - H_i) + C_{il}(H_l - H_i) + \cdots = Q_i$$ (12.2.6)

which is of the same nature as Eq. (12.2.2).

*Boundary conditions.* If a node of the grid falls on a prescribed head boundary, then the corresponding $H_j$ is known. A finite difference equation is not written for this node, and each time $H_j$ appears in another equation, the term $C_{ij}H_j$ is transferred to the right-hand side. If the side of a polygon falls on a prescribed flux boundary, then this prescribed flux is substituted for the corresponding term of the contour integral and then transferred to the right-hand side of the equation. More precisely, if $\partial h/\partial n$ is prescribed along AB, then

$$\int_{AB} \left( T_x \frac{\partial h}{\partial x} n_x + T_y \frac{\partial h}{\partial y} n_y \right) ds = \int_{AB} (T_x n_x^2 + T_y n_y^2) \frac{\partial h}{\partial n} ds$$

which can be evaluated. Note that, when a grid is designed over a domain in finite differences, its nodes should fall on the prescribed head boundaries and its sides on the prescribed flux boundaries.

### 12.2.4. Integrated Finite Differences: Special Cases

(a)  *Rectangles or squares.*   If the principle directions of anisotropy of the transmissivity $x$ and $y$ are parallel to the sides of the grid, then the contour integral in Eq. (12.2.3) can be written (see Fig. 12.5) as

$$\int_{\Gamma_i} \left( T_x \frac{\partial h}{\partial x} n_x + T_y \frac{\partial h}{\partial y} n_y \right) ds = \int_{AB} T_x \frac{\partial h}{\partial x} dy + \int_{BB'} T_y \frac{\partial h}{\partial y} dx$$

$$+ \int_{B'A'} T_x \frac{\partial h}{\partial x} dy + \int_{A'A} T_y \frac{\partial h}{\partial y} dx$$

Then, if the nodes are the centers of the rectangles (intersection of the diagonals), Taylor's first-order series expansion can be written with only two adjacent nodes and gives, for instance, for $M \in AB$:

$$\left( \frac{\partial h}{\partial x} \right)_m = \frac{H_e - H_c}{x_e - x_c}$$

If we denote by

$$T_{xec} = \frac{1}{|AB|} \int_{AB} T_x ds$$

the average directional transmissivity along AB, where $|AB|$ is the length of segment AB, then

$$\int_{AB} T_x \frac{\partial h}{\partial x} dy = T_{xec} \frac{y_b - y_a}{x_e - x_c} (H_e - H_c)$$

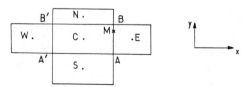

**Fig. 12.5.** Rectangular grid.

and similarly, we obtain in the end

$$T_{ync}\frac{x_b - x_{b'}}{y_n - y_c}(H_n - H_c) + T_{xec}\frac{y_b - y_a}{x_e - x_c}(H_e - H_c) + T_{ysc}\frac{x_a - x_{a'}}{y_s - y_c}(H_s - H_c)$$

$$+ T_{xwc}\frac{y_{b'} - y_{a'}}{x_w - x_c}(H_w - H_c) = Q_c \qquad (12.2.7)$$

If all these rectangles have the same size ($a$ along $x$, $b$ along $y$), Eq. (12.2.7) reduces to

$$T_{ync}\frac{a}{b}(H_n - H_c) + T_{xec}\frac{b}{a}(H_e - H_c) + T_{ysc}\frac{a}{b}(H_s - H_c) + T_{xwc}\frac{b}{a}(H_w - H_c) = Q_c$$

If the medium is isotropic ($T_x = T_y = T$) and the grid is made up of squares ($b = a$), this expression reduces to Eq. (12.2.2), which we have established earlier with the two simpler methods. In the case of rectangles it is important to note the ratio of the coefficients of two unknowns in different directions, e.g., $H_n$ and $H_e$ in the linear system. It is

$$\frac{T_{ync}}{T_{xec}}\left(\frac{a}{b}\right)^2$$

If the anisotropy ratio is close to 1 and $a/b$ is close to 10, the ratio of two coefficients of the matrix of the linear system will be close to 100. Depending on the method used to invert the matrix, its size, and the accuracy of the computer (16-, 32-, or 60-bit words), it is often found that such a large ratio creates a numerical difficulty by round-off errors. The calculated solution may then be unreliable. With 32-bit words a ratio $a/b$ of 5 is often a maximum.

(b) *Nested squares.* It is often necessary to have more precision in one area of the domain than in others, which means that one should be able to vary the mesh size in the grid. One way to do it is to use rectangles of variable size (Fig. 12.6a), but this procedure increases the total number of meshes unnecessarily and is limited by the $a/b$ ratio quoted above. Another way is to use nested square meshes (Fig. 12.6b). Provided that two adjacent squares have at most a factor of 2 of difference in size, there is no limitation to the relative size of the smallest and largest squares. When two adjacent squares are of the same size, the general expression [e.g., $T_{ync}(H_n - H_c)$] is used. When adjacent squares are of different size (Fig. 12.7), the three-node Taylor's series expansion of Eq. (12.2.4) must be used to evaluate the contour integral. If $a$ is the size of the large square in Fig. 12.7, we find from Eq. (12.2.4) that

$$\left(\frac{\partial h}{\partial x}\right)_m = \frac{2}{3a}[(H_j - H_i) + (H_k - H_i)]$$

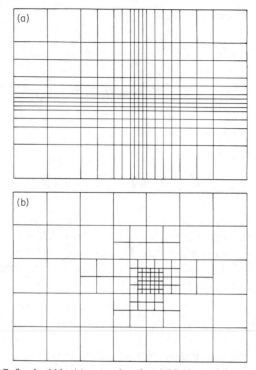

**Fig. 12.6.** Refined grid by (a) rectangles of variable size and (b) nested squares.

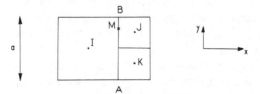

**Fig. 12.7.** Adjacent nested squares.

and as, along AB, $n_x = 1$ and $n_y = 0$,

$$\int_{AB} \left( T_x \frac{\partial h}{\partial x} n_x + T_y \frac{\partial h}{\partial y} n_y \right) ds = \frac{2 T_{xijk}}{3} [(H_j - H_i) + (H_k - H_i)]$$

where

$$T_{xijk} = \frac{1}{a} \int_{AB} T_x \, ds$$

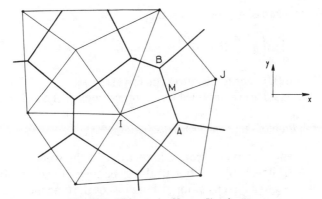

**Fig. 12.8.** Thiessen (or Voronoï) polygons.

is the average directional transmissivity along AB. As the fluxes crossing AB must be identical when evaluated in meshes I, J or K, the above expression is decomposed into $(2T_{xijk}/3)(H_j - H_i)$ for exchanges between I and J, and into $(2T_{xijk}/3)(H_k - H_i)$ for exchanges between I and K.

Note that the ratio of two coefficients of the matrix of the linear system in an isotropic medium is at most $\frac{2}{3}$ and independent of the size of the squares: there is indeed no limitation in the mesh size for nested squares.

(c) *Thiessen polygons.* One of the first finite-difference models built (Tyson and Weber, 1964) used as a grid Voronoï polygons, also known as Thiessen polygons (Fig. 12.8). Given a set of nodes, which can be selected arbitrarily, Voronoï polygons are built as the union of the mediators of each segment successively joining all adjacent nodes. For such polygons, and only if the medium is isotropic ($T_x = T_y = T$), the contour integral can be evaluated with a Taylor series expansion limited to two adjacent nodes:

$$\int_{AB} \left( T\frac{\partial h}{\partial x}n_x + T\frac{\partial h}{\partial y}n_y \right) ds = \int_{AB} T\frac{\partial h}{\partial n} ds$$

if $n$ is the outer normal to AB. But

$$\left( \frac{\partial h}{\partial n} \right)_m = \frac{H_j - H_i}{|IJ|} \qquad M \in AB$$

and, if we use

$$T_{ij} = \frac{1}{|AB|} \int_{AB} T\,ds,$$

the average transmissivity along AB, then

$$\int_{AB} \left( T\frac{\partial h}{\partial x}n_x + T\frac{\partial h}{\partial y}n_y \right) ds = T_{ij}\frac{|AB|}{|IJ|}(H_j - H_i)$$

where $|AB|$ and $|IJ|$ refer to distances (positive).

Similar expressions are obtained for all other sides of the polygons.

### 12.2.5. Estimation of Average Transmissivities

Finite-difference equations require knowledge of the value of the average transmissivity (isotropic or directional) along the sides of each mesh. In some models these are given directly as input, but most of the time the input data are the average transmissivities (isotropic or directional) over the area of each mesh. These can, for example, be obtained by kriging (Section 11.4.7) from the local measurements obtained by pumping tests. We shall also see in Section 12.6 that these transmissivities are often adjusted by calibration of the model.

To calculate the required contour transmissivities, we shall limit the discussion to the special cases (rectangles or squares in anisotropic media as well as Thiessen polygons in isotropic media), but it could be extended to the general case.

(a)  *Rectangles or squares (Fig. 12.5).*  We had to evaluate integrals such as $\int_{AB} T_x \partial h/\partial x\, ds$ when we used Taylor's expansion between C and E to write

$$\left( \frac{\partial h}{\partial x} \right)_m = \frac{H_e - H_c}{x_e - x_c}$$

This derivative was assumed uniform between C and E, but if there is a constant transmissivity $T_{xc}$ in C, and $T_{xe}$ in E, the gradient can no longer be uniform between these two blocks that are in series. By virtue of mass balance, we can write along the interface AB (see Section 6.3.b)

$$T_{xc}\left( \frac{\partial h}{\partial x} \right)_m^c = T_{xe}\left( \frac{\partial h}{\partial x} \right)_m^e \quad \text{and} \quad \left( \frac{\partial h}{\partial y} \right)_m^c = \left( \frac{\partial h}{\partial y} \right)_m^e$$

where the upper indices c and e mean that the derivatives are evaluated in the medium respectively to the left or to the right of AB. By also writing a Taylor expansion separately in each medium, we obtain

$$h_c = h_m + (x_c - x_m)\left( \frac{\partial h}{\partial x} \right)_m^c + (y_c - y_m)\left( \frac{\partial h}{\partial y} \right)_m^c$$

$$h_e = h_m + (x_e - x_m)\left( \frac{\partial h}{\partial x} \right)_m^e + (y_e - y_m)\left( \frac{\partial h}{\partial y} \right)_m^e$$

Thus

$$(h_e - h_c) = (x_e - x_k)\left(\frac{\partial h}{\partial x}\right)_m^e - (x_c - x_k)\left(\frac{\partial h}{\partial x}\right)_m^c$$

$$= \left(\frac{\partial h}{\partial x}\right)_m^c \times [(x_e - x_k)\frac{T_{xc}}{T_{xe}} - (x_c - x_e)]$$

where $x_k = x_m$ is the coordinate along $x$ of AB.

Assuming again that $h_e = H_e$ and $h_c = H_c$, we get

$$\left(\frac{\partial h}{\partial x}\right)_m^c = \frac{T_{xe}}{T_{xc}(x_e - x_k) + T_{xe}(x_k - x_c)}(H_e - H_c)$$

Let us evaluate the contour integral along AB in the homogeneous medium in C (the same result would be obtained if we evaluated it in E); then $T_x = T_{xc}$ and

$$\int_{AB} T_x \frac{\partial h}{\partial x} dy = T_{xc}\left(\frac{\partial h}{\partial x}\right)_m^c (y_b - y_a)$$

$$= \frac{T_{xc}T_{xe}(y_b - y_a)}{T_{xc}(x_e - x_k) + T_{xe}(x_k - x_c)}(H_e - H_c)$$

If we compare this expression with that given in Eq. (12.2.7), we see that the "average" transmissivity defined earlier is the harmonic mean:

$$T_{xec} = \frac{T_{xc}T_{xe}(x_e - x_c)}{T_{xc}(x_e - x_k) + T_{xe}(x_k - x_c)}$$

The same expression could be established for all directions.

For squares or rectangles, all of the same size, $K$ is the middle of CE and

$$T_{xec} = \frac{2T_{xc}T_{xe}}{T_{xc} + T_{xe}}$$

$$T_{ync} = \frac{2T_{yc}T_{yn}}{T_{yc} + T_{yn}} \qquad \text{etc.}$$

For isotropic media these expressions hold when we substitute $T$ for $T_x$ or $T_y$, for example

$$T_{sc} = \frac{2T_s T_c}{T_s + T_c}$$

(b) *Nested squares (Fig. 12.7).* A similar calculation between squares I and J and I and K gives:

$$\int_{AB} \left( T_x \frac{\partial h}{\partial x} \right) dy = \frac{4 T_{xi} T_{xj} T_{xk}}{T_{xi} T_{xj} + T_{xi} T_{xk} + 4 T_{xj} T_{xk}} (H_j - H_i + H_k - H_i)$$

A slightly simpler approximate expression, where the fluxes between each of the squares are decomposed, is

$$\int_{AB} \left( T_x \frac{\partial h}{\partial x} \right) dy = \frac{2 T_{xi} T_{xj}}{T_{xi} + 2 T_{xj}} (H_j - H_i) + \frac{2 T_{xi} T_{xk}}{T_{xi} + 2 T_{xk}} (H_k - H_i)$$

(c) *Theissen polygons* (*Fig.* 12.8). Similarly, with an isotropic medium, one finds

$$\int_{AB} T \left( \frac{\partial h}{\partial x} n_x + \frac{\partial h}{\partial y} n_y \right) ds = \frac{|AB| T_i T_j}{|IM| T_j + |MJ| T_i} (H_j - H_i)$$

where $|AB|$, $|IM|$, and $|MJ|$ are distances.

### 12.2.6. *Finite Differences in a Transient State*

So far we have only used the steady-state flow equation, (12.2.1). If we want to solve the transient-flow equation, this only adds one term to the right-hand side of the equation:

$$\frac{\partial}{\partial x} \left( T_x \frac{\partial h}{\partial x} \right) + \frac{\partial}{\partial y} \left( T_y \frac{\partial h}{\partial y} \right) = S \frac{\partial h}{\partial t} + q$$

where $S$ is the storage coefficient in a confined aquifer or the specific yield in an unconfined aquifer. If we use the integrated finite-difference approach (Section 12.2.3), this will add the term

$$\iint_{D_i} S \frac{\partial h}{\partial t} \, dx \, dy$$

to the right-hand side of the finite-difference equation.

If $H_i$ is the finite-difference head at node I, we will assume that $\partial h / \partial t$ over $D_i$ can be approximated by $\partial H_i / \partial t$.

Then we define the average storage coefficient as

$$S_i = \frac{1}{|D_i|} \iint_{D_i} S \, dx \, dy$$

where $|D_i|$ is the area of $D_i$.

The new term to add to the finite difference equation is simply $|D_i| S_i (\partial H_i / \partial t)$.

The general equation, Eq. (12.2.6), would then become

$$C_{ij}(H_j - H_i) + C_{ik}(H_k - H_i) + \cdots = Q_i + |D_i| S_i \frac{\partial H_i}{\partial t} \qquad (12.2.8)$$

For the sake of simplicity we shall now use a matrix notation and write Eq. (12.2.8) as

$$MH = S \frac{\partial H}{\partial t} + Q \quad \text{or} \quad \frac{\partial H}{\partial t} = S^{-1} MH - S^{-1} Q \qquad (12.2.9)$$

If there are $p$ nodes in the grid, where a finite difference equation similar to Eq. (12.2.8) is written, $M$ is a $p \times p$ matrix, the coefficients of which are the $C_{ij}$ or their sum on the diagonal; $S$ is a diagonal $p \times p$ matrix with $|D_i| S_i$ as coefficients on the diagonal; $Q$ is a column vector with $Q_i$ as coefficients; $H$ is a column vector of the $p$ unknowns $H_i$; and $\partial H / \partial t$ is the derivative of $H$, i.e., a column vector of the $p$ derivatives of the unknowns $\partial H_i / \partial t$.

There are two basic methods for solving the differential system of Eq. (12.2.9): the integral and the differential method.

(a) *Integral method.* Direct integration of the differential system leads to the solution

$$H^t = (H^0 - M^{-1} Q) \exp(S^{-1} Mt) + M^{-1} Q \qquad (12.2.10)$$

where the exponential of a matrix is defined as

$$e^A = I + A + \frac{1}{2} A^2 + \cdots + \frac{A^n}{n!} + \cdots$$

and where $I$ is the identity matrix and $H^0$ is the vector of the initial conditions, i.e., vector $H^t$ for $t = 0$; $Q$ is assumed independent of time.

It is then possible to approximate the matrix exponential operator by a matrix polynomial operator and, in principle, to solve Eq. (12.2.10) for any time $t$. In practice, large time steps are used, because of the error involved in the polynomial approximation and also because $Q$ generally varies with time and can only be considered constant for a given time step.

This method was successfully used by Emsellem and Ledoux (1971), but it is not widely used.

(b) *Differential method.* This is commonly used and consists in approximating the time derivative by a finite difference $\partial H / \partial t = (H^{t+\Delta t} - H^t)/\Delta t$, where the upper indices represent the time at which vector $H$ is considered and $\Delta t$ is the time step of the approximation. This is, in fact, a first-order Taylor series expansion and can be written formally in three different ways.

(1) *Explicit approximation.*  Taylor's expansion is written

$$H^{t+\Delta t} = H^t + \Delta t \frac{\partial H^t}{\partial t} + \frac{\Delta t^2}{2} \frac{\partial^2 H^t}{\partial t^2} + \cdots \qquad (12.2.11)$$

To the first order, and taking into account Eq. (12.2.9),

$$\frac{H^{t+\Delta t} - H^t}{\Delta t} = \frac{\partial H^t}{\partial t} = S^{-1} M H^t - S^{-1} Q^t$$

Note that we have written Eq. (12.2.9) at time $t$ for both sides of the equation. Rearranging, we have

$$H^{t+\Delta t} = H^t + \Delta t (S^{-1} M H^t - S^{-1} Q^t) \qquad (12.2.12)$$

Given $H^t$, $H^{t+\Delta t}$ is thus obtained *explicitly*, i.e., simply by multiplying vector $H^t$ by a matrix and adding a few terms. When we look at the implicit approximation, the simplicity of Eq. (12.2.12) will become clear. Note that $S^{-1}$, the inverse of a diagonal matrix, is simply a diagonal matrix having $1/|D_i|S_i$ as coefficient on the diagonal. The solution of Eq. (12.2.9) is thus obtained time step by time step. Given the initial conditions $H^0$, $H^1$ is calculated, then $H^2$ etc. The length of the time steps $\Delta t$ may vary, as well as the source/sink term $Q^t$, during the simulation.

In general, small time steps are used at the beginning of a simulation or each time $Q^t$ changes significantly (see Section 12.6).

There is, however, a limitation on the magnitude of the time step. If $\Delta t > \Delta t_c$, which is called the critical time step, the explicit approximation becomes unstable. This can easily be understood from Eq. (12.2.12). If, at time $t$, a small approximation error $\varepsilon^t$ was made in the evaluation of $H^t$, then at time $t + \Delta t$ this error is multiplied by $(\Delta t\, S^{-1} M)$. If the norm of this matrix is larger than 1, the errors are amplified from one time step to the next and very soon the results are meaningless. Let us write explicitly one equation of the linear system Eq. (12.2.12):

$$H_i^{t+\Delta t} = H_i^t + \frac{\Delta t}{|D_i|S_i} [C_{ij}(H_j^t - H_i^t) + C_{ik}(H_k^t - H_i^t) + \cdots - Q_i^t]$$

As the $C_{ij}$ are positive, it is clear that the largest coefficient of the matrix $(\Delta t\, S^{-1} M)$ is

$$\frac{\Delta t \sum_j C_{ij}}{|D_i|S_i}$$

where the summation over $j$ is extended to all the neighboring nodes of node $i$.

As this coefficient must be smaller than 1 for all equations of the linear system, the critical time step $\Delta t_c$ is

$$\Delta t_c = \min_i \frac{|D_i| S_i}{\sum_j C_{ij}}$$

When the explicit approximation is used, it is first necessary to evaluate $\Delta t_c$ and then to keep $\Delta t < \Delta t_c$. Note that, in general, $\Delta t_c$ depends on the area of the smallest mesh of the grid.

(2) *Implicit approximation.*   Taylor's expansion is written

$$H^t = H^{t+\Delta t} - \Delta t \frac{\partial H^{t+\Delta t}}{\partial t} + \frac{\Delta t^2}{2} \frac{\partial^2 H^{t+\Delta t}}{\partial t^2} + \cdots \qquad (12.2.13)$$

In the same way as before, to the first order and taking Eq. (12.2.9) into account,

$$\frac{H^{t+\Delta t} - H^t}{\Delta t} = \frac{\partial H^{t+\Delta t}}{\partial t} = S^{-1} M H^{t+\Delta t} - S^{-1} Q^{t+\Delta t}$$

Here Eq. (12.2.9) has been written at time $t + \Delta t$ on both sides of the equation. Rearranging,

$$\left(\frac{1}{\Delta t} S - M\right) H^{t+\Delta t} = \frac{1}{\Delta t} S H^t - Q^{t+\Delta t} \qquad (12.2.14)$$

Now the solution of this linear system is no longer straightforward: the matrix $[(1/\Delta t)S - M]$ needs to be inverted (see Section 12.4). Given the initial conditions $H^0$, $H^1$ can be computed by solving Eq. (12.2.14), then $H^2$, etc. But at each time step a linear system must be solved, which takes a considerable amount of computer time compared to the explicit approximation. The advantage of the implicit approximation is that there is no stability criterion: the method is stable for any length of the time step. But, of course, as for any first-order approximation, the shorter the time step, the better the precision (see Section 12.6).

(3) *Crank–Nicholson's approximation.*   Let us subtract Taylor's implicit expansion [Eq. (12.2.13)] from the explicit one [Eq. (12.2.11)]. We find

$$H^{t+\Delta t} - H^t = H^t - H^{t+\Delta t} + \Delta t \left(\frac{\partial H^t}{\partial t} + \frac{\partial H^{t+\Delta t}}{\partial t}\right)$$

$$+ \frac{\Delta t^2}{2}\left(\frac{\partial^2 H^t}{\partial t^2} - \frac{\partial^2 H^{t+\Delta t}}{\partial t^2}\right) + \cdots$$

We see that the second-order terms almost cancel out, so that the first-order approximation is almost correct to the third order; it becomes

$$\frac{H^{t+\Delta t} - H^t}{\Delta t} = \frac{1}{2}\left(\frac{\partial H^t}{\partial t} + \frac{\partial H^{t+\Delta t}}{\partial t}\right)$$

Using Eq. (12.2.9), we could replace $\partial H/\partial t$ by $S^{-1}(MH - Q)$ at each time. In order to gain generality, let us do this substraction again, but this time, multiplying Eq. (12.2.13) by $\alpha$ and Eq. (12.2.11) by $(1 - \alpha)$. We obtain

$$\frac{H^{t+\Delta t} - H^t}{\Delta t} = (1 - \alpha)\frac{\partial H^t}{\partial t} + \alpha\frac{\partial H^{t+\Delta t}}{\partial t}$$

The parameter $\alpha$ can vary between 0 and 1. For $\alpha = 0$, the approximation is fully explicit. For $\alpha = 1$, the approximation is fully implicit. For $\alpha = \frac{1}{2}$, the approximation is called "Crank–Nicholson," but other values of $\alpha$ between 0 and 1 can be used. Using Eq. (12.2.9) and rearranging,

$$\left(\frac{1}{\Delta t}S - \alpha M\right)H^{t+\Delta t} = \left[\frac{1}{\Delta t}S + (1 - \alpha)M\right]H^t - (1 - \alpha)Q^t - \alpha Q^{t+\Delta t}$$

$$(12.2.15)$$

As for the implicit approximation, a linear system needs to be solved at each time step if $\alpha \neq 0$. But it can also be shown that for $\alpha \leq \frac{1}{2}$, the method is unstable for time steps larger than the critical one defined for the explicit approximation. In practice one always uses an $\alpha$ a little larger than 0.5, e.g., 0.55 or 0.6.

Narasimhan and Neuman (1977) have also proposed an explicit–implicit scheme where $\alpha$ varies from one mesh to the next in the domain. If a local stability criterion is met for a given length of the time step, $\alpha$ is set to zero and the equation of that mesh is solved explicitly. Then, for all the meshes where the stability criterion is not met $\alpha$ is set to 1 and the system of equations is solved implicitly.

(4) *Gir's approximation.* So far, the interpolation used for $H$ between $t$ and $t + \Delta t$ has always been linear. Gir's approximation consists in using a parabolic approximation in time:

$$H_i = at^2 + bt + c$$

At each time step, the three coefficients $a$, $b$, and $c$, for each mesh are adjusted by imposing three conditions on the parabola: it passes through the two previous time steps, $H_i^{t-\Delta t}$ and $H_i^t$, and its derivative at time $t + \Delta t$ is equal to that given by Eq. (12.2.9) written at $t + \Delta t$. For instance, if $\Delta t$ is the same for the two consecutive time steps, one finds

$$H^{t+\Delta t} = -\tfrac{1}{3}H^{t-\Delta t} + \tfrac{4}{3}H^t + \tfrac{2}{3}\Delta t(S^{-1}MH^{t+\Delta t} - S^{-1}Q)$$

Rearranging, one would have to solve a linear system at each time step. Gir's approximation is stable for all time steps as well as second-order correct. As one needs to know $H$ for two consecutive time steps, another method (e.g., implicit) must be used for the first time step.

### 12.2.7. Nonlinear Problem

A very common nonlinear problem in groundwater modeling is that of unconfined aquifers. We have seen, in Section 5.1.d, that in the flow equations [Eqs. (5.1.1), (5.1.2), or (5.1.3)] the transmissivity is a function of the saturated thickness of the aquifer, i.e., of the hydraulic head $h$.

This can be handled in numerical models by changing iteratively the value of the coefficients of matrix $M$ in Eq. (12.2.9). This method is explained in Section 12.4.2.3 both for steady and transient states.

One must also check that the elevation of the hydraulic head in an unconfined aquifer model does not fall below the substratum or rise above the ground surface. In the first case the aquifer actually becomes dry and the corresponding mesh in the domain should be taken out, which introduces a no-flow boundary condition. In practice, it is simpler to accept that the model includes a calculated head, which can fall below the substratum, and to give the mesh a *positive* transmissivity, which is very small, e.g., $10^{-2}$ or $10^{-3}$ times the normal value of the transmissivity in the domain. For all practical purposes, this mesh becomes a no-flow boundary, but if infiltration takes place or if the head rises again (in transient conditions), the mesh can again act as a portion of the aquifer. In the second case, when the head rises above the ground surface, this means that an outlet is created (spring, flow in a river, etc.). The corresponding mesh in the model becomes a prescribed head boundary, where the prescribed head is the elevation of the ground (elevation of the spring, the river bed in the mesh, etc.). Then it becomes necessary to check that the flow in this mesh remains an outflow. When this is no longer the case in a transient state, the head falls below the ground surface and the mesh should no longer be a prescribed head boundary. It can only remain a prescribed head boundary if there is enough surface water available in that mesh to ensure an inflow of water into the model (e.g., water coming from upstream in a river). See Section 12.2.10 for a discussion of how a prescribed head boundary is in practice applied to a river.

Another nonlinear problem is that of dewatering in a confined aquifer. As soon as the hydraulic head in a confined aquifer falls below the elevation of the confining bed, (1) the storage coefficient $S$ in the mesh must be changed into the specific yield $\omega_d$ and (2) the transmissivity may become a function of the saturated thickness of the aquifer, i.e., of the head.

Note that the transmissivity in an unconfined aquifer is, by definition,

$$T = \int_{\sigma}^{h} K \, dz$$

where $\sigma$ is the substratum and $h$ the head. The variation of $T$ with $h$ can sometimes be disregarded if the distribution of $K(z)$ is such that highly permeable material lies at the bottom of the aquifer and only low-permeability material lies at the top.

The flow equations in multiphase flow (Section 9.1.e) and in the unsaturated zone (Section 9.2.1) are also highly nonlinear. The method for solving them is described in Section 12.4.2.3.

### 12.2.8. *Multilayered Systems*

In large sedimentary basins one frequently finds a succession of pervious and impervious (or semipervious) layers which form, respectively, aquifers and aquitards or aquicludes. Aquitards are layers where water cannot be withdrawn through wells but which are pervious enough for significant leakage to occur toward the adjacent aquifers. Aquicludes are less pervious layers, for which leakage is insignificant during a pumping test in an adjacent aquifer but through which leakage can be significant over a large regional area [see Javandel and Witherspoon (1969), Neuman and Witherspoon (1969a,b)]. Such systems are called multilayered systems.

Modeling multilayered systems is easy. One makes the assumptions (1) that flow is essentially parallel to the layering in the aquifers, (2) essentially orthogonal to it in the aquitards or aquicludes, and (3) that the leakage flux can be introduced as a source term in the flow equation of the aquifers. This last assumption has actually been demonstrated in Section 5.3.9.

The model will represent each aquifer in the system by a two-dimensional layer of meshes. In order to make the problem tractable, the mesh size is made identical in the two aquifers covering each other. Only for nested square meshes is it feasible to have overlying meshes with a difference in size of one rank (Fig. 12.9).

There are two methods for evaluating the leakage flux between two superimposed meshes through an aquitard.

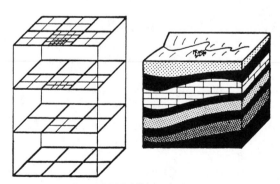

**Fig. 12.9.** Multilayered systems.

*Apply Darcy's law directly* and write the leakage flux as

$$F = -\frac{K'}{e'}(H_b - H_t) \tag{12.2.16}$$

where $H_t$ and $H_b$ are the heads in the top and bottom meshes, respectively. This flux $F$ is then integrated over the mesh size. In finite differences this means multiplying $F$ by the area of the mesh. In finite elements (see Section 12.3), since $H_b$ and $H_t$ vary over the domain of integration $D_i$, this integration is done either after substituting the expression of $H$ in Eq. (12.2.16) as for the "consistent" transient formulation in Eq. (12.3.10) or by using the nodal value of $H_i$ in $D_i$ as for the "lumped" formulation. With Galerkin's formulation, this integration is also weighted by a basis function $N_i$ as in Eq. (12.3.17).

The resulting source term is added to the right-hand side of Eq. (12.2.6) in finite differences, or Eq. (12.3.7) in finite elements, for the system of equations of layer t and subtracted from that of layer b (remember that the source term of the flow equation is negative for a source, positive for a sink).

The resulting source term is then transferred to the left-hand side of the equation, which adds a new unknown to the linear system, for instance,

$$C_{ij}(H_j - H_i) + C_{ik}(H_k - H_i) + \cdots + C_{it}(H_t - H_i) = Q_i$$

where $H_t$ is the head in the mesh at the top of mesh I.

This only increases the size of the linear system that must be solved and makes each layer dependent on the behavior of the others (which is what happens in reality). One can have two such terms simultaneously, one for an underlying and one for an overlying aquifer.

This formulation is strictly valid for a steady state; in a transient state it assumes that the steady-state flux through the aquitard is reached instantly and disregards all storage of water in the aquitards.

To extend its validity, one adds half of the storage coefficient of the aquitard to each of the storage coefficients of the underlying and the overlying aquifers: in this manner, the storage of water in the aquitard is accounted for. But the validity of the assumption of steady-state flow through the aquitard can only be checked by

$$\exp\left(-\frac{\pi^2 K' \Delta t}{S_s' e'^2}\right) \ll 0.5 \tag{12.2.17}$$

where $K'$ is the hydraulic conductivity, $S_s'$ the specific storage coefficient [see Eq. (5.3.8)], $e'$ the thickness, and $\Delta t = $ length of the time step of the transient calculation of the aquitard.

If the assumption of steady-state flow in the aquitard is not valid, then it is possible to use an analytical solution of the one-dimensional flow equation in the aquitard and to evaluate analytically the leakage flux $F$ at the limit of the adjacent aquifers. But this analytical expression involves a convolution and the calculations are rather lengthy:

$$F^t(t) = -\int_0^t \frac{\partial H_b}{\partial t} f^b(t - \tau)\, d\tau + \int_0^t \frac{\partial H_t}{\partial t} f^t(t - \tau)\, d\tau$$

with

$$f(t) = \frac{K'}{e'}\left[1 + 2\sum_{n=1}^{\infty}(\alpha)^n \exp\left(-\frac{n^2\pi^2 K't}{S_s' e'^2}\right)\right]$$

with $\alpha = -1$ for $f^b$ and $+1$ for $f^t$. $F^t$ is the leakage flux in the top layer.

However, the convolution integrals can be calculated by recurrence without the need to store the value of $H$ as a function of time [see Marsily et al. (1978), Trescott et al. (1976), Herrera and Yates (1977) and Hennart et al. (1981). Equation (12.2.17) comes from this analytical solution.

### 12.2.9. Three-Dimensional Systems

In cases where a two-dimensional or multilayered approximation is not valid—i.e., when the true components of the flux in three dimensions have to be evaluated—then a three-dimensional network must be built. In finite differences, cubes or parallepipeds will be used. In finite elements, tetrahedra (linear elements) or hexahedra (bilinear elements) will be used. Just as for multilayered systems, this involves additional terms in the discretized form of the flow equation, representing the fluxes in, e.g., six directions for a cube (north, south, east, west, top and bottom around the central cube). The differences are now that (1) the hydraulic conductivity $K$ and the specific storage coefficient $S_s$ must be used instead of $T$ and $S$ in two-dimensions; (2) the contour integrals of $T$ along the border line of a mesh now become a surface integral of $K$ over the side of the cube; (3) the surface integral of $S$ now becomes a volume integral of $S_s$; and (4) the source term must be defined per unit volume and then integrated inside the volume of each element.

The resulting linear equation, however, is identical to Eq. (12.2.6) or Eq. (12.3.7), but the $C_{ij}$ are given by these surface integrals.

Note that if the medium is anisotropic, the ratio of the horizontal–vertical mesh size must be adjusted so that the resulting $C_{ij}$ are of the same order of magnitude in the three directions. Otherwise the linear system cannot be solved accurately because of round-off errors (see Section 12.2.4).

Using a three-dimensional model is extremely costly since the number of meshes rapidly becomes extremely large. Freeze (1971) has calculated the flow in three-dimensions on a watershed, including both the saturated and unsaturated zone. In practice, three-dimensional modelling is only applied to local problems (e.g., dams, dewatering of an excavation).

It is often much better to study the flow on several two-dimensional cross-sections of the medium rather than on a three-dimensional model. For such modeling one considers a unit thickness of the medium, orthogonal to the plane of the cross-section. Consequently one uses $K$ and $S_s$ instead of $T$ and $S$, which we generally use in the two-dimensional equations.

On such cross-sections it is often necessary to locate the position of the free surface. In a steady state this can be done iteratively by first imposing the condition $h = z$ on a surface chosen *a priori* (see Section 6.3.d) and then checking that the second condition ($K \, \partial h / \partial n$ prescribed) is also fulfilled. If not, the position of the surface is modified. The finite-elements method can handle such problems with a moving surface more accurately than the finite-differences method (see Neuman and Witherspoon (1970, 1971)).

In a transient state it is also possible to move a free surface in the same fashion, but this does not accurately represent the physical processes of drainage/imbibition of an unsaturated medium. It is therefore much better to solve a complete saturated–unsaturated flow problem, using Richard's equation expressed with the head as the unknown (see Section 9.2.1), and then determine the position of the free surface in the domain as the place where $p = 0$ (or $h = z$) [see Freeze (1971)].

### 12.2.10. *Representation of Rivers*

For unconfined aquifers in temperate climates, rivers act as sinks or sources for the aquifer. They can be represented by prescribing the head in each node of the model where a river flows. This head is then the elevation of the water in the river.

In practice, river beds are very often covered by a silt layer, and the flux exchanged between the river and the aquifer creates a difference of head between the two. In Section 6.3.c, we have shown that this may be represented by a Fourier boundary condition. When modeling an aquifer, one therefore prefers to prescribe the head in the river for a mesh overlying the aquifer and linked to it by an exchange coefficient similar to the one used to represent leakage in a steady state through an aquitard,

$$C_{ir} = \frac{K'a'^2}{e'}$$

where $K'$ is the hydraulic conductivity of the silt layer on the river bed, $e'$ the thickness of this low conductivity layer between the river and the acquifer, and $a'^2$ the area of the river bed in contact with the aquifer in the mesh.

The term $C_{ir}(H_r - H_i)$, where $H_r$ is the prescribed head in the river, is then added to the left-hand side of the flow equation, e.g., Eq. (12.2.6). Generally, neither $K'$ nor $e'$ nor $a'^2$ is measured. The coefficient $C_{ir}$ is adjusted so that the difference in head $H_r - H_i$ observed in reality is reproduced by the model when the flow balance is respected (e.g., total flow drained by the river).

Even if there is no significant head difference between river and aquifer, such a representation is still used with a large $C_{ir}$. This makes it very easy to calculate the flow exchanged between the river and the aquifer and (if necessary) to limit this flow to a prescribed figure. This may be necessary when the river recharges the aquifer. If the head in the aquifer is drawn down substantially in the vicinity of a river, it may in reality reach a point where the river and the aquifer are disconnected: either the medium becomes unsaturated or a recharge mound with a vertical gradient of 1 is created beneath the river. In either case the recharge rate is no longer a function of the head in the aquifer and becomes a constant prescribed figure. In groundwater modeling, such a representation of rivers by an overlying mesh with a prescribed head, an exchange coefficient, and a prescribed limiting recharge rate is called a drain with a limiting flux.

In arid zones, where rivers are generally dry and carry water only during floods, the recharge through the river bed is a prescribed flux, which is a function of the total volume of each flood. This recharge may take a very long time to reach the aquifer if the unsaturated zone is thick. Methods for estimating both this flux and the transient time have been suggested by Besbes *et al.* (1978).

### 12.2.11. *Estimation of Regional Recharge*

For an unconfined aquifer the distributed term $q$ essentially represents recharge to the aquifer. In Section 1.3, we showed how infiltration (recharge) can be roughly estimated from rainfall and potential evapotranspiration data, using a simple reservoir model to represent the storage of water in the root zone. Although this type of model is very rough, it it the only one that can be applied in practice to regional groundwater modeling. In more sophisticated reservoir models, the rate at which evapotranspiration withdraws water from the reservoir is made a function of its saturation, and so are infiltration and runoff. In areas where runoff becomes significant it is therefore interesting to be able to calculate simultaneously all the components of the water balance at the ground surface (runoff, infiltration, and evapotranspiration) and then to check these calculations by modeling surface-water flow and comparing the calculated and measured flow at the river gauges. These models are called coupled surface-water–groundwater models. The coupling also takes account of the infiltration into or drainage from the aquifers by the rivers. Such a model has been made by Girard *et al.* (1981) and Ledoux *et al.* (1984) using a nested square mesh both for the aquifer and for the runoff in the surface layer. Such meshes are very appropriate for representing the river network with small elements.

If the unsaturated zone between the ground surface and the aquifer is not too thick (e.g., a few meters), the infiltration beneath the root zone is transferred very rapidly to the aquifer: recharge is equal to infiltration. However, if this unsaturated zone is very thick (e.g., tens of meters) considerable delay and damping are introduced, and recharge at the aquifer surface is, in fact, a "convolution" of the infiltration at the ground surface by a "transfer function" representing the vertical flow through the unsaturated zone. One could, in principle, try to solve the unsaturated flow equation in order to represent this transfer, but this would prove much too complex, costly, and difficult (because of the lack of data on the properties of the unsaturated zone) to be applicable to regional groundwater modeling.

Besbes and Marsily (1984) have shown how a simple linear convolution can be estimated from rainfall and piezometric data. Morel-Seytoux (1984) has also shown how this linear convolution is related to the actual nonlinear unsaturated flow equation.

### 12.2.12. *Representation of Wells*

In regional groundwater modeling, the dimension of a well is generally much too small to be accurately described by the grid of the model (e.g., the diameter of the well may be 0.5 m when the mesh size is 200 × 200 m). In a given mesh there may often be several wells.

In the partial differential equation representing the flow in the aquifer the sink term $q$ for such a well would be

$$q = Q_0 \, \delta(x_0, y_0)$$

where $Q_0$ is the flow rate of the well, $x_0$ and $y_0$ are coordinates of the well, and $\delta$ is the Dirac function at location $(x_0, y_0)$, i.e.,

$$\delta(x, y) = \infty \quad \text{if} \quad (x, y) = (x_0, y_0)$$

$$\delta(x, y) = 0 \quad \text{if} \quad (x, y) \neq (x_0, y_0)$$

$$\iint_D \delta \, dx \, dy = 1 \quad \forall D \subset (x_0, y_0)$$

In the corresponding discretized form of the flow equation the integrated source/sink term is

$$Q_i = \iint_{D_i} q \, dx \, dy = Q_0 \quad \text{if} \quad q = Q_0 \, \delta(x_0, y_0)$$

Thus, $Q_i$ will be the algebraic sum of the integral of the distributed source term (representing recharge) and of the actual flow rates in the various wells in the mesh.

If there is only one well, e.g., in a square mesh, it is still possible to estimate the order of magnitude of the hydraulic head in the well, given the value of the calculated head at the node of the mesh. This calculation is based on Dupuit's steady state expression (Section 7.3.a).

The flow equation for mesh I in a steady state would be written, as in Eq. (12.2.6);

$$\sum_j C_{ij}(H_j - H_i) = Q_i$$

The summation over $j$ may extend both horizontally and vertically in multilayered systems; $Q_i$ is the integrated source term, including the flow rate $q_i$ in the well. If no withdrawal had occured in this well, the computed head $\bar{H}_i$ would have been given by:

$$\sum_j C_{ij}(H_j - \bar{H}_i) = Q_i - q_i$$

This assumes that the head in the adjacent meshes $H_j$ is not modified by the term $q_i$. In other words, $\bar{H}_i - H_i$ is the additional drawdown created by the well, which can be calculated by the numerical model, between the center of mesh I and the adjacent meshes J, i.e., at a distance of $a$ if $a$ is the size of the square mesh. What we want, in fact, is the actual drawdown for a well of radius $r_0$ between this well and the adjacent meshes at a distance $a$. This is estimated by Dupuit's formula as

$$s = \frac{q_i}{2\pi T_i} \ln \frac{a}{r_0}$$

where $T_i$ is the transmissivity in mesh I.

From these two expressions one can derive the head $h_i$ in the well:

$$h_i = H_i - q_i \left( \frac{1}{2\pi T_i} \ln \frac{a}{r_0} - \frac{1}{\sum_j C_{ij}} \right)$$

where $h_i$ is the head in the well bore and $H_i$ the head calculated by the model. This expression is approximate and does not account for quadratic head losses, which must be subtracted if they are significant.

In a transient state the same expression is used, assuming that the logarithmic profile of Dupuit's formula is valid at the scale of the mesh. This is true as soon as Jacob's logarithmic expression can be used at distance $a$.

## 12.3. Finite Elements

The method of finite elements constitutes a very flexible and powerful technique for integrating a partial differential equation over space. It involves three major steps.

(1)   The domain is decomposed into a set of "elements," which in two dimensions generally are triangles or quadrilaterals but can be more complex.

(2)   On each element the unknown function $h(x, y)$ is decomposed on a set of known basis functions $b_k(x, y)$ as

$$h(x, y) = \sum_1^m a_k b_k(x, y)$$

The unknowns are then the $a_k$ coefficients in each element.

(3)   Some kind of integral equation is written to ensure that $h(x, y)$ approximately satisfies the partial differential equation in question or mass balance.

We shall limit this presentation to two examples: linear interpolation on triangles and linear isoparametric elements with the method of Galerkin. A good description of the method can be found in Remson *et al.* (1971), Strang and Fix (1973), Zienkiewicz (1977), Pinder and Gray (1977), Mitchell and Wait (1977), Dhatt and Touzot (1981), and Wang and Anderson (1982).

### 12.3.1.   *Linear Finite Elements on Triangles*

We shall start with the steady-state equation, Eq. (12.2.1). The domain is decomposed into a set of triangular elements. Let IJK (Fig. 12.10) be one of them. On each triangle the unknown function $h(x, y)$ is supposed to be linear:

$$h(x, y) = a_0 + a_1 x + a_2 y \tag{12.3.1}$$

The unknowns are $a_o$, $a_1$, and $a_2$. When writing a balance equation, the first idea would be to use the integrated form [Eq. (12.2.3)] of the flow equation (see Section 12.2.3) on the area of triangle IJK. But this would lead us nowhere, since a linear expression for $h$ such as Eq. (12.3.1) would satisfy $\text{div}(\mathbf{T}\,\text{grad}\,h) = 0$ if the tensor $\mathbf{T}$ is constant over the triangle. Thus, the balance equation can never be satisfied over a linear element unless the source term $q \equiv 0$.

Instead we shall use a polygon surrounding each node I of the grid. These polygons must constitute a partitioning of the domain, i.e., the union of the polygons is equal to the domain itself and their intersection is $\phi$. One usually considers the union of the medians of each triangle (Fig. 12.11). Let $D_i$ be this

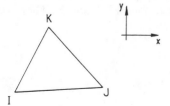

**Fig. 12.10.**  Triangle for linear finite elements.

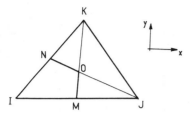

Fig. 12.11. Polygon of integration for a finite-element mesh.

polygon and $\Gamma_i$ its perimeter. The integrated flow equation writes, as in Eq. (12.2.3),

$$\int_{\Gamma_i}\left(T_x\frac{\partial h}{\partial x}n_x + T_y\frac{\partial h}{\partial y}n_y\right)ds = \iint_{D_i} q\,dx\,dy \qquad (12.3.2)$$

We shall assume initially that $x$ and $y$ are the principal directions of the tensor $\mathbf{T}$ in the triangle IJK. Let MO and ON be the two sides of $\Gamma_i$ inside IJK.

We need to calculate the contour integral along MON. This could be done directly, but as $h$ inside IJK is a linear function of the coordinates, we know that $\mathrm{div}(\mathbf{T}\,\mathbf{grad}\,h) = 0$, i.e.,

$$\int_{\mathrm{MON}}\left(T_x\frac{\partial h}{\partial x}n_x + T_y\frac{\partial h}{\partial y}n_y\right)ds + \int_{\mathrm{NIM}}\left(T_x\frac{\partial h}{\partial x}n_x + T_y\frac{\partial h}{\partial y}n_y\right)ds = 0$$

Given Eq. (12.3.1) and the coordinates of I, J, and K, it is simpler to calculate the second integral. Very simple algebra gives

$$\int_{\mathrm{NI}}\left(T_x\frac{\partial h}{\partial x}n_x + T_y\frac{\partial h}{\partial y}n_y\right)ds = \tfrac{1}{2}T_x a_1(y_i - y_k) - \tfrac{1}{2}T_y a_2(x_i - x_k)$$

$$\int_{\mathrm{IM}}\left(T_x\frac{\partial h}{\partial x}n_x + T_y\frac{\partial h}{\partial y}n_y\right)ds = \tfrac{1}{2}T_x a_1(y_j - y_i) - \tfrac{1}{2}T_y a_2(x_j - x_i)$$

Therefore

$$\int_{\mathrm{MON}}\left(T_x\frac{\partial h}{\partial x}n_x + T_y\frac{\partial h}{\partial y}n_y\right)ds = \tfrac{1}{2}T_x a_1(y_j - y_k) - \tfrac{1}{2}T_y a_2(x_j - x_k) \quad (12.3.3)$$

Instead of using the unknowns $a_0$, $a_1$, and $a_2$ in Eq. (12.3.1), one usually prefers to introduce the values of the head at nodes, I, J, and K. They verify:

$$H_i = a_0 + a_1 x_i + a_2 y_i$$
$$H_j = a_0 + a_1 x_j + a_2 y_j \qquad (12.3.4)$$
$$H_k = a_0 + a_1 x_k + a_2 y_k$$

The head at a node I must, of course, be the same for all triangles having I as their apex.

Solving Eq. (12.3.4), we obtain

$$a_1 = \frac{(y_i - y_k)(H_i - H_j) - (y_i - y_j)(H_i - H_k)}{(x_i - x_j)(y_i - y_k) - (x_i - x_k)(y_i - y_j)}$$

$$a_2 = \frac{(x_i - x_j)(H_i - H_k) - (x_i - x_k)(H_i - H_j)}{(x_i - x_j)(y_i - y_k) - (x_i - x_k)(y_i - y_j)}$$

(12.3.5)

and introducing these values in Eq. (12.3.3),

$$\int_{\text{MON}} \left( T_x \frac{\partial h}{\partial x} n_x + T_y \frac{\partial h}{\partial y} n_y \right) ds$$

$$= \frac{1/2}{(x_i - x_j)(y_i - y_k) - (x_i - x_k)(y_i - y_j)}$$

$$\times \{ [T_x(y_i - y_k)(y_j - y_k) + T_y(x_i - x_k)(x_j - x_k)](H_i - H_j)$$

$$- [T_x(y_i - y_j)(y_j - y_k) + T_y(x_i - x_j)(x_j - x_k)](H_i - H_k) \} \quad (12.3.6)$$

or

$$\int_{\text{MON}} \left( T_x \frac{\partial h}{\partial x} n_x + T_y \frac{\partial h}{\partial y} n_y \right) ds = C_{ij}(H_i - H_j) + C_{ik}(H_i - H_k)$$

We return to the balance equation, Eq. (12.3.2) and build the total contour integral over $\Gamma_i$ by adding similar terms calculated in each of the triangles with I as its apex. We finally obtain

$$C_{ij}(H_i - H_j) + C_{ik}(H_i - H_k) + C_{il}(H_i - H_l) + \cdots = Q_i \quad (12.3.7)$$

where $Q_i$ is the integral of the source term over $D_i$.

Equations like Eq. (12.3.7) can be written for each node I where the head is not prescribed.

Note that this expression is very similar to Eq. (12.2.6), which we obtained in integrated finite differences over a polygon of any shape. The main differences are

(1)  The domain of integration $D_i$ is *not* the elementary mesh of the approximation. As we have used triangular elements here, $D_i$ is constituted by portions of all the triangles having I as apex. If the parameter values (e.g., $T$) are given on the elementary triangles, then $T$ varies inside $D_i$, contrary to finite differences.

(2)  Finite differences compute "average" heads over a polygon affected to a central node without making any assumption on the form of the variation of

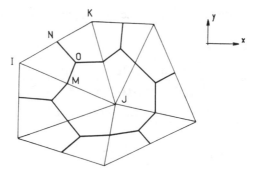

**Fig. 12.12.** Boundary conditions for a finite-element mesh.

this head from one node to the next. On the contrary, finite elements precisely define the variation of the head within one element, linearly in this case. Values at the nodes are only calculated for convenience, but $H$ is defined everywhere. From one triangular element to the next the head varies continuously.

(3) However, the fluxes along the side IJ of a triangle are discontinuous: they are different when evaluated in the two triangles having IJ in common. Linear finite elements do not conserve mass on an elementary triangle; this only happens on a polygon surrounding a node.

*Boundary conditions.* If the head is prescribed along a boundary, the nodes of the triangles falling on this boundary will be prescribed. Equations such as Eq. (12.3.7) will not be written for these nodes.

If a flux is prescribed along a side of a triangle, e.g., IK (Fig. 12.12), then the polygons $D_i$ and $D_k$ will have IN and NK as sides. For polygon $D_i$ the flux along NI will be evaluated using the prescribed boundary condition and introduced as a known term into the total contour integral of $D_i$, in Eq. (12.3.7). Note that the flux along MON will still be evaluated inside IJK by Eq. (12.3.6) without any change.

Contrary to finite differences, the boundary must follow the sides of the elements both for prescribed heads and prescribed flux conditions.

*Anisotropy.* In Eq. (12.3.6) we have assumed that $x$ and $y$ are the principal directions of the anisotropy tensor of the transmissivity inside triangle IJK. If this were not the case, Eq. (12.3.6) would still apply if $x$ and $y$ now form a local coordinate system inside IJK, parallel to the principal directions of anisotropy.

Let $X$, $Y$ be the general coordinate system and $\theta$ the angle between the two. The rotation $\theta$ of the axis (Fig. 12.13) gives

$$x = X \cos \theta + Y \sin \theta$$
$$y = -X \sin \theta + Y \cos \theta$$

(12.3.8)

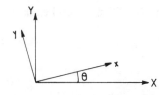

**Fig. 12.13.** Rotation of axis for taking into account the anisotropy.

which can be introduced into Eq. (12.3.5). For instance, $C_{ij}$ becomes

$$
\begin{aligned}
C_{ij} = {} & T_x[(X_k - X_i)\sin\theta + (Y_i - Y_k)\cos\theta] \\
& \times [(X_k - X_j)\sin\theta + (Y_j - Y_k)\cos\theta] \\
& + T_y[(X_i - X_k)\cos\theta + (Y_i - Y_k)\sin\theta] \\
& \times [(X_j - X_k)\cos\theta + (Y_j - Y_k)\sin\theta]
\end{aligned}
$$

where $T_x$ and $T_y$ are still the local transmissivities in the local axis of anisotropy. Such corrections can be introduced into each triangle and, if necessary, with a different angle $\theta$.

*Transient state.* In transient state, the following term must be added to the right-hand side of Eq. (12.3.2) or Eq. (12.3.7):

$$
\iint_{D_i} S\frac{\partial h}{\partial t}\,dx\,dy \tag{12.3.9}
$$

There are two methods for evaluating this term: the lumped and the consistent formulations.

*Lumped approximation.* One assumes that $\partial h/\partial t$ within $D_i$ can be approximated by $\partial H_i/\partial t$. If $S_i$ is the integral of $S$ over $D_i$,

$$
S_i = \iint_{D_i} S\,dx\,dy
$$

then the term $S_i(\partial H_i/\partial t)$ is added to the right-hand side of Eq. (12.3.7). The discretization of this term can be made exactly as for finite differences, i.e., explicit, implicit, etc. (see Section 12.2.6.d).

*Consistent Formulation.* In reality, the linear expression of the head over each triangle [Eq. (12.3.1)] makes it possible to evaluate Eq. (12.3.9) more rigorously. Let MONI (Fig. 12.11) be the portion of $D_i$ inside the triangle IJK. We can write

$$
\iint_{\text{MONI}} S\frac{\partial h}{\partial t}\,dx\,dy = \iint_{\text{MONI}} S_{ijk}\frac{\partial}{\partial t}(a_0 + a_1 x + a_2 y)\,dx\,dy \tag{12.3.10}
$$

where $S_{ijk}$ is the storage coefficient of element IJK. But in Eq. (12.3.5) we have evaluated $a_1$ and $a_2$ as functions of $H_i$, $H_j$, and $H_k$. Then $a_0$ can be taken from Eq. (12.3.4) as

$$a_0 = H_i - a_1 x_i - a_2 y_i \qquad (12.3.11)$$

Thus Eq. (12.3.10) becomes

$$\iint\limits_{\text{MONI}} S\frac{\partial h}{\partial t}\,dx\,dy = S_{ijk}\left[\frac{\partial H_i}{\partial t}\iint\limits_{\text{MONI}} f(x,y)\,dx\,dy \right.$$

$$\left. + \frac{\partial H_j}{\partial t}\iint\limits_{\text{MONI}} f'(x,y)\,dx\,dy + \frac{\partial H_k}{\partial t}\iint\limits_{\text{MONI}} f''(x,y)\,dx\,dy\right]$$

where $f$, $f'$, and $f''$ are linear functions of $x$ and $y$, with coefficients depending on the coordinates of IJK. These functions and their integrals over MONI can be evaluated analytically using Eqs. (12.3.5) and (12.3.11). This term and the equivalent ones for all triangles having I as apex will be added to the right-hand side of Eq. (12.3.7).

In principle, the explicit, implicit, Crank–Nicholson's, and Gir's approximations can also be used to solve the resulting consistent equation. However, using the explicit approximation would be of no interest, because it is no longer possible to solve explicitly for $H_i^{t+\Delta t}$ on the left-hand side as in Eq. (12.2.12). Each equation will now involve several unknowns at time $t + \Delta t$: $H_i^{t+\Delta t}$, $H_j^{t+\Delta t}$, $H_k^{t+\Delta t}$, etc., and the solution thus requires the inversion of the matrix of the system at each time step, as in the implicit approximation.

As the implicit or Crank–Nicholson's approximations (with $\alpha > 0.5$) are unconditionally stable and do not require more computational effort in this case, they are systematically preferred to the explicit approximation in consistent finite elements.

### 12.3.2. Linear Isoparametric Finite Elements using Galerkin's Approximation

(a) *Basic element.* In two dimensions the basic element is a quadrilateral IJKL (Fig. 12.14). For nonlinear elements the sides of this element could represent polynomials of higher degree, e.g., parabolas, but we shall restrict this presentation to linear elements for which the sides are straight lines.

One usually defines a linear transformation of the coordinate system $(x, y)$ for each element so that IJKL becomes a square $ijkl$ in the new system $(\xi, \eta)$ (Fig. 12.14).

This transformation is defined by

$$x = N_i(\xi,\eta)x_I + N_j(\xi,\eta)x_J + N_k(\xi,\eta)x_K + N_l(\xi,\eta)x_L$$
$$y = N_i(\xi,\eta)y_I + N_j(\xi,\eta)y_J + N_k(\xi,\eta)y_K + N_l(\xi,\eta)y_L \qquad (12.3.12)$$

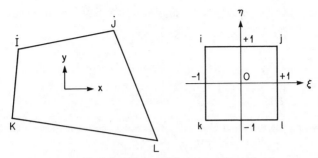

**Fig. 12.14.** Quadrilateral element and linear transformation in a square.

In the system $(\xi, \eta)$ the functions $N_i$ are bilinear functions, called the chapeau functions:

$$N_i = \tfrac{1}{4}(1 + \xi\xi_i)(1 + \eta\eta_i) \tag{12.3.13}$$

For instance, for apex $i(\xi_i = -1, \eta_i = +1)$;

$$N_i = \tfrac{1}{4}(1 - \xi)(1 + \eta)$$

This function $N_i$ is equal to 1 in $i$ and to 0 in $j$, $k$, and $l$. It varies linearly with $\eta$ and $\xi$ along the sides of the square and bilinearly inside the square. In Fig. 12.15, we have drawn the contour line of a chapeau function around a node $i$ in the $(\xi, \eta)$ plane, assuming the value of $N$ to be on an axis orthogonal to the plane $(\xi, \eta)$. The name "chapeau" (meaning hat) comes from the shape of this function.

When $(\xi, \eta)$ describes the square $ijkl$, it is easy to see that $(x, y)$ from Eq. (12.3.12) describes IJKL.

(b)  *Basis functions.*  On the element IJKL the unknown $h(x, y)$ (the head

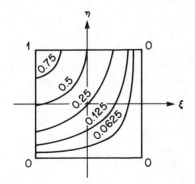

**Fig. 12.15.** Contour line of a chapeau function.

here) will be approximated by the sum of four bilinear basis functions:

$$h(x, y) = H_I N_I(x, y) + H_J N_J(x, y) + H_K N_K(x, y) + H_L N_L(x, y) \quad (12.3.14)$$

where $H_I, \ldots, H_L$ will be the values of the head at the nodes $I, \ldots, L$ (the unknowns of the problem) and the bilinear basis functions $N_I, \ldots, N_L$ will again be the chapeau functions defined above, i.e., $N_I(I) = 1$, $N_I(J) = N_I(K) = N_I(L) = 0$, and $N_I$ varies bilinearly in $x$ and $y$. More precisely,

$$N_I(x, y) = N_i(\xi, \eta) \quad (12.3.15)$$

with $(x, y)$ given by Eq. (12.3.12) from $(\xi, \eta)$ and $N_i$ defined in Eq. (12.3.13).

(c) *Integral equation.* Instead of integrating the flow equation exactly over a given domain $D_i$ around each node I, as we did for integrated finite differences or linear triangular finite elements, the Galerkin formulation requires the integration of this equation with a weighting factor. In a steady state and assuming $x$ and $y$ to be the principal directions of anisotropy, we write

$$\frac{\partial}{\partial x}\left(T_x \frac{\partial h}{\partial x}\right) + \frac{\partial}{\partial y}\left(T_y \frac{\partial h}{\partial y}\right) - q = 0$$

and then

$$\iint_{D_i} W_i(x, y)\left[\frac{\partial}{\partial x}\left(T_x \frac{\partial h}{\partial x}\right) + \frac{\partial}{\partial y}\left(T_y \frac{\partial h}{\partial y}\right) - q\right] dx\, dy = 0 \quad (12.3.16)$$

where $W_i$ is a weighting function. In other words, the flow equation will be satisfied "on the average" over $D_i$ but as a weighted average. Several types of weighting functions could be used, but in Galerkin's formulation the weighting functions $W_i$ are again the same chapeau functions as the ones used as basis functions, and the domain of integration $D_i$ is made up of the four quadrilaterals surrounding each node. Equations like (12.3.16) can be written for each node of the mesh where the head $h$ is not prescribed.

Note that with Galerkin's formulation $D_i$ is no longer a polygon over which mass balance is satisfied. But it can be shown that (1) mass balance is satisfied globally for the entire domain, and (2) around each node one can find a domain included in $D_i$ for which mass balance is satisfied. These domains form a partitioning of the entire domain, but their actual shape is a function of the nodal value of the head (Goblet, 1981) and cannot be defined *a priori*.

(d) *Calculation of the integral.* We can integrate Eq. (12.3.16) by parts.

Substituting $N_I$ for $W_I$, we get

$$\int_{\Gamma_i} N_I(x, y) \left( T_x \frac{\partial h}{\partial x} n_x + T_y \frac{\partial h}{\partial y} n_y \right) ds - \iint_{D_i} \left( \frac{\partial N_I}{\partial x} T_x \frac{\partial h}{\partial x} + \frac{\partial N_I}{\partial y} T_y \frac{\partial h}{\partial y} \right) dx\, dy$$

$$= \iint_{D_i} q N_I\, dx\, dy \qquad (12.3.17)$$

where $n_x$ and $n_y$ are the direction cosines of the outer normal of $\Gamma_i$. However, by definition of the chapeau function, $N_I \equiv 0$ over $\Gamma_i$, so that the contour integral cancels out for all nodes *inside* the domain. We shall see later how to deal with the boundary conditions. We now calculate the integrals over $D_i$ separately for each quadrilateral having I as a node. We then use Eq. (12.3.14) for the head in each quadrilateral. For instance, for IJKL, the left-hand side of Eq. (12.3.17) would become

$$-H_I \iint_{IJKL} \left[ \frac{\partial N_I}{\partial x} T_x \frac{\partial N_I}{\partial x} + \frac{\partial N_I}{\partial y} T_y \frac{\partial N_I}{\partial y} \right] dx\, dy$$

$$-H_J \iint_{IJKL} \left[ \frac{\partial N_I}{\partial x} T_x \frac{\partial N_J}{\partial x} + \frac{\partial N_I}{\partial y} T_y \frac{\partial N_J}{\partial y} \right] dx\, dy \qquad (12.3.18)$$

$$-H_K(\text{similar term}) - H_L(\text{similar term})$$

Finally, Eq. (12.3.17) will take the form

$$-C_I H_I - C_J H_J - \cdots (9 \text{ terms, in general}) = Q_I \qquad (12.3.19)$$

where $Q_I$ is the weighted integral of the source term over $D_i$. Note that $C_I$ is the sum of the integrals given in Eq. (12.3.18) for all four quadrilaterals like IJKL. To calculate these integrals, which are only functions of the coordinates and directional transmissivities, several methods can be used.

*Analytical integration.*    It is then useful to integrate in the $(\xi, \eta)$ coordinate system. We shall assume that $T_x$ and $T_y$ are constant over each quadrilateral. We must then evaluate integrals such as

$$\iint_{IJKL} \frac{\partial N_I}{\partial x} \frac{\partial N_J}{\partial x} dx\, dy$$

$$= \int_{-1}^{+1} \int_{-1}^{+1} \left( \frac{\partial N_i}{\partial \xi} \frac{\partial \xi}{\partial x} + \frac{\partial N_i}{\partial \eta} \frac{\partial \eta}{\partial x} \right) \left( \frac{\partial N_j}{\partial \xi} \frac{\partial \xi}{\partial x} + \frac{\partial N_j}{\partial \eta} \frac{\partial \eta}{\partial x} \right) \det(J)\, d\xi\, d\eta$$

where $\det(J)$ is the determinant of the Jacobian of the change of coordinates from $(x, y)$ to $(\xi, \eta)$:

$$\det(J) = \begin{vmatrix} \dfrac{\partial x}{\partial \xi} & \dfrac{\partial y}{\partial \xi} \\ \dfrac{\partial x}{\partial \eta} & \dfrac{\partial y}{\partial \eta} \end{vmatrix}$$

These derivatives,

$$\frac{\partial N_i}{\partial \xi} \qquad \frac{\partial x}{\partial \xi} \qquad \frac{\partial \xi}{\partial x}$$

can be explicitly calculated from Eqs. (12.3.12) and (12.3.13) and by inversion of the Jacobian matrix.

*Numerical integration.* We use a Gaussian integration of the form

$$\iint\limits_{ijkl} G(\xi, \eta)\, d\xi\, d\eta = \sum_{n=1}^{m} \lambda_n G(M_n)$$

The points $M_n$ and the weights $\lambda_n$ are the Gaussian points and weight of *ijkl*, and are known. The number of points to use is a function of the degree of the polynomial expression $G$: with $m$ points, the integration is exact for a polynomial expression of degree $2m - 1$. Four points are used in general (Zienkiewicz, 1977; Dhatt and Touzot 1981).

One can take for four points

$$(\lambda, \xi, \eta) = (1, \pm 1/\sqrt{3}, \pm 1/\sqrt{3})$$

or for seven points

$$(\lambda, \xi, \eta) = [\tfrac{8}{7}, 0, 0] \quad \text{and} \quad [\tfrac{20}{63}, 0, \pm\sqrt{\tfrac{14}{15}}] \quad \text{and} \quad [\tfrac{20}{36}, \pm\sqrt{\tfrac{3}{5}}, \pm\sqrt{\tfrac{3}{5}}]$$

*Anisotropy.* If the principal directions of anisotropy $(x, y)$ of the transmissivity tensor **T** inside IJKL are not the true coordinate system $(X, Y)$ of the entire domain, one must first change the $(X, Y)$ system locally into $(x, y)$ by a rotation, exactly as for triangular elements; see Eq. (12.3.8).

*Mixed elements.* It is also possible to use linear triangular elements and bilinear quadrilaterals simultaneously in Galerkin's approximation, making the mesh more flexible. The chapeau functions to use for a triangle are then $(1 - \xi - \eta)$, $(\xi)$, and $(\eta)$, if the three apexes of the triangle are located in $(0, 0)$, $(0, 1)$, and $(1, 0)$ in the $(\xi, \eta)$ diagram (Fig. 12.16).

If a numerical integration is performed over the triangle, the Gaussian points and weights are, for $(\lambda, \xi, \eta)$,

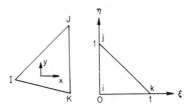

**Fig. 12.16.** Triangle in the $(\xi, \eta)$ system.

Three points:

$$[\tfrac{1}{6}; \tfrac{1}{2}; \tfrac{1}{2}]; \; [\tfrac{1}{6}, 0, \tfrac{1}{2}]; \; [\tfrac{1}{6}, \tfrac{1}{2}, 0]$$

Four points:

$$[-\tfrac{27}{96}, \tfrac{1}{3}, \tfrac{1}{3}]; \; [\tfrac{25}{96}, \tfrac{1}{5}, \tfrac{1}{5}]; \; [\tfrac{25}{96}, \tfrac{3}{5}, \tfrac{1}{5}]; \; [\tfrac{25}{96}, \tfrac{1}{5}, \tfrac{3}{5}].$$

*Boundary conditions.*    We have seen that for a prescribed head boundary the head on the nodes falling on this boundary will be known, and no equation will be written for them. For a prescribed flux boundary the contour integral in Eq. (12.3.17) is not zero but can easily be calculated knowing the flux on the boundary and the chapeau function $N_I$. This term is then known and can be transferred to the right-hand side of Eq. (12.3.19).

*Transient state.*    In transient state the term $S \, \partial h / \partial t$ is added inside the brackets in Eq. (12.3.16). We then have to evaluate in Eq. (12.3.17) or Eq. (12.3.19) terms such as

$$\iint\limits_{D_i} N_I S \frac{\partial h}{\partial t} \, dx \, dy \qquad (12.3.20)$$

This can be done in two ways, as for the triangular elements. One can either use the consistent formulation, i.e., substitute $\sum_I H_I N_I$ [Eq. (12.3.14)] for $h$ in Eq. (12.3.20), and then evaluate the integrals of $N_I S$ over each quadrilateral, or use the lumped formulation, i.e., assume that $\partial H_I / \partial t$ can be used to represent $\partial h / \partial t$ inside the integral (see Neuman, 1975b).

This lumped approximation has the advantage of making it possible to use the explicit formulation. It is, however, less accurate than the consistent formulation.

*Matrix assembly.*    Instead of calculating successively for each node I, the coefficients $C_i$, $C_j, \ldots$ of *each line* of the system's global matrix (Eqs. (12.3.11) or (12.3.19)), it is more efficient to first determine all of these coefficients for *each element* of the mesh, by calculating for each element successively the integrals given in Eq. (12.3.18). Once this is done, the coefficients for each node $\dot{I}$ (each line of the matrix) are calculated by adding the relevant term of each element which has $\dot{I}$ as its apex. This is called the system's matrix assembly.

### 12.3.3. Higher-Order Elements

Instead of using linear or bilinear basis functions, it is possible to use higher-order functions in finite elements to increase the precision and/or decrease the mesh size. There are at least two possibilities.

(1)   One can use higher-order polynomials as basis functions on the elements (e.g., quadratic or cubic) that will increase the number of unknowns on each element. One will therefore use more nodes on each element (e.g., nine nodes on a quadrilateral—the four corners, four nodes in the middle of each side, and one in the center for quadratic functions—or 16 nodes for a cubic function). But with such elements the interpolation function still does not have a continuous derivative from one element to the next.

(2)   One can use hermitian elements. The number of nodes per element is not increased, but instead $h$ and its derivatives at the nodes are taken as unknowns. If cubic hermitian polynomials are used, $h$ is continuous, $\partial h / \partial n$ is still discontinuous, but $\partial h / \partial x$ and $\partial h / \partial y$ are identical in each quadrilateral at a node: if the elements are rectangles, $\partial h / \partial x$ and $\partial h / \partial y$ are continuous. Higher-order hermitian polynomials can ensure continuity of the first- and second-order derivatives. This has proved very efficient for solving the transport equation (Section 12.5.3); see Van Genuchten (1977).

All these techniques are described by Pinder and Gray (1977).

## 12.4. Solving Large Linear Systems

Except for the explicit formulation in a transient state in finite differences, all numerical techniques that have been presented end up with a large linear system of equations to be solved.

The matrices of these systems are generally sparse, i.e., they have only a few nonzero coefficients per line (five in finite differences on rectangles or squares, nine in finite elements on quadrilaterals, etc.). These matrices are also symmetrical. This is imposed by mass balance in finite differences (see Section 12.2.3) and can easily be checked on Eq. (12.3.18) for finite elements.

The simplest method of solving the linear system would be to compute its inverse matrix; however, this matrix would no longer be sparse and would require the storage of $p \times p$ coefficients if there are $p$ unknowns. This becomes unfeasible for $p$ larger than 100, approximately. The time also becomes prohibitive and the accuracy of direct matrix inversion insufficient for larger $p$. Therefore, other methods must be used. Here we will briefly introduce a few of the most common ones.

### 12.4.1. *Direct Methods*

Direct methods are all more or less based on Gaussian elimination, which means that the linear system is solved without ever computing the inverse matrix. The simplest method of elimination is triangularization, but several decomposition methods are also used.

(a)  *Triangularization.*  If $MH = Q$ is the linear system to solve, we will write it

$$
\begin{bmatrix} m_{11} \cdots m_{1p} \\ \vdots \quad \ddots \\ m_{p1} \cdots m_{pp} \end{bmatrix}
\begin{bmatrix} h_1 \\ \vdots \\ h_p \end{bmatrix}
=
\begin{bmatrix} q_1 \\ \vdots \\ q_p \end{bmatrix}
\tag{12.4.1}
$$

In the first equation, we write $h_1$ as a function of $h_2 \cdots$ and $q_1$. This expression of $h_1$ is then introduced into the second and the following equations. The second equation is then no longer a function of $h_1$. This equation is again written to give $h_2$ as a function of $h_3 \cdots h_j$ and $q_i$, and this expression is introduced into the third and following lines. This third equation is then no longer a function of $h_1$ and $h_2$, and so on until the last equation, which can then be solved, thus giving the value of $h_p$. By back-substitution, $h_{p-1}, \ldots, h_2, h_1$ are then calculated. This method is called triangularization because at each step $M$ is transformed into a triangular matrix by linear combinations of its lines.

(b)  *Decomposition.*  We shall only give Choleski's algorithm, which is applicable to symmetrical positive definite matrices such as the matrices presented earlier. For Choleski's algorithm the coefficients $m_{ii}$ must be positive.

This may require the solving of the system $-MH = -Q$ if the $m_{ii}$ are negative. We shall assume below that the $m_{ii}$ are positive in $MH = Q$.

One looks for a lower triangular matrix $R$ such that $M = RR^T$ (T indicates transposition). The coefficients $r_{ij}$ of matrix $R$ are calculated by recurrence with

$$
r_{11} = \sqrt{m_{11}}
$$

$$
r_{j1} = m_{j1}/r_{11}, \qquad j = 2, \ldots, p
$$

$$
r_{ii} = \sqrt{m_{ii} - \sum_{k=1}^{i-1} r_{ik}^2}
$$

$$
r_{ji} = \frac{1}{r_{ii}} \left( m_{ij} - \sum_{k=1}^{i-1} r_{ik} r_{jk} \right), \qquad i \le j
$$

Once $R$ is calculated, the solution of $RR^T H = Q$ is straightforward. If $V = R^T H$, then $RV = Q$, where $V$ is a vector. We then solve the triangular system $RV = Q$ directly, whence $V$, and then we again solve the triangular system $R^T H = V$, whence $H$.

A very large number of other direct methods is available. Furthermore, with the development of array processors, new and better direct methods are being developed at present. One should know, however, that due to round-off errors, direct methods are sometimes less accurate than the indirect methods, which we shall now introduce.

### 12.4.2. Iterative Methods

Initially, iterative methods were widely used for solving linear systems in groundwater modeling. Their advantage is that they require much less core storage than the direct methods, since the matrix $M$ is never stored in the computer, but its coefficients are recalculated at each line. They may be less efficient in terms of CPU time if a large core is available, but they are less sensitive to round-off errors, especially for computers with 16-bit words. They are therefore still very much in use, especially with micro- or minicomputers. We shall briefly present two iterative algorithms, known as point successive over-relaxation (PSOR) and alternate directions implicit (ADI), but we shall first define an iterative algorithm using Jacobi's decomposition.

Let us decompose matrix $M$ as the sum of three matrices:

$$M = \quad L \quad + \quad D \quad + \quad U$$

$$\underbrace{\text{lower}}_{\text{triangular}} \quad \text{diagonal} \quad \underbrace{\text{upper}}_{\text{triangular}}$$

where $L$ and $U$ are, respectively, the lower and the upper triangular matrices, strictly below and above the diagonal, and $D$ is the diagonal of $M$. If $M = (m_{ij})$, then

$$L = (m_{ij}) \quad j < i$$

$$U = (m_{ij}) \quad j > i$$

$$D = (m_{ii})$$

To solve the system $MH = Q$ we can write identically

$$(L+D+U)H=Q \quad \text{or} \quad DH=Q-(L+U)H \quad \text{or} \quad H=D^{-1}[Q-(L+U)H]$$

As $D$ is diagonal, $D^{-1} = (1/m_{ii})$. From this expression, an iterative algorithm is built by recurrence by

$$^0H = \text{arbitrary initial value given to } H$$

$$^{n+1}H = D^{-1}[Q - (L + U)^n H] \tag{12.4.2}$$

The indices $0$, $n$, and $n + 1$ refer here to the iteration number of the calculation. If the iterative algorithm converges, then $^nH \to H$ as $n \to \infty$ $\forall \, ^0H$, $H$ being the solution of the system. For strictly diagonally dominant matrices (i.e., $|m_{ii}| \geq \sum_{k \neq i} |m_{ik}| \, \forall \, i$, with a strict inequality at least for one $i$), Jacobi's algorithm converges and the only problem is to stop the iterations at some point. One generally computes the difference between two successive estimates,

$$E = \,^{n+1}H - \,^nH$$

and stops the iterations when $\|E\| < \varepsilon$, where $\varepsilon$ is a prescribed figure and $\| \; \|$ is a norm (e.g., $\sum_i e_i^2$, or $\sum_i |e_i|$, or $\max_i e_i$). Note that $E$ *is not* the difference between $^{n+1}H$ and the true solution $H$, but only the difference between two successive estimates of $H$. Therefore $\varepsilon$ must be *much smaller* than the required precision for $H$.

Here we have used a matrix notation of Jacobi's iterative algorithm. One must realize that, in practice, in programming, matrices $D^{-1}$, $L$, and $U$ are never actually stored. Each line of Eq. (12.4.2) is indeed written using the notations of Eq. (12.4.1):

$$^{n+1}h_i = \frac{1}{m_{ii}} \left( q_i - \sum_{j \neq i} m_{ij}\,^nh_j \right), \qquad i = 1, \ldots, p \qquad (12.4.3)$$

Generally, there are only a few nonzero terms in the summation over $j$ (four in rectangles in finite differences, eight in quadrilaterals in finite elements), so that only the nonzero $m_{ij}$ are stored or calculated. The FORTRAN codes for iterative algorithms are thus extremely simple to write.

However, Jacobi's method is very inefficient. The two following methods, PSOR and ADI, are much better.

(a) *Frankel–Young's method or point successive over-relaxation (PSOR)*. We start from the same decomposition of matrix $M$ as Jacobi's, but we write the identity

$$(D + L)H = Q - UH$$

The corresponding iterative algorithm, known as the Gauss–Seidel, algorithm is written

$$^0H = \text{arbitrary value}$$
$$^{n+1}H = (D + L)^{-1}(Q - U\,^nH) \qquad (12.4.4)$$

Note that $(D + L)$ is a lower triangular matrix, so that solving Eq. (12.4.4) does not involve any matrix inversion and is done by recurrence:

$$^{n+1}h_1 = \frac{1}{m_{11}} \left( q_1 - \sum_{j>1} m_{1j} {}^n h_j \right)$$

$$^{n+1}h_2 = \frac{1}{m_{22}} \left( q_2 - m_{21} {}^{n+1}h_1 - \sum_{j>2} m_{2j} {}^n h_j \right)$$

$$\vdots$$

$$^{n+1}h_i = \frac{1}{m_{ii}} \left( q_i - \sum_{j<i} m_{ij} {}^{n+1}h_j - \sum_{j>i} m_{ij} {}^n h_j \right)$$

Frankel–Young's method, or over-relaxation, consists in accelerating the rate at which this algorithm converges. One first writes, as before,

$$*h_i = \frac{1}{m_{ii}} \left( q_i - \sum_{j<i} m_{ij} {}^{n+1}h_j - \sum_{j>i} m_{ij} {}^n h_j \right)$$

and then $d_i = *h_i - {}^n h_i$; $d_i$ is the magnitude of the variation of $h_i$ during this iteration. The over-relaxation consists in amplifying this variation by a coefficient $\rho$ and writing

$$^{n+1}h_i = {}^n h_i + \rho d_i$$

This can be written

$$^{n+1}h_i = (1 - \rho)^n h_i + \frac{\rho}{m_{ii}} \left[ q_i - \sum_{j<i} m_{ij} {}^{n+1}h_j - \sum_{j>i} m_{ij} {}^n h_j \right]$$

or in matrix form,

$$^0H = \text{arbitrary}, \qquad ^{n+1}H = (D + \rho L)^{-1}\{[(1 - \rho)D - \rho U]^n H + \rho Q\}$$

$$(12.4.5)$$

where $\rho$ is the over-relaxation coefficient. It can be shown that the algorithm converges if $\rho < 2$. But $\rho$ must be larger than 1 to accelerate the convergence. In order to stop the iterations, a test similar to the one given for Jacobi's algorithm must be used. The over-relaxation coefficient $\rho$ is a complex function of the geometry of the mesh and properties of the medium. It is better to underestimate $\rho$ than to overestimate it. In practice, the optimal value of $\rho$ can be obtained by trial and error (value for which the minimum number of iterations is required to obtain a given precision) or by the following expression:

$$\rho = \frac{2}{1 + \sqrt{1 - \mu^2}}$$

where $\mu$ is the largest eigenvalue of matrix $D^{-1}(L + U)$. For finite-difference equations written with squares in a homogeneous medium and for a rectangular domain of $n$ lines and $m$ columns, $\mu$ is given by

$$\mu = \tfrac{1}{2}\left(\cos\frac{\pi}{n+1} + \cos\frac{\pi}{m+1}\right)$$

If the domain is not rectangular, $n$ and $m$ are taken as the average number of lines and columns.

If the medium is not homogeneous, $\mu$ can be estimated iteratively by the Rayleigh ratio of the following iterative Jacobi's algorithm:

$$^{0}A = \text{arbitrary vector}$$

$$^{n+1}A = D^{-1}(L + U)^{n}A$$

$$\mu = \lim_{n \to \infty} \frac{(^{n+1}A,\,{}^{n}A)}{(^{n}A,\,{}^{n}A)}$$

where $(A, B) = \sum_i a_i b_i$ is the scalar product.

However, this procedure of estimating $\mu$ and $\rho$ is quite lengthy, and the value of $\rho$ is generally fixed empirically.

Note that these iterative algorithms converge for any $^{0}H$, but the closer $^{0}H$ is to the solution, the faster the convergence. One tries to select $^{0}H$ as close as possible to the expected solution. In particular, in a transient state (implicit or Crank–Nicholson), one always takes $^{0}H^{t+\Delta t} = H^{t}$, i.e., the solution calculated at the previous time step.

When the iterations are stopped, it is always preferable to compute $(MH - Q)$ and see if the residual errors, which are given in terms of flow rate per mesh, are acceptable given the precision on the source term $Q$.

(b)  *Alternate directions or alternate directions implicit (ADI) in transient state.*   Alternate directions are a decomposition method of matrix $M$ that can only be used for rectangular meshes in finite differences, where there are only five nonzero coefficients per line in matrix $M$. This matrix is decomposed into the sum of two matrices.

For the sake of simplicity, we shall present the method on the example of a square grid as given in Eq. (12.2.2):

$$T_{nc}(H_n - H_c) + T_{ec}(H_e - H_c) + T_{sc}(H_s - H_c) + T_{wc}(H_w - H_c) = Q_c$$

where $n$ and $s$ correspond to the direction of the columns and $e$ and $w$ correspond to the direction of the lines.

We can write Eq. (12.2.2) as the sum of columns and lines equations:
Column equation:

$$T_{nc}(H_n - H_c) + T_{sc}(H_s - H_c) = A_1 H$$

Line equation:

$$T_{ec}(H_e - H_c) + T_{wc}(H_w - H_c) = A_2H$$

In matrix form, this is $(A_1 + A_2)H = Q$. The alternate directions algorithm is then

$$^0H = \text{arbitrary}$$

$$^{n+1/2}H = (A_1 + \rho_1 I)^{-1}[(\rho_1 I - A_2)^n H + Q] \tag{12.4.6}$$

$$^{n+1}H = (A_2 + \rho_2 I)^{-1}[\rho_2 I - A_1)^{n+1/2}H + Q]$$

where $\rho_1$ and $\rho_2$ are acceleration factors and can vary from one iteration to the next (they only need to be positive), $n + \frac{1}{2}$ is an intermediate step in the iteration, and $I$ is the identity matrix. In other words, the linear system is first solved column by column, then line by line. The only problem is to solve a linear system like $(A_1 + \rho_1 I)$ or $(A_2 + \rho_2 I)$. These systems only have three nonzero coefficients per line. They are solved by Gaussian elimination, which is very simple for such matrices. If we want to solve the tridiagonal system $AX = B$, where each line of matrix $A$ has three coefficients, we denote them $a_{ii-1}$, $a_{ii}$, $a_{ii+1}$ and then the two-step elimination algorithm, known as Thomas' algorithm, becomes for the forward step

$$
\left.\begin{aligned}
w &= a_{11} \\
f_1 &= a_{12}/w \\
g_1 &= b_1/w
\end{aligned}\right\}
\quad \text{and then} \quad
\left.\begin{aligned}
w &= a_{ii} - a_{ii-1}f_{i-1} \\
f_i &= a_{ii+1}/w \\
g_i &= (b_i - a_{ii-1}g_{i-1})/w
\end{aligned}\right\}
\quad i = 2,\ldots,p
$$

Only $f$ and $g$ need to be stored; $w$ is only a dummy variable. If $p$ is the dimension of matrix $A$, then, the backward step is,

$$x_p = g_p$$

$$x_i = g_i - f_i x_{i+1} \qquad i = p-1,\ldots,1$$

The alternate directions method is often used with rectangular grids and in transient state in the implicit approximation. Sometimes the number of iterations is limited to one per time step [i.e., one resolution per column, one per line. See Pinder and Bredehoeft (1968)]. But this does not ensure convergence of the solution and can only be used if the time steps are very small: otherwise it is necessary to check the convergence with an error criterion $\|E\| < \epsilon$, as for the Jacobi or PSOR method.

The difficulty with the ADI method is to select the optimal $\rho$ for each iteration. It is also less flexible and less easy to program than PSOR, especially if internal boundary conditions exist inside the domain (e.g., rivers). As they are only more efficient than PSOR for the set of optimal $\rho$ terms (compared to an optimal $\rho$ for PSOR), very often PSOR is the preferred algorithm for iterative solution of linear systems in groundwater modeling.

(c) *Other iterative methods.* There are many other iterative methods than PSOR and ADI. See, for instance, Varga (1962). A very efficient method, called the incomplete Choleski conjugate gradient algorithm, can be found in Gambolati and Perdon (1984).

(d) *Nonlinear problems.* Iterative methods are well suited to solving nonlinear problems, i.e., when the coefficients of matrix $M$ are functions of the solution $H$.

In a steady state the first simple thing that comes to mind is to change the values of the coefficients of the matrix $M$ at each iteration of the solution as vector $H$ changes. But this method, which sometimes works, is very dangerous: there is absolutely no proof that a nonlinear iterative algorithm converges, and very often it does not. The correct method is to solve the iterative system for a fixed value of the parameter $M_0$, assuming a given initial value for the unknown $H_0$. One then obtains a first approximation $H_1$ of $H$. The coefficients are changed and matrix $M_1$ is determined and, by solving the system iteratively, $H_2$ is calculated, etc. If $H_n \to H$ as $n \to \infty$, the nonlinear system has been solved. But for each solution one knows that the iterative system will converge. In practice, the number of iterations is kept small for the first $H_i$ terms and then increased as $H_i$ converges.

In a transient state, nonlinearities can be solved assuming that the parameters of matrix $M$ between $t$ and $t + \Delta t$ are those estimated at time $t$, no matter if one solves an implicit equation. If this is not precise enough, the predictor–corrector method consists in predicting $H^{t+\Delta t}$ by an approximate method (e.g., the explicit, using matrix $M$ at time $t$, even if the time step is larger than the critical time step). This approximate $H^{t+\Delta t}$ is used to determine the coefficients of matrix $M$ at time $t + \Delta t$. Then the correct $H^{t+\Delta t}$ is calculated (e.g., by the implicit scheme) using this updated matrix. If this is still not precise enough, it is also possible to iterate on the value of the coefficients, as explained above for a steady state.

This can also be done if direct methods of solution are used.

## 12.5. Solving the Transport Equation

The transport equation (for solute or heat) was established in Chapter 10. It has the form

$$\text{div}(\mathbf{D} \, \text{grad} \, C - UC) = \omega_c \frac{\partial C}{\partial t} \qquad (12.5.1)$$

The numerical solution of this equation raises two problems. (1) The principal directions of the dispersion tensor **D** are in principle parallel and orthogonal to the direction of the Darcy velocity in the medium. As the direction of this velocity may change with time and space, the numerical method must be able to incorporate this variation. (2) The derivatives on the left-hand side of Eq. (12.5.1) are of order 1 and 2: if a second-order correct approximation of the first derivative is used, the associated error may be of the same order of magnitude as the true second-order derivative. This is called "numerical dispersion;" the numerical solution adds a purely numerical dispersion to the real phenomenon of hydrodynamic dispersion and this increases the total dispersion of the solute in the medium erroneously. Now, if third-order correct approximations are used, the numerical solution may become unstable: the numerical solution is often a compromise between stability and numerical dispersion.

A large number of numerical methods have been proposed in the literature and are still being developed. Here we shall briefly present three methods most commonly used at present: finite differences, method of characteristics, and finite elements.

### 12.5.1. *Finite Differences*

Simple finite differences can be used in one or in two dimensions with square or rectangular meshes if the direction of the velocity always remains parallel to one of the axes of the mesh.

(a) *In one dimension.* Let us first assume that the dispersion coefficient $D$, the Darcy velocity $U$, and the kinematic porosity $\omega_c$ are constant. The transport equation is

$$D\frac{\partial^2 C}{\partial x^2} - U\frac{\partial C}{\partial x} = \omega_c \frac{\partial C}{\partial t} \qquad (12.5.2)$$

The convective term $U(\partial C/\partial x)$ can be taken as a centered or backward difference.

Centered difference:

$$\frac{\partial C}{\partial x} = \frac{C(x + \Delta x) - C(x - \Delta x)}{2\,\Delta x}$$

Backward difference:

$$\frac{\partial C}{\partial x} = \frac{C(x) - C(x - \Delta x)}{\Delta x}$$

where $\Delta x$ is the size of the spatial discretization.

The centered differences are correct to the third order (thus without numerical dispersion), but they can be unstable if the numerical Peclet number $P_e = U \Delta x/D = \Delta x/\alpha$ is too large ($\alpha$ is the dispersivity assuming that $D = \alpha U$). Price et al. (1966) have shown that the stability is ensured if $P_e < 2$.

The backward differences are unconditionally stable if the velocity is directed from $x - \Delta x$ to $x$. However, they introduce a numerical dispersion, which can be corrected as suggested by Lantz (1971). We will treat it together with the time difference, which can also introduce numerical dispersion. Using Taylor's series expansion, one can write:

Backward differences:

$$\left(\frac{\partial C}{\partial x}\right)_x = \frac{C_x - C_{x+\Delta x}}{\Delta x} + \tfrac{1}{2}\Delta x \left(\frac{\partial^2 C}{\partial x^2}\right)_x + O(\Delta x^2) \qquad (12.5.3)$$

Sum of backward and forward differences:

$$\left(\frac{\partial^2 C}{\partial x^2}\right)_x = \frac{C_{x+\Delta x} + C_{x-\Delta x} - 2C_x}{\Delta x^2} + O(\Delta x) \qquad (12.5.4)$$

Explicit time difference

$$\left(\frac{\partial C}{\partial t}\right)_t = \frac{C_{t+\Delta t} - C_t}{\Delta t} - \tfrac{1}{2}\Delta t \left(\frac{\partial^2 C}{\partial t^2}\right)_t + O(\Delta t^2) \qquad (12.5.5)$$

Implicit time difference

$$\left(\frac{\partial C}{\partial t}\right)_{t+\Delta t} = \frac{C_{t+\Delta t} - C_t}{\Delta t} + \tfrac{1}{2}\Delta t \left(\frac{\partial^2 C}{\partial t^2}\right)_{t+\Delta t} + O(\Delta t^2) \qquad (12.5.6)$$

where $\Delta t$ is the size of the time step.

The second-order time derivative can be evaluated by differentiating Eq. (12.5.2):

$$\frac{\partial^2 C}{\partial t^2} = \frac{1}{\omega_c} \frac{\partial}{\partial t} \left( D \frac{\partial^2 C}{\partial x^2} - U \frac{\partial C}{\partial x} \right)$$

With a permutation of the order of derivation, disregarding third-order derivatives and using Eq. (12.5.2) again, we obtain

$$\frac{\partial^2 C}{\partial t^2} = -\frac{U}{\omega_c} \frac{\partial}{\partial x} \left(\frac{\partial C}{\partial t}\right) = -\frac{U}{\omega_c^2} \frac{\partial}{\partial x} \left( D \frac{\partial^2 C}{\partial x^2} - U \frac{\partial C}{\partial x} \right)$$

$$= \frac{U^2}{\omega_c^2} \frac{\partial^2 C}{\partial x^2} + \text{third-order derivatives} \qquad (12.5.7)$$

Equation (12.5.7) is then introduced into Eq. (12.5.6) or in Eq. (12.5.5). These are taken into Eq. (12.5.2) together with Eq. (12.5.3); all terms in $\partial^2 C/\partial x^2$ are grouped and (12.5.4) is then used. One gets, in explicit,

$$\left(D - \tfrac{1}{2}U\,\Delta x + \tfrac{1}{2}\frac{U^2\,\Delta t}{\omega_c}\right)\left(\frac{C_{x+\Delta x} + C_{x-\Delta x} - 2C_x}{\Delta x^2}\right)_t - U\left(\frac{C_x - C_{x-\Delta x}}{\Delta x}\right)_t$$

$$= \omega_c \frac{C_x^{t+\Delta t} - C_x^t}{\Delta t}$$

and in implicit,

$$\left(D - \tfrac{1}{2}U\,\Delta x - \tfrac{1}{2}\frac{U^2\,\Delta t}{\omega_c}\right)\left(\frac{C_{x+\Delta x} - C_{x-\Delta x} - 2C_x}{\Delta x^2}\right)_{t+\Delta t} - U\left(\frac{C_x - C_{x-\Delta x}}{\Delta x}\right)_{t+\Delta t}$$

$$= \omega_c \frac{C_x^{t+\Delta t} - C_x^t}{\Delta t}$$

Lantz's correction for numerical dispersion is to use an apparent dispersion coefficient given by, in explicit,

$$D^* = D - \tfrac{1}{2}U\,\Delta x + \tfrac{1}{2}\frac{U^2\,\Delta t}{\omega_c}$$

and in implicit,

$$D^* = D - \tfrac{1}{2}U\,\Delta x - \tfrac{1}{2}\frac{U^2\,\Delta t}{\omega_c}$$

and to solve,

$$D^*\frac{C_{x+\Delta x} + C_{x-\Delta x} - 2C_x}{\Delta x^2} - U\frac{C_x - C_{x-\Delta x}}{\Delta x} = \omega_c \frac{C^{t+\Delta t} - C^t}{\Delta t}$$

Alternatively, in explicit, one can choose the time step $\Delta t$ or the mesh size $\Delta x$ so that

$$\tfrac{1}{2}U\,\Delta x - \tfrac{1}{2}\frac{U^2}{\omega_c}\Delta t = 0, \quad \text{i.e.,} \quad U\,\Delta t = \omega_c\,\Delta x$$

which means that during each time step the convective flux of water entering into a mesh, $U\,\Delta t$, is strictly equal to the volume of mobile water in that mesh $\omega_c\,\Delta x$.

If $U$, $D$, or $\Delta x$ vary in the model, the correction can be made mesh by mesh, but this is rather complex.

If the centered Crank–Nicholson's approximation is used in the time domain, there is no numerical dispersion for that term.

Note that this one-dimensional discretization can also be used in radial coordinates [see for instance Sauty (1977, 1978a)].

(b) *In two dimensions.* Backward differences are also used for the two components of the velocity. Note that if the velocity is not parallel to one of the axes, a five-point finite difference scheme cannot handle the anisotropy of the dispersion tensor. One must therefore assume that **D** is isotropic, which is far from correct, or use a nine-point finite-difference scheme. However, these methods are not very efficient and finite elements should be preferred.

In any case, with backward differences (in the direction of the velocity) and the time derivative, a correction for numerical dispersion can be introduced. One finds that the apparent dispersion coefficient becomes, e.g., using the explicit method (Goblet, 1981),

$$D_{xx}^* = D_{xx} - \frac{U_x \Delta x}{2} + \frac{U_x^2 \Delta t}{2\omega_c}$$

$$D_{yy}^* = D_{yy} - \frac{U_y \Delta y}{2} + \frac{U_y^2 \Delta t}{2\omega_c}$$

$$D_{xy}^* = D_{yx}^* = D_{xy} + \frac{U_x U_y \Delta t}{2\omega_c}$$

where $D_{xx}$, $D_{yy}$, and $D_{xy}$ are components of the dispersion tensor and $\Delta x$, $\Delta y$, and $\Delta t$ the discretization step in space and time.

But these corrections do not solve all the difficulties and for large $\Delta x$, $\Delta y$, or $\Delta t$ the higher order terms start to play a role. Van Genuchten (1977) has suggested more elaborate schemes using centered differences both in space and time.

In general, finite differences are not very well suited to solving the transport equation apart from very simple one-dimensional problems.

Note that the "mixing-cell" approach proposed, e.g., by Simpson and Duckstein (1975) can be seen in one dimension as a transport equation that only has the convective term and that is discretized with a backward difference. The hydrodynamic dispersion is introduced by the numerical dispersion: if $\Delta x$ and $\Delta t$ are chosen so that the order of magnitude of this dispersion is correct, the mixing-cell model provides a good description of transport in one dimension.

### 12.5.2. *Method of Characteristics*

This method has been widely used and can be applied to finite differences or finite elements, in two or three dimensions. The basic idea is to decouple the convective part and the dispersive part in the transport equation and to solve them successively. Two different ways of treating the dispersion term have been proposed: the particle in cells (PIC) and the discrete parcel random walk (DPRW).

(a) *Particle in cells* (*PIC*).  This method was introduced by Garder *et al.* (1964) and has been presented with application, e.g., by Pinder and Cooper (1970) Bredehoeft and Pinder (1973) and Konikow and Bredehoeft (1974, 1978). The flow equation is first solved with a numerical model, and the velocity in each mesh is determined. Then at each location in the aquifer where there is a source term of solute a large number of "particles" is introduced at each time step (e.g., hundreds). These "particles" are assumed to represent the solute transported with the flow. If the solute is initially distributed over an area of the aquifer, a regular number of particles (e.g., 4 or 5) is introduced in each mesh of this particular area at the beginning of the simulation.

To represent the convective transport, one keeps track of the coordinates of each particle and moves them in the aquifer with the velocity of the flow. If $x_i^t$ are the coordinates (in two or three dimensions) of particle $i$ at time $t$, which falls inside mesh $j$, then its coordinates at time $t + \Delta t$ will be

$$\mathbf{x}_i^{t+\Delta t} = \mathbf{x}_i^t + \mathbf{V}_j^t \Delta t$$

where $\mathbf{V}_j^t$ is the pore velocity vector in mesh $j$ (i.e., $\mathbf{U}/\omega_c$, the Darcy velocity divided by the kinematic porosity) at time $t$ if $\mathbf{V}$ is not constant. The time step $\Delta t$ is in general adjusted so that the maximum traveled distance of a particle during a time step is smaller than one half of the mesh size.

In the PIC method each "particle" represents a concentration of solute. Initially this concentration is given by the dilution of the flux of solute in the mesh where this solute is injected. For instance, if the volume of water in this mesh (area × thickness × porosity) is $10^4 \, m^3$ and if $10^3 \, kg$ of elements are injected per time step $\Delta t$, the concentration in the mesh will be $0.1 \, kg/m^3$. Alternatively, the initial concentration in the mesh may be known. Then each particle present in a mesh is given the value of the concentration of that mesh.

The particles are then moved with the convective velocity $\mathbf{V}$. A new concentration in each mesh is then calculated by taking the average of the concentration of each particle present in each mesh. This value is given to the node in the center of the mesh for that time step.

To represent the dispersive part of the transport equation, the concentrations calculated in each mesh, as above, are now modified at each time step by solving a purely dispersive equation:

$$\text{div } \mathbf{D} \text{ grad } C = \omega_c \frac{\partial C}{\partial t}$$

This equation (which is identical to the simple flow equation) is solved numerically using finite differences for the time step $\Delta t$. Then the new concentration of each particle in a given mesh is the sum of its previous value (at the end of the convective step) and the change in concentration calculated in that mesh during the dispersive step; this new figure is "labeled" on each of these particles.

Note that the coordinates of the particles are kept constant during this dispersive part of the transport.

Then this whole procedure is repeated for a new time step: the particles are moved convectively, new concentrations are determined in each mesh, these are modified by dispersion, etc.

Although this technique is rather simple, it poses quite a few problems:

(1)   Accurate bookkeeping of a large number of particles (several hundreds) is necessary, giving their coordinates and concentration. As new particles are added at the sources while others may leave the model at the boundaries, this is not a very easy job.

(2)   The method introduces some numerical dispersion when the concentrations are calculated from the number of particles.

(3)   There may be a net loss of solute at the front of the system in a mesh where dispersion introduces solute but where there are no particles. This can be corrected by introducing new particles, but this makes the method even more complex.

(4)   Finally, the method does not converge systematically when the number of particles is increased. For instance, if 2, 5, or 10 times as many particles are used to solve an identical problem, the concentration may oscillate from one solution to the next. As the number of particles always becomes small toward the front of the solute plume, the method is not very accurate.

The second method of characteristics (DPRW) is preferable, although it does not solve all these problems.

(b)   *The discrete parcel random walk (DPRW).*   This method was described by Ahlstrom *et al.* (1977) and Prickett *et al.* (1981). It differs from the PIC method in two ways:

(1)   Each particle represents a mass of solute, not a concentration. To determine the concentration in a mesh, one divides the sum of the mass of the particles by the volume of water in the mesh. This is more satisfactory as masses are additive quantities, whereas concentrations are not. As long as particles are not "lost," mass balance is always conserved.

(2)   The dispersive part of transport is not represented by solving the dispersion equation but only by giving an additional displacement to each particle at the end of each convective displacement, without modifying the mass of the particle. This dispersive displacement is random: in each direction (longitudinal and transversal) it is determined for each particle by randomly sampling a Gaussian distribution with zero mean and a variance equal to $2D_i \Delta t$, $D_i$ being the dispersion coefficient in the longitudinal or transversal directions. This random walk can be seen intuitively as a Brownian motion, which is known to be responsible for molecular diffusion. If the number of

particles is large enough, these random walks will indeed correctly represent Fickian dispersion.

Another approach that has been used is to represent dispersion by randomly sampling in a distribution of velocity for the convective transport instead of using a unique "average" convective velocity. This can be done either by defining this distribution of velocity at each point or by generating different "realizations" of convective velocity fields and then averaging the purely convective displacement of each particle in these realizations.

The advantages of DRPW are that it does not involve numerical dispersion since concentrations are never calculated (except to plot the results). It also conserves mass and, moreover, it is possible to take into account reactions during transport, e.g., radioactive decay: the mass of each particle is simply decreased as a function of time. The difficulties are due to the large number of particles, their bookkeeping when they enter or leave the model, and the determination of their displacement. In particular, when the velocity varies a great deal in the medium, it must be changed during a time step each time a particle enters into a different medium. Otherwise a particle entering, for instance, mesh 2 with the velocity calculated in mesh 1, which was much greater than that of mesh 2, can be "stuck" in mesh 2 if its distance of displacement is calculated using $V_1 \, \Delta t$. With $V_2$ it would never have moved so far in the mesh and would not have become "stuck."

The DPRW is, however, unstable when the number of particles is increased: the solution oscillates and must be smoothed. Therefore it is not very accurate.

### 12.5.3. Finite Elements

Solving the transport equation using isoparametric elements and Galerkin's procedure is straightforward. The calculations presented in Section 12.3.2 to integrate div($T \, \mathbf{grad} \, h$) are used to integrate div($D \, \mathbf{grad} \, C - UC$). The integrals given in Eq. (12.3.18) will now also involve the chapeau functions $N_I$ instead of only their derivative, but this does not create any major difficulties.

It can be shown that Galerkin's procedure is equivalent to a centered approximation for a regular mesh, so that numerical dispersion from the first-order space derivative is minimum. The stability criterion given by Price *et al.* (1966) also applies, i.e., $P_e = U \, \Delta x / D = \Delta x / \alpha < 2$, $\Delta x$ being the size of the mesh in the direction of the velocity. Otherwise oscillations of the solution will be observed at the front of the plume of solute.

To increase the stability, Huyakorn (1976) has suggested the use of a sort of "backward difference" scheme in finite elements, which he has called an upstream weighting function. This was extended to two dimensions and to the lumped transient approximation. However, it increases the numerical dispersion, it is difficult to extend to anisotropic dispersion (the numerical Peclet numbers are different in the longitudinal and transverse directions), and it is lengthy to use (Goblet, 1981). It is more appropriate to use smaller meshes or higher-order elements. For quadratic elements, the stability criterion becomes $P_e < 4$ (Christie *et al.*, 1976) and for second-order hermitian elements, $P_e < 4.64$ (Jenson and Finlayson, 1978).

If Crank–Nicholson's (or Gir's) scheme is used in the time domain, it is also a centered approximation that limits numerical dispersion.

Stability also requires that the current number $C_0 = (U/\omega_c)(\Delta t/\Delta x)$ be on the order of 1, but in practice it should stay smaller than $\frac{1}{3}$ (Neuman, 1980). This relates to the displacement distance of the convective front with respect to the mesh size and is a constraint on the time step.

The choice of a mesh appropriate to the flow system greatly increases the stability and precision of the finite-element method. In practice, one tries to use a mesh that more or less follows the flow lines of the system. Away from the zone of displacement of a sharp front, the criterion $P_e < 2$ can be relaxed; values up to 20 have been used without major instabilities: as time increases in the simulation and fronts become less sharp, the criteria on both the Peclet and current numbers can be progressively relaxed.

It is also possible to use apparent dispersion coefficients smaller than the actual ones, to correct the numerical dispersion approximately, but these corrections are difficult to evaluate if the mesh is irregular and the medium properties vary in space.

Note that in finite elements the anisotropy of the dispersion tensor can easily be accounted for as shown in Section 12.3.1.

Recently, procedures incorporating some features of the method of characteristics into the finite element formulation, i.e., moving or deformable meshes, have been suggested by Varoglu and Finn (1978), Neuman (1980), Cady and Neuman (1986), and Farmer (1986).

### 12.5.4. Determination of the Velocity

If the velocity is not prescribed (e.g., uniform velocity field), it must be calculated with an ordinary flow model, generally in a steady state. In finite differences, the components of the velocity are calculated between each node by applying Darcy's law. In linear finite elements the components of the velocity will be given by differentiating the bilinear basis functions to get the derivatives of the head and multiplying them by the hydraulic conductivity.

However, if the solute transport can modify the mass per unit volume of the fluid, the flow and transport equations are coupled and must be solved successively at each time step (see Section 10.1.1.d).

## 12.6. Use of a Model

In the preceding sections, we have shown how to set up the equations of a groundwater flow or transport problem, how to represent the boundary conditions, the source/sink terms and some special featues (e.g., leakage, wells) and finally, how to solve the equations. In this section we shall look at the data collecting, the choice of the parameters, the fitting of the model, and, finally, how to use the model for prediction. We shall essentially discuss the problem of regional groundwater flow modeling, which must, in any case, precede any transport modeling.

### 12.6.1. Data Collecting

The geometry of the aquifer(s) must first be described. Besides mapping the outrcrops, this entails making a synthesis of all the well logs, especially in sedimentary basins with multilayered systems. The continuity and thickness of each layer (aquifer and aquitard) have to be estimated. For unconfined aquifers the elevation of the bedrock (or underlying aquitard) must be known. All these data can be estimated in space with kriging from the borehole measurements. Surface geophysical measurements (seismic or electric) can also be used.

Then the transmissivities must be estimated. Data may come from pumping tests, specific capacity, slug tests, borehole flow meter, hydraulic conductivity measurements on cores, or only from the thickness of layers and description of material, which makes it possible to correlate areas where pumping tests have been made with areas where they are lacking. Again, kriging can be used to interpolate this information. In general, log $T$ will be used and averages will be estimated directly in each of the meshes of the model. If there is a systematic drift in the transmissivity of the aquifer (thickness or conductivity or both), and if this drift is approximately known from the borehole geological data, it can be imposed in the universal kriging approach by writing $m(x) = ap(x)$, where $p(x)$ is this prescribed drift. In universal kriging, the local optimal value of $a$ will be determined automatically.

Whether it is kriging or not, an automatic procedure capable of estimating the value of a parameter in the mesh of a model from local measurements helps to save time in setting up a model.

Storage coefficients, or specific yield, generally come from long-term pumping tests. Short-term pumping tests (less than a month) generally underestimate the specific yield, but there is, in general, less variability of these coefficients than of transmissivity. Storage coefficents can be roughly estimated simply from thickness and compressibility of layers as a function of the geologic description.

Interpretation of natural fluctuations of water levels in wells will generally give a good estimation of the diffusivity $T/S($ or $T/\omega_d)$. In an unconfined aquifer one will select one or several flow lines with several piezometers and try to interpret the decay of the piezometric level over several months in one dimension, following the rise of the wet season. Analytic solutions given in Section 8.5.a will be used. Alternatively, in the vicinity of a river, the fluctuations of the piezometric level, caused by the fluctuations of the water level in the river, can be used.

For confined aquifers natural piezometric fluctuations are linked to barometric variations, earth tide, and sometimes seismic events. The first two can be interpreted in terms of diffusivity (see Ardity, 1978). Diffusivity can help to assess both transmissivity and storativity, depending on which one is best known.

Leakage factors of aquitards, as well as their storativity, will come from long pumping tests with several observation holes inside the aquitards above and below them (see Neuman and Witherspoon, 1972). Estimations can also come from thicknesses of layers, geologic descriptions, and tests on cores. Environmental tracers (e.g., $^{14}C$, $^{36}Cl$) can provide information concerning the velocity of transfer through an aquitard or an aquifer, but there are still large uncertainties about these tracers, especially for very-long-term transfers (*in situ* radiogenic sources, interactions with the medium, etc.).

Extraction from the aquifers through the wells has to be determined, which may prove rather difficult. Nevertheless, this may be as important for the calibration of the model as estimating transmissivities. Apart from exhaustive surveys and inquiries, indirect estimations can be based on agricultural or industrial production given the water requirements for each product, areas of irrigation surfaces (e.g., from satellite images), energy consumption for pumping (electricity or oil), and population and livestock density (domestic water supply). A monthly extraction rate must often be estimated from the annual consumption. Frequently, these data must be known over several years.

Natural discharge at springs or into rivers must be measured, e.g., by gauging the river in different sections, at low flow, when most of the water comes from groundwater.

Direct evapotranspiration in low lands where the water level in the aquifer is close to the surface will be evaluated from empirical formulas (see Section 1.3 and Appendix 1).

Recharge is also of paramount importance (see Sections 1.3 and 12.2.11). In semiarid or arid zones it may sometimes be more important to assess the variability of the recharge than the actual recharge in a given year: one often finds that in such climates 80% or more of the average recharge over a long period may come from one "catastrophic" event, occurring every 20 or 30 years. Recharge from rivers to an aquifer is often difficult to evaluate. In some cases differential gauging can be used, or, if the river is "polluted" by a good tracer, the dilution of this tracer can help estimate the recharge.

Piezometric heads will be required for calibration of the model. Piezometric maps in a steady state (if there is one) or at several dates in transient state are necessary and can be built by kriging. There is often a problem of precision in these maps if the piezometers or wells are not precisely topographically leveled. Some maps may be used as initial conditions. Piezometric records, over several years if there are significant variations, are necessary at several locations.

The nature and position of the boundary conditions must be determined often by looking at the piezometric maps (see Section 6.3). For long-term planning of aquifer development, it is essential that natural boundaries of the aquifers be used. Taking artificial limits at some distance in an aquifer and imposing an arbitrary condition (prescribed head or prescribed flux) will, in general, lead to significant errors in the long term.

### 12.6.2. *Choice of Parameters*

This essentially concerns the size of the mesh and of the time step. It must be realized that these parameters are completely independent of the quality and availability of the data described in Section 12.6.1: they are only related to the required precision in the solution of a discretized partial differential equation. As all numerical models globally ensure mass balance, if the choice of the mesh or of the time steps is poor, the global water balance in the system will still be correct (input, output, storage); only the heads or velocities will be approximated.

In order to define the mesh, one starts from the "center(s)" of the model, i.e., the areas where the head and velocities have to be known with precision. These areas generally coincide with areas of development (wells, drainage, etc.) where hydraulic gradients are variable. Given the type of problem, the smallest mesh may represent a complete well field or, on the contrary, each well will be included in a different mesh. Having then defined the position of the actual boundaries of the aquifer, one must choose the rest of the mesh in order (1) to fill the gap between the boundaries and the "center(s)"; (2) gradually increase the meash size, starting from these centers; (3) keep small meshes in areas where the head gradient varies significantly (e.g., around rivers acting as drains or sources), whereas large meshes can be used in areas of uniform head gradient; and (4) keep the total number of meshes below a limit fixed by the computer resources available (core storage and amount of CPU time). An average number is a few thousands (from 500 to 10,000).

In finite differences, nested squares, as shown in Section 12.2.4 or Section 11.8.2, are ideally suited to this purpose both in the plane and, in the case of multilayered systems, in the vertical direction (see Section 12.2.8). They eliminate the need for submodels, which are used to increase the precision on the "centers." A "regional" model with coarse meshes would give the global behavior of the aquifer. A local submodel would then represent each "center" using prescribed head boundary conditions on its outer artificial limits. These prescribed heads would come from the "regional" model. This technique is feasible but very cumbersome.

If the natural limits of the aquifer are really impossible to reach because they are extremely far away, one must sometimes use artificial limits and boundary conditions, e.g., prescribed heads along a contour line of the piezometric map or no-flow along a flow line. In such cases it is recommended that one check the influence of the boundary conditions on the long-term predictions of the model: change prescribed head to prescribed flux, and vice versa. If the resulting changes are insignificant, the model is valid.

Note that one can sometimes use different meshes for the calibration and the prediction phase of a model. There is, for instance, no point in having a finely discretized "center" in the calibration phase if this center represents a well field that does not yet exist. One must be able to change the discretization inside the model easily without changing the numbering and other characteristics of all the other meshes.

Finally, in case of uncertainty concerning the choice of the mesh, tests can always be made with a finer mesh to check that the results are not greatly modified.

The choice of the time step in a transient state follows roughly the same rules. Small time steps must be used (if precision is required) each time there is a change in the slope of $\partial h/\partial t$. In particular, each time a well (or well field) is started, stopped, or its flow rate modified, small time steps must be used. They can start at an hour, a day, a week, etc. depending on the mesh size, the parameters, and the period of interest. A good order of magnitude for the suitable time step is given by the "critical time step" $\Delta t_c$ of the explicit formulation (see Section 12.2.6), even if an implicit or Crank–Nicholson formulation is used. Each time a source term, boundary condition, etc. is changed the time step to use is a fraction $(\frac{1}{10}, \frac{1}{2})$ or a multiple $(2, 10)$ of the critical time step. In general, the time steps are then increased as a geometric progression of ratio 1.2 to 1.5 ($\sqrt{2}$ is often a good choice) until the next change in the system requires one to start with small time steps again. Very often, the geometric progression is stopped at a given length (e.g., a week, a month, a year) in order to obtain outputs at regular intervals and use constant time steps.

One can compromise and use constant time steps all through the simulation that are larger than the critical time step but short compared to the length of the simulation. In this case a rule of thumb is that five time steps without significant changes in the source/sink functions or boundary conditions are required before the calculated head can be considered accurate.

Note that an explicit formulation should be used each time the time step is shorter than $\Delta t_c$ and an implicit or Crank–Nicholson one in other cases. If iterative methods of solution are used, a constant time step is helpful because one can try to optimize the relaxation factor. A compromise can also be found between the length of the time step and the number of iterations: in general, the larger the former, the larger the latter.

It is always possible to check the suitability of the time steps by running the simulation again with twice as many time steps of half the length to see if the results are identical.

### 12.6.3. Calibration

This is the most important phase in the construction of the model. Most of the time, the procedure used is called "trial and error." All the data and parameters are introduced as described above. Then, one tries to calculate the head for a period with available observed heads. Finally, one compares the two. The model is said to be calibrated when the comparison is favorable, i.e., the differences between the two are considered negligible.

If this is not achieved, the parameters are modified in the appropriate direction in order to improve the fitting, i.e., decrease this difference. Most of the time the transmissivities are the least known parameters and thus are considerably modified during these trials. If kriging has been used for estimating $T$, the standard deviation of the estimation error provides a useful guide to keep the modifications of $T$ within the confidence interval, although this is not an absolute rule. Often the initial estimations of leakage factors, storativity, recharge and discharge, and boundary conditions must also be modified: the model helps to find a compromise between the various estimations of not-so-well-known independent parameters. The final maps of parameters, after calibration, are therefore much more reliable than before. However, calibration does not, in general, have a unique solution: different sets of parameters may fit the model. Only the hydrogeologist's own experience will tell which set has the most likelihood.

Concerning the observed heads, two types of calibration are possible: steady or transient state. If a real steady state can be found in an aquifer, it should certainly be simulated because then one of the unknown parameters drops out, the storativity. In this case one compares the observed and the computed head maps. If kriging has been used for estimating the observed head, the associated estimation error can also be used: it is only necessary that the calculated heads fall inside the confidence interval of the observed heads. If no true steady state can be found, it is always useful to select an approximate steady state (e.g., "average" head or low-water head) and to use it for calibration before continuing with transient fitting. An artificial source/sink term can always be estimated and added to the recharge/discharge term to represent $S \, \partial h/\partial t$.

If possible, transient fitting should always be made after the initial steady-state trials. The comparison can be made on several head maps at different times or only on the piezometric records at selected observation wells. It requires an evaluation of the variation with time of recharge and discharge but, in general, it produces a much better calibration with much less uncertainty than steady-state fitting.

In many cases the comparison can also be made on the calculated discharge from the aquifer, e.g., in a river, a spring, or an underground excavation. This greatly improves how representative the model is, because if no flow rates are available the linearity of the flow equation makes it possible to compensate for an error on $T$ by an error on $Q$ in a steady state without modifying the head.

Even if the fitting of the model is good, one must remember that the parameters estimated by fitting in an area with little data will always be questionable and that additional measurements must be made prior to any development in such an area.

Automatic fitting of a model has been called "the inverse problem." There is a great deal of literature on this topic, although very few models have actually been fitted with automatic methods so far.

An important aspect of automatic fitting is how to define the gradient of the objective function in the parameter space. When this objective function is the integral (in time and/or space) of the squares of the differences between observed and computed heads, Chavent et al. (1975) have shown that these gradients can easily be calculated with the help of an adjoint state equation. Neuman (1980) has also applied this technique. Solving the inverse problem is nowadays seen as a multicriterion problem. One must fit the model (i.e., minimize the objective function) while maximizing the likelihood or plausibility of the parameters, especially the transmissivities. Many, but not all, of the recently developed techniques rely on kriging to incorporate this likelihood [see, for instance Emsellem and Marsily (1971), Cooley (1977, 1979, 1982), Neuman (1973a, 1980), Neuman and Yakowitz (1979), Neuman et al. (1980), Yeh et al. (1983), Marsily et al. (1984), Carrera (1984), and Townley and Wilson (1985).]

## 12.6.4. Predicting with a Model

Once a model is calibrated, it can be used for predictions. These may range from extrapolations of existing conditions to the determination of the influence of new developments or works. Very often these predictions will require forecasts of the natural recharge of the aquifer. One may use an "average" annual recharge or simulate passed climatic series (e.g., the last 20 years or selected episodes of floods or droughts).

Another possibility, if the model is linear, is to apply the principle of superposition and to determine the drawdown due to a development in the aquifer directly (see Section 7.1.b). This drawdown will then be subtracted from the piezometric map of the record of any of the passed 20 years, for instance, in order to give a range for the possible influence of the development.

Another very helpful consequence of the linearity of the flow equation is the possibility of using linear programming in conjunction with a model in order to optimize both the location and the flowrate of wells in an aquifer to meet a given goal. In general, the objective function will depend on the total flowrate extracted from the aquifer (e.g., maximize the extraction, maximize the economic return from the extraction, minimize the pumping costs). The drawdown and flowrates per well will be limited by physical reasons (e.g., maximum flow per well, maximum economical drawdown) and the relations between extraction in one well and drawdown anywhere else in the aquifer will be linear and additive. This is, by definition, a linear programming problem.

The drawdown at any location for a unit extraction rate in a well is usually called an influence coefficient. It can be calculated easily with a model, either in a steady state or as a function of time in a transient state. A whole matrix of these coefficients is calculated: (1) locations where wells could be drilled and (2) locations where the drawdown should be limited [which can be the same location as (1)]. The result of the linear programming optimization is (1) the location of the "optimal" wells within the possible locations and (2) the "optimal" extraction rate in these wells.

This method is well described by Illangasekare and Morel-Seytoux (1982), Marsily et al. (1978), and Maddock (1972). Additional developments can be found in Morel-Seytoux et al. (1981) and Hubert (1984).

# Formulas for Estimating the Potential Evapotranspiration

### 1. Thornthwaite's Formula

The potential evapotranspiration ($ET_p$) per month or ten days is given by:

$$ET_p = 16(10\theta/I)^a \times F(\lambda)$$

Here, $ET_p$ is given in millimeters per month.

$\theta$      mean temperature of the period in question (°C) measured under shelter,

$a$      $6.75 \times 10^{-7} I^3 - 7.71 \times 10^{-5} I^2 + 1.79 \times 10^{-2} I + 0.49239$

$I$      annual thermal index, sum of twelve monthly thermal indexes $i$,

$i$      $(\theta/5)^{1.514}$

$F(\lambda)$      correction coefficient, function of the latitude and the month, given by Table A.1.1.

### 2. Turc's Formula

Turc prefers different formulas according to whether the mean relative humidity is above or below 50%. If $U_m > 50\%$ (usual in temperate zones)

$$ET_p \quad (mm/10\,d) = 0.13 \frac{\theta}{\theta + 15}(R_g + 50)$$

If $U_m < 50\%$

$$ET_p \quad (mm/10\,d) = 0.13 \frac{\theta}{\theta + 15}(R_g + 50)\left[1 + \frac{50 - U_m}{70}\right]$$

$\theta$      mean temperature of the period in question (°C) measured under shelter,

$R_g$      overall solar radiation $\simeq I_{ga}(0.18 + 0.62\,h/H)$

**Table A.1.1.**

Correction Coefficient $F(\lambda)$ Depending on the Latitude and the Month[a]

| Lat. N. | J | F | M | A | M | J | J | A | S | O | N | D |
|---|---|---|---|---|---|---|---|---|---|---|---|---|
| 0 | 1.04 | 0.94 | 1.04 | 1.01 | 1.04 | 1.01 | 1.04 | 1.04 | 1.01 | 1.04 | 1.01 | 1.04 |
| 5 | 1.02 | 0.93 | 1.03 | 1.02 | 1.06 | 1.03 | 1.06 | 1.05 | 1.01 | 1.03 | 0.99 | 1.02 |
| 10 | 1.00 | 0.91 | 1.03 | 1.03 | 1.08 | 1.06 | 1.08 | 1.07 | 1.02 | 1.02 | 0.98 | 0.99 |
| 15 | 0.97 | 0.91 | 1.03 | 1.04 | 1.11 | 1.08 | 1.12 | 1.08 | 1.02 | 1.01 | 0.95 | 0.97 |
| 20 | 0.95 | 0.90 | 1.03 | 1.05 | 1.13 | 1.11 | 1.14 | 1.11 | 1.02 | 1.00 | 0.93 | 0.94 |
| 25 | 0.93 | 0.89 | 1.03 | 1.06 | 1.15 | 1.14 | 1.17 | 1.12 | 1.02 | 0.99 | 0.91 | 0.91 |
| 26 | 0.92 | 0.88 | 1.03 | 1.06 | 1.15 | 1.15 | 1.17 | 1.12 | 1.02 | 0.99 | 0.91 | 0.91 |
| 27 | 0.92 | 0.88 | 1.03 | 1.07 | 1.16 | 1.15 | 1.18 | 1.13 | 1.02 | 0.99 | 0.90 | 0.90 |
| 28 | 0.91 | 0.88 | 1.03 | 1.07 | 1.16 | 1.16 | 1.18 | 1.13 | 1.02 | 0.98 | 0.90 | 0.90 |
| 29 | 0.91 | 0.87 | 1.03 | 1.07 | 1.17 | 1.16 | 1.19 | 1.13 | 1.03 | 0.98 | 0.90 | 0.89 |
| 30 | 0.90 | 0.87 | 1.03 | 1.08 | 1.18 | 1.17 | 1.20 | 1.14 | 1.03 | 0.98 | 0.89 | 0.88 |
| 31 | 0.90 | 0.87 | 1.03 | 1.08 | 1.18 | 1.18 | 1.20 | 1.14 | 1.03 | 0.98 | 0.89 | 0.88 |
| 32 | 0.89 | 0.86 | 1.03 | 1.08 | 1.19 | 1.19 | 1.21 | 1.15 | 1.03 | 0.98 | 0.88 | 0.87 |
| 33 | 0.88 | 0.86 | 1.03 | 1.09 | 1.19 | 1.20 | 1.22 | 1.15 | 1.03 | 0.97 | 0.88 | 0.86 |
| 34 | 0.88 | 0.85 | 1.03 | 1.09 | 1.20 | 1.20 | 1.22 | 1.16 | 1.03 | 0.97 | 0.87 | 0.86 |
| 35 | 0.87 | 0.85 | 1.03 | 1.09 | 1.21 | 1.21 | 1.23 | 1.16 | 1.03 | 0.97 | 0.86 | 0.85 |
| 36 | 0.87 | 0.85 | 1.03 | 1.10 | 1.21 | 1.22 | 1.24 | 1.16 | 1.03 | 0.97 | 0.86 | 0.84 |
| 37 | 0.86 | 0.84 | 1.03 | 1.10 | 1.22 | 1.23 | 1.25 | 1.17 | 1.03 | 0.97 | 0.85 | 0.83 |
| 38 | 0.85 | 0.84 | 1.03 | 1.10 | 1.23 | 1.24 | 1.25 | 1.17 | 1.04 | 0.96 | 0.84 | 0.83 |
| 39 | 0.85 | 0.84 | 1.03 | 1.11 | 1.23 | 1.24 | 1.26 | 1.18 | 1.04 | 0.96 | 0.84 | 0.82 |
| 40 | 0.84 | 0.83 | 1.03 | 1.11 | 1.24 | 1.25 | 1.27 | 1.18 | 1.04 | 0.96 | 0.83 | 0.81 |
| 41 | 0.83 | 0.83 | 1.03 | 1.11 | 1.25 | 1.26 | 1.27 | 1.19 | 1.04 | 0.96 | 0.82 | 0.80 |
| 42 | 0.82 | 0.83 | 1.03 | 1.12 | 1.26 | 1.27 | 1.28 | 1.19 | 1.04 | 0.95 | 0.82 | 0.79 |
| 43 | 0.81 | 0.82 | 1.02 | 1.12 | 1.26 | 1.28 | 1.29 | 1.20 | 1.04 | 0.95 | 0.81 | 0.77 |
| 44 | 0.81 | 0.82 | 1.02 | 1.13 | 1.27 | 1.29 | 1.30 | 1.20 | 1.04 | 0.95 | 0.80 | 0.76 |
| 45 | 0.80 | 0.81 | 1.02 | 1.13 | 1.28 | 1.29 | 1.31 | 1.21 | 1.04 | 0.94 | 0.79 | 0.75 |
| 46 | 0.79 | 0.81 | 1.02 | 1.13 | 1.29 | 1.31 | 1.32 | 1.22 | 1.04 | 0.94 | 0.79 | 0.74 |
| 47 | 0.77 | 0.80 | 1.02 | 1.14 | 1.30 | 1.32 | 1.33 | 1.22 | 1.04 | 0.93 | 0.78 | 0.73 |
| 48 | 0.76 | 0.80 | 1.02 | 1.14 | 1.31 | 1.33 | 1.34 | 1.23 | 1.05 | 0.93 | 0.77 | 0.72 |
| 49 | 0.75 | 0.79 | 1.02 | 1.14 | 1.32 | 1.34 | 1.35 | 1.24 | 1.05 | 0.93 | 0.76 | 0.71 |
| 50 | 0.74 | 0.78 | 1.02 | 1.15 | 1.33 | 1.36 | 1.37 | 1.25 | 1.06 | 0.92 | 0.76 | 0.70 |

Lat. S.

| Lat. S. | J | F | M | A | M | J | J | A | S | O | N | D |
|---|---|---|---|---|---|---|---|---|---|---|---|---|
| 5 | 1.06 | 0.95 | 1.04 | 1.00 | 1.02 | 0.99 | 1.02 | 1.03 | 1.00 | 1.05 | 1.03 | 1.06 |
| 10 | 1.08 | 0.97 | 1.05 | 0.99 | 1.01 | 0.96 | 1.00 | 1.01 | 1.00 | 1.06 | 1.05 | 1.10 |
| 15 | 1.12 | 0.98 | 1.05 | 0.98 | 0.98 | 0.94 | 0.97 | 1.00 | 1.00 | 1.07 | 1.07 | 1.12 |
| 20 | 1.14 | 1.00 | 1.05 | 0.97 | 0.96 | 0.91 | 0.95 | 0.99 | 1.00 | 1.08 | 1.09 | 1.15 |
| 25 | 1.17 | 1.01 | 1.05 | 0.96 | 0.94 | 0.88 | 0.93 | 0.98 | 1.00 | 1.10 | 1.11 | 1.18 |
| 30 | 1.20 | 1.03 | 1.06 | 0.95 | 0.92 | 0.85 | 0.90 | 0.96 | 1.00 | 1.12 | 1.14 | 1.21 |
| 35 | 1.23 | 1.04 | 1.06 | 0.94 | 0.89 | 0.82 | 0.87 | 0.94 | 1.00 | 1.13 | 1.17 | 1.25 |
| 40 | 1.27 | 1.06 | 1.07 | 0.93 | 0.86 | 0.78 | 0.84 | 0.92 | 1.00 | 1.15 | 1.20 | 1.29 |
| 42 | 1.28 | 1.07 | 1.07 | 0.92 | 0.85 | 0.76 | 0.82 | 0.92 | 1.00 | 1.16 | 1.22 | 1.31 |
| 44 | 1.30 | 1.08 | 1.07 | 0.92 | 0.83 | 0.74 | 0.81 | 0.91 | 0.99 | 1.17 | 1.23 | 1.33 |
| 46 | 1.32 | 1.10 | 1.07 | 0.91 | 0.82 | 0.72 | 0.79 | 0.90 | 0.99 | 1.17 | 1.25 | 1.35 |
| 48 | 1.34 | 1.11 | 1.08 | 0.90 | 0.80 | 0.70 | 0.76 | 0.89 | 0.99 | 1.18 | 1.27 | 1.37 |
| 50 | 1.37 | 1.12 | 1.08 | 0.89 | 0.77 | 0.67 | 0.74 | 0.88 | 0.99 | 1.19 | 1.29 | 1.41 |

[a] Thornthwaite's formula, from Brochet and Gerbier (1974).

$h$     actual amount of sunshine in hours per day,

$H$     maximum possible amount of sunshine (astronomical length of the day),

$I_{g_a}$     direct solar radiation at the top of the atmosphere,

$I_{g_a}$ and $H$ are tabulated according to the latitude and the date on Tables A.1.2 and A.1.3.

**Table A. 1.2.**

Monthly $I_{g_a}$ Values in Small Calories per cm$^2$ of Horizontal Surface Area and per Day[a]

| Latitude North | 30° | 40° | 50° | 60° |
|---|---|---|---|---|
| January | 508 | 364 | 222 | 87.5 |
| February | 624 | 495 | 360 | 215 |
| March | 764 | 673 | 562 | 432 |
| April | 880 | 833 | 764 | 676 |
| May | 950 | 944 | 920 | 880 |
| June | 972 | 985 | 983 | 970 |
| July | 955 | 958 | 938 | 908 |
| August | 891 | 858 | 800 | 728 |
| September | 788 | 710 | 607 | 487 |
| October | 658 | 536 | 404 | 262 |
| November | 528 | 390 | 246 | 111 |
| December | 469 | 323 | 180 | 55.5 |

[a] From Brochet and Gerbier (1974)

**Table A.1.3.**

Length of the Astronomical Day $H$ (mean monthly values in hours per day)[a]

| Latitude North | 30° | 40° | 50° | 60° |
|---|---|---|---|---|
| January | 10.45 | 9.71 | 8.58 | 6.78 |
| February | 11.09 | 10.64 | 10.07 | 9.11 |
| March | 12.00 | 11.96 | 11.90 | 11.81 |
| April | 12.90 | 13.26 | 13.77 | 14.61 |
| May | 13.71 | 14.39 | 15.46 | 17.18 |
| June | 14.07 | 14.96 | 16.33 | 18.73 |
| July | 13.85 | 14.68 | 15.86 | 17.97 |
| August | 13.21 | 13.72 | 14.49 | 15.58 |
| September | 12.36 | 12.46 | 12.63 | 12.89 |
| October | 11.45 | 11.15 | 10.77 | 10.14 |
| November | 10.67 | 10.00 | 9.08 | 7.58 |
| December | 10.23 | 9.39 | 8.15 | 6.30 |

[a] From Brochet and Gerbier (1974)

## 3. Penman's Formula

$$E = \frac{1}{L} \frac{R_n(F'\theta/\gamma)}{1 + (F'\theta/\gamma)} + E_a \frac{1}{1 + (F'\theta/\gamma)}$$

$L$      latent heat of water evaporation (59 cal/cm$^2$ for 1 mm of equivalent water),

$R_n$      net radiation, evaluated from the formula

$$R_n = I_{ga}(1 - a)(0.18 + 0.62h/H) - \sigma\theta^4(0.56 - 0.08\sqrt{e})(0.10 + 0.9h/H)$$

$E_a$      evaporating power of the air $= (e_w - e)0.26(1 + 0.4V)$,

$\gamma$      psychometric constant ($\gamma \simeq 0.65$),

$a$      albedo of the evaporating surface (generally $a = 0.25$),

$I_{ga}$      direct solar radiation at the top of the atmosphere,

$h$      actual amount of sunshine,

$H$      astronomical length of the day,

$\theta$      air temperature under shelter (K),

$\sigma$      $1.19 \times 10^7$ cal/cm$^2 \times$ d $\times$ K,

$e$      tension of the water vapor measured under shelter, in mbar

$e_w$      maximum tension of the water vapor in mbar for the temperature $\theta$,

$V$      mean wind velocity measured at 10 m above the evaporating surface (m/s),

$F'\theta$      slope of the curve of maximum water vapor tension

# *Appendix 2*

## Commonly Used Physical Quantities

### 1. Measurement Units of the International System (SI)

#### 1.1. *Basic Units*

There are seven basic units in the SI system given in Table A.2.1 with their dimensions and abbreviations. In front of each unit, a prefix, given in Table A.2.2, can be added to scale the unit. These prefixes are attached to the basic symbols. For example, 1000 A = 1 kA.

#### 1.2. *Geometric Units*

The following lists give the various units, their abbreviations, and conversion factors. See also, table A.2.3 for conversion factors.

*Length.*   The basic unit is the meter (m).

*Metric units*      Micrometer (mm)    $10^{-6}$ m,
                    Ångström (Å)       $10^{-10}$ m,

*English units*

| | |
|---|---|
| inch | 0.0254 m, |
| foot | 0.3048 m, |
| yard | 0.9143 m, |
| mile | $1.609 \times 10^3$ m, |
| nautical mile | $1.8532 \times 10^3$ m. |

*Surface area.*   The basic unit is the meter squared ($m^2$).

*Metric units*

| | |
|---|---|
| hectare (ha) | 10,000 $m^2$, |
| are (a) | 100 $m^2$, |

**Table A.2.1**

Basic SI Units

| Unit | Dimension | Abbreviation |
|------|-----------|--------------|
| meter | length | m |
| kilogram | mass | kg |
| second | time | s |
| ampere | electric current intensity | A |
| kelvin | temperature | K |
| candela | luminous intensity | cd |
| mole | quantity of matter | mol. |

**Table A.2.2**

SI Prefixes

| Prefix | Symbol | Factor by which the unit is multiplied | |
|--------|--------|------------------------------|---|
| exa | E | 1 000 000 000 000 000 000 | $= 10^{18}$ |
| peta | P | 1 000 000 000 000 000 | $= 10^{15}$ |
| tera | T | 1 000 000 000 000 | $= 10^{12}$ |
| giga | G | 1 000 000 000 | $= 10^{9}$ |
| mega | M | 1 000 000 | $= 10^{6}$ |
| kilo | k | 1 000 | $= 10^{3}$ |
| hecto | h | 100 | $= 10^{2}$ |
| deca | da | 10 | $= 10$ |
| deci | d | 0.1 | $= 10^{-1}$ |
| centi | c | 0.01 | $= 10^{-2}$ |
| milli | m | 0.001 | $= 10^{-3}$ |
| micron | $\mu$ | 0.000 001 | $= 10^{-6}$ |
| nano | n | 0.000 000 001 | $= 10^{-9}$ |
| pico | p | 0.000 000 000 001 | $= 10^{-12}$ |
| femto | f | 0.000 000 000 000 001 | $= 10^{-15}$ |
| atto | a | 0.000 000 000 000 000 001 | $= 10^{-18}$ |

*English units*

$$\text{square foot (ft}^2) \quad 9.29 \times 10^{-2} \text{ m}^2,$$
$$\text{acre} \quad 4.047 \times 10^{3} \text{ m}^2.$$

*Volume.*   The basic unit is the meter cubed ($m^3$).

*Metric units*

The liter (0.001 $m^3$) must not be used instead of the $dm^3$ when extremely precise results are desired.

*English units*

| | |
|---|---|
| liquid ounce | $2.95412 \times 10^{-5}$ m$^3$, |
| ft$^3$ | $2.832 \times 10^{-2}$ m$^3$, |
| US gal | $3.785 \times 10^{-3}$ m$^3$, |
| UK gal | $4.546 \times 10^{-3}$ m$^3$, |
| barrel of oil | $0.156$ m$^3$. |

*Plane angle.* The basic unit is the radian (rad). The degree is equal to $\pi/180$ rad, and the grad is equal to $\pi/200$ rad.

*Solid angle.* The basic unit is the steradian (sr). It is the solid angle with its apex in the center of a sphere and subtending an area on the surface of the sphere equal to that of a square with side the length of the radius of the sphere.

### 1.3. *Units of Mass and Matter*

*Mass*

*Metric units.* The basic unit is the kg.

| | |
|---|---|
| metric ton (t) | 1,000 kg (also written Mg) |
| quintal (q) | 100 kg, |

*English units*

| | |
|---|---|
| ounce (oz) | $2,835 \times 10^{-2}$ kg, |
| pound (lb) | 0.4536 kg, |
| ton (short) (tn.s) | $0.907 \times 10^3$ kg, |
| ton (long) (tn.1) | $1.016 \times 10^3$ kg. |

*Mass per unit volume.* The basic unit is the kilogram per cubic meter (kg/m$^3$).

*Quantity of matter.* The basic unit is the mole (mol). This is the quantity of matter in a system containing the same amount of elementary entities as there are atoms in 0.012 kg of carbon 12 (the nature of the entities must be specified).

*Concentration.* The basic unit is the mol per cubic meter (mol/m$^3$). Volumetric concentrations in kg/m$^3$ and massic concentrations in kg/kg are also used. (See also the introduction of Chapter 10 for other concentration units.)

### 1.4. *Mechanical Units*

*Velocity.* The basic unit is the meter per second (m/s). In navigation, 1 knot = 0.514444 m/s.

**Table A.2.3**

SI–English, English–SI Conversion Table

| Units | English → SI | | SI → English | |
|---|---|---|---|---|
| Length | 1 inch | = 2.54 cm | 1 cm | = 0.3937 inch |
| | 1 ft | = 0.3048 m | 1 m | = 3.281 ft |
| | 1 mi | = 1.609 km | 1 km | = 0.6215 mi |
| Area | 1 inch$^2$ | = 6.4516 cm$^2$ | 1 cm$^2$ | = 0.155 inch$^2$ |
| | 1 ft$^2$ | = 0.0929 m$^2$ | 1 m$^2$ | = 10.76 ft$^2$ |
| | 1 acre | = 0.4047 ha | 1 ha | = 2.471 acres |
| | 1 mi$^2$ | = 0.4047 × 10$^4$ m$^2$ | 1 km$^2$ | = 0.3861 mi$^2$ |
| | | = 2.590 km$^2$ | | |
| Volume | 1 US fl oz | = 29.54 cm$^3$ | 1 cm$^3$ | = 3.385 × 10$^{-2}$ Fl.oz |
| | 1 ft$^3$ | = 2.832 × 10$^{-2}$ m$^3$ | 1 l | = 3.531 × 10$^{-2}$ ft$^3$ |
| | | = 28.32 liter | 1 l | = 0.2642 US gal |
| | 1 US gal | = 3.785 × 10$^{-3}$ m$^3$ | 1 l | = 0.2200 UK gal |
| | | = 3.785 liter | 1 m$^3$ | = 264.2 US gal |
| | 1 UK gal | = 4.546 × 10$^{-3}$ m$^3$ | 1 m$^3$ | = 220.0 UK gal |
| | | = 4.546 liter | 1 m$^3$ | = 28.38 US bushel |
| | 1 US bushel | = 3.524 × 10$^{-2}$ m$^3$ | 1 m$^3$ | = 6.41 oil barrel |
| | | = 35.24 liter | | |
| | 1 oil barrel | = 0.156 m$^3$ | | |
| | | = 156 liter | | |
| Flow rate | 1 cubic ft/s | = 2.832 × 10$^{-2}$ m$^3$/s | 1 m$^3$/s | = 35.311 ft$^3$/s |
| | | = 28.32 liter/s | 1 l/s | = 0.0353 ft$^3$/s |
| | 1 US gal/min | = 6.309 × 10$^{-5}$ m$^3$/s | 1 l/s | = 10$^{-3}$ m$^3$/s |
| | | = 6.309 × 10$^{-2}$ liter/s | 1 l/s | = 15.85 US gal/min |
| | 1 UK gal/min | = 7.576 × 10$^{-5}$ m$^3$/s | 1 l/s | = 13.20 UK gal/min |
| | | = 7.576 × 10$^{-2}$ liter/s | | |

**Table A.2.3** (*Continued*)

| Units | English → SI | | SI → English | |
|---|---|---|---|---|
| Mass | 1 oz | $= 28.35$ g | 1 g | $= 3.257 \times 10^{-2}$ oz |
| | 1 lb$_m$ | $= 0.4536$ kg | 1 kg | $= 2.205$ lb$_m$ |
| | 1 s. ton | $= 907$ kg | 1 metric ton | $= 1{,}000$ kg |
| | 1 l. ton | $= 1{,}016$ kg | 1 metric ton | $= 1.103$ s. ton |
| | | | 1 metric ton | $= 0.984$ l. ton |
| Mass per unit volume | 1 lb$_m$/ft$^3$ | $= 16.02$ kg/m$^3$ | 1 kg/m$^3$ | $= 6.242 \times 10^{-2}$ lb$_m$/ft$^3$ |
| | | $= 16.02$ g/liter | | $= 1$ g/liter |
| Force | 1 lb$_f$ | $= 4.448$ N | 1 N | $= 0.2248$ lb$_f$ |
| Stress and pressure | 1 lb$_f$/foot$^2$ | $= 47.88$ Pa | 1 Pa | $= 2.089 \times 10^{-2}$ lb$_f$/ft$^2$ |
| | 1 psi | $= 6.895 \times 10^3$ Pa | 1 Pa | $= 1.450 \times 10^{-4}$ psi |
| | 1 atm | $= 1.013 \times 10^5$ Pa | 1 Pa | $= 10^{-5}$ bars |
| | 1 bar | $= 10^5$ Pa | 1 MPa | $= 10$ bars |
| | | $= 0.1$ MPa | | |
| Work or energy | 1 ft lb$_f$ | $= 1.356$ J | 1 J | $= 0.7374$ ft lb$_f$ |
| | 1 calorie | $= 4.185$ J | 1 J | $= 0.2389$ calorie |
| | 1 BTU | $= 1.055 \times 10^3$ J | 1 J | $= 9.479 \times 10^{-4}$ BTU |
| Temperature | $x°\text{F} = \frac{5}{9}(x - 32)°\text{C}$ | | $x°\text{C} = \frac{9}{5}x + 32°\text{F}$ | |
| | $- 459.69°\text{F} = 0$ K | | $- 273.15°\text{C} = 0$ K | |
| Hydraulic conductivity | 1 ft/s | $= 0.3048$ m/s | 1 m/s | $= 3.281$ ft/s |
| | 1 US gal/day ft$^2$ | $= 4.720 \times 10^{-7}$ m/s | 1 m/s | $= 2.119 \times 10^6$ US gal/day ft$^2$ |
| Transmissivity | 1 ft$^2$/s | $= 9.290 \times 10^{-2}$ m$^2$/s | 1 m$^2$/s | $= 10.76$ ft$^2$/s |
| | 1 US gal/day ft | $= 1.438 \times 10^{-7}$ m$^2$/s | 1 m$^2$/s | $= 6.954 \times 10^6$ US gal/day ft |
| Intrinsic permeability | 1 ft$^2$ | $= 9.290 \times 10^{-2}$ m$^2$ | 1 m$^2$ | $= 10.76$ ft$^2$ |
| | | $= 9.412 \times 10^{10}$ darcy | 1 m$^2$ | $= 1.013 \times 10^{12}$ darcy |
| | 1 darcy | $= 0.987 \times 10^{-12}$ m$^2$ | | |

*Acceleration.*   The basic unit is the meter per second squared (m/s²). The acceleration due to gravity is $g = 9.80665$ m/s².

*Angular velocity.*   The basic unit is the radian per second (rad/s).

*Frequency.*   The basic unit, the hertz (Hz), is equal to 1 cycle/s.

*Force.*   The basic unit is the newton (N). It is the force which gives a body with a mass of 1 kg an acceleration of 1 m/s², dimension [MLT⁻²].[†] The dyne is equal to $10^{-5}$ N. Gravity produces a force of 9.80665 N on a mass of 1 kg.

*Moment of a force.*   The basic unit is newton meter (N m).

*Stress and pressure.*   The basic unit is the pascal (Pa) produced by a force of 1 N applied over an area of 1 m², dimension [ML⁻¹T⁻²]. The megapascal (MPa), i.e., $10^6$ Pa, is used more frequently. The bar ($10^5$ Pa) and the millibar ($10^2$ Pa) are also used.

Do not use

| | |
|---|---|
| standard atmosphere | 1.0133 bar $= 1.0133 \times 10^5$ Pa, |
| mm of mercury | $1.33322 \times 10^2$ Pa, |
| m of water at 4°C | $9.80638 \times 10^3$ Pa, |
| dyne/cm² | 0.1 Pa, |
| psi (pound per square inch) | $6.895 \times 10^3$ Pa, |
| kg/cm² | $0.981 \times 10^5$ Pa. |

*Dynamic viscosity* ($\mu$).   The basic unit is the pascal second (Pa s) dimension [ML⁻¹T⁻¹]. The poise is equal to 0.1 Pa s.

*Kinematic viscosity* ($v = \mu/\rho$).   The basic unit is the meter squared per second (m²/s). The stokes is equal to $10^{-4}$ m²/s.

## 1.5. Energy Units

*Work or quantity of heat.*   The basic unit is the joule (J). It is the work done by a force of 1 N moving its point of application through 1 m (dimension [ML²T⁻²]). The erg is equal to $10^{-7}$ J, the kWh to $3.6 \times 10^6$ J, the calorie (small) to 4.185 J. The calorie is the energy required to increase the temperature of 1 g of water by 1°C. The kilocalorie (or large calorie) is $4.185 \times 10^3$ J, the therm ($10^3$ kilocalories) is $4.185 \times 10^6$ J, and the BTU (British Thermal Unit) is $1.055 \times 10^3$ J.

---

[†] Dimensions are given in brackets with capital letters. M = mass, L = length, T = time, K = temperature.

*Power*

| watt (W) | 1 J/s, dimension $[ML^2T^{-3}]$, |
|---|---|
| 1 kW | $10^3$ W; 1 MW = $10^6$ W; 1 GW = $10^9$ W, |
| horsepower | 736 W, |
| erg/s | $10^{-7}$ W, |
| BTU/s | $1.055 \times 10^3$ W. |

*Thermodynamic temperature.* The kelvin (K) is the basic unit. The celsius degree (°C) is the same unit of temperature, but the Celsius scale has its zero at 273.15 K.

*Heat conductivity* $\lambda$. The basic unit is W/m K $[MLT^{-3}K^{-1}]$. The kcal/s m °C = $4.18 \times 10^3$ W/m K is also used.

*Massic heat capacity* $c$. The basic unit is J/kg K $[L^2T^{-2}K^{-1}]$. The kcal/kg °C = $4.18 \times 10^3$ J/kg K is also used.

*Volumetric heat capacity* $\rho c$. The basic unit is J/m³ K $[ML^{-1}T^{-2}K^{-1}]$. The kcal/m³ °C = $4.18 \times 10^3$ J/m³ K is also used.

*Heat diffusivity* $\lambda/\rho c$. The basic unit is m²/s $[L^2T^{-1}]$.

## 1.6. Electric Units

*Intensity of electric current.* The basic unit is the ampere (A).

*Electric charge.* The basic unit is the coulomb (C). It is the quantity of electricity transported in one second by a current of 1 ampere.

*Electrical potential.* The basic unit is the volt (V). It is the difference in potential that dissipates a power of 1 W for a constant current of 1 A.

*Electric resistance.* The basic unit is the ohm ($\Omega$). It is the resistance of a conductor where 1 A circulates under a difference of potential of 1 V. The siemens (S) is the conductance (the inverse of the resistance) of a conductor with a resistance of 1 $\Omega$.

*Electric capacitance.* The basic unit is the farad (F). It is the capacitance of a condensor that becomes charged with 1 C under a difference of potential of 1 V.

*Electric inductance.* The basic unit is the henry (H). It is the electric inductance of a closed circuit in which a potential difference of 1 V is produced when the current going through the circuit varies uniformly at a rate of 1A/s.

*Magnetic flux.* The basic unit is the weber (Wb). It produces a potential difference of 1 V through a circuit of one single coil, if it is reduced to zero in 1 s by uniform decrease.

*Magnetic flux density.* The basic unit is the tesla (T). It is the uniform magnetic flux density that produces a total magnetic flux of 1 Wb across a normal plane surface of 1 m² over which it is uniformly distributed.

*Magnetic motive force.* The basic unit is the ampere (A). It is the magnetic motive force that corresponds to a current of 1 A in a single coil.

*Intensity of the magnetic field.* The basic unit is the ampere per meter (A/m). It is the intensity of the magnetic field created at the center of a circuit with a diameter of 1 m by the passing of a current of 1 A through the circuit, which is constituted by a conducting wire of negligible cross-section area.

## 1.7. Radiological Units

*Radionuclear activity.* The basic unit is the becquerel (Bq). It corresponds to one disintegration per second of a radioactive body; the curie (Ci) is equal to 37 GBq (gigabecquerel).

*Half-life.* This is the length of time needed for one-half of the initial mass of the radioactive element to disappear by radioactive decay.

*Quantity of x or γ radiation.* The basic unit is the coulomb per kilogram (C/kg). It is the quantity of x or γ radiation that is such that the corpuscular

Table A.2.4

| Temperature (°C) | Specific Weight (kN/m³), $\gamma = \rho g$ | Mass per unit volume, $\rho$ (kg/m³) | Dynamic viscosity, $\mu$ ($10^3$/Pa s) | Kinematic viscosity, $v(= \mu/\rho)$ ($10^{-6}$ m²/s) | Latent heat of vaporization (J/g) |
|---|---|---|---|---|---|
| 0 | 9.805 | 999.8 | 1.781 | 1.785 | 2500.3 |
| 5 | 9.807 | 1000.0 | 1.518 | 1.519 | 2488.6 |
| 10 | 9.804 | 999.7 | 1.307 | 1.306 | 2476.9 |
| 15 | 9.798 | 999.1 | 1.139 | 1.139 | 2465.1 |
| 20 | 9.789 | 998.2 | 1.002 | 1.003 | 2453.0 |
| 25 | 9.777 | 997.0 | 0.890 | 0.893 | 2441.3 |
| 30 | 9.764 | 995.7 | 0.798 | 0.800 | 2429.6 |
| 40 | 9.730 | 992.2 | 0.653 | 0.658 | 2405.7 |
| 50 | 9.689 | 988.0 | 0.547 | 0.553 | 2381.8 |
| 60 | 9.642 | 983.2 | 0.466 | 0.474 | 2357.6 |
| 70 | 9.589 | 977.8 | 0.404 | 0.413 | 2333.3 |
| 80 | 9.530 | 971.8 | 0.354 | 0.364 | 2308.2 |
| 90 | 9.466 | 965.3 | 0.315 | 0.326 | 2282.6 |
| 100 | 9.399 | 958.4 | 0.282 | 0.294 | 2256.7 |

emission associated with it in 1 kg of air produces ions in the air that transport a quantity of electricity (of either sign) equal to 1 C; the roentgen is equal to $2.58 \times 10^{-4}$ C/kg.

*Absorbed dose of ionizing radiation.* The basic unit is the gray (Gy). It corresponds to an absorbed energy equal to 1 J/kg; the rad is equal to 0.01 Gy.

*Effective biological dose.* The basic unit is the sievert (Sv). It is the dose caused by an ionizing radiation that has an effect equal to that of x or $\gamma$ radiation of 200 to 250 kV; the rem is equal to 0.01 Sv. The International Commission for Radiological Protection (ICRP) recommended dose limits are

0.05 Sv/yr for workers in the nuclear industry
0.005 Sv/yr for members of the public, from all possible sources
0.001 Sv/yr for members of the public, for long term exposure.

## 1.8. Optical Units

*Luminous intensity.* The basic unit is the candela (cd). It is the luminous intensity in a given direction from a source that emits monochromatic rays of a frequency of 540 THz and has an energy intensity in this direction of 1/683 watt per steradian.

Properties of Pure Water

| Absolute vapor pressure (kPa) | Compressibility ($10^{-10}$ Pa$^{-1}$) | Young's modulus ($10^6$ kPa) | Coefficient of volume trial heat expansion, $\alpha = d(\ln \rho)/dT$ ($10^{-6}$ K) | Specific mass heat, $C$ (J/kg K) | Heat conductivity, $\lambda$ (W/m K) | Heat diffusivity, $\lambda/\rho c$ ($10^{-8}$ m$^2$/s) |
|---|---|---|---|---|---|---|
| 0.61 | 5.098 | 2.02 | −68 | 4217.4 | 0.564 | 13.4 |
| 0.87 | 4.928 | 2.06 | 16 | — | — | — |
| 1.23 | 4.789 | 2.10 | 88 | 4191.9 | 0.578 | 13.8 |
| 1.70 | 4.678 | 2.15 | 151 | — | — | — |
| 2.34 | 4.591 | 2.18 | 207 | 4181.6 | 0.598 | 14.2 |
| 3.17 | 4.524 | 2.22 | 257 | — | — | |
| 4.24 | 4.475 | 2.25 | 303 | 4178.2 | 0.607 | 14.6 |
| 7.38 | 4.422 | 2.28 | 385 | 4178.3 | 0.628 | 15.2 |
| 12.33 | 4.417 | 2.29 | 458 | 4180.4 | — | — |
| 19.92 | 4.450 | 2.28 | 523 | 4184.1 | 0.652 | 15.8 |
| 31.16 | 4.515 | 2.25 | 584 | 4189.3 | — | — |
| 47.34 | 4.610 | 2.20 | 641 | 4196.1 | 0.669 | 16.4 |
| 70.10 | 4.734 | 2.14 | 696 | 4204.8 | — | — |
| 101.33 | 4.890 | 2.07 | 750 | 4215.7 | 0.671 | 16.6 |

*Luminance.*   The basic unit is the candela per square meter $(cd/m^2)$. It is the luminance of a source with an area of $1 m^2$ emitting with a luminous intensity of 1 cd. The stilb (sb) is equal to $10^4 cd/m^2$.

*Luminous flux.*   The basic unit is the lumen (lm). It is the luminous flux emitted in the solid angle of 1 sr by a uniform point source placed at the apex of the solid angle and having a luminous intensity of 1 cd.

*Illumination.*   The basic unit is the lux (lx). It is the illumination of an area which receives with normal incidence the flux of $1 lm/m^2$ uniformly distributed. The phot (ph) is equal to $10^4$ lux.

*Vergence of optical systems.*   The basic unit is the diopter ($\delta$). It is the refractive power of an optical system with a focal distance of 1 m in a medium with a refractive index of 1.

## 2. Values of Common Hydrogeological Quantities

### 2.1. *Properties of pure water*

These are given in Table A.2.4.

### 2.2. *Properties of Ice at $-5°C$*

| | |
|---|---|
| Mass per unit volume, $\rho$ | $917 kg/m^3$ |
| Latent heat of fusion | $334 \times 10^3 J/kg$ |
| Specific heat, $c$ | $2075 J/kg K$ |
| Heat conductivity, $\lambda$ | $2.3 W/m K$ |

### 2.3. *Properties of Saltwater (NaCl):*

**Table A.2.5**

Properties of Seawater at $34 kg/m^3$

| Temperature °C | Mass per unit volume, $\rho$ $(kg/m^3)$ | Specific heat, c $(J/kg K)$ | Kinematic viscosity, $\nu$ $(10^6 m^2/s)$ |
|---|---|---|---|
| 0 | 1027.32 | 3989 | 1.8 |
| 5 | 1026.91 | 3992 | 1.6 |
| 10 | 1026.19 | 3995 | 1.4 |
| 15 | 1025.22 | 3997 | 1.2 |
| 20 | 1024.02 | 4000 | 1.1 |
| 25 | 1022.61 | 4002 | 0.94 |

Table A.2.6

Properties of Saltwater at Various Concentrations

| Concentration of Nacl (kg/m³) | Mass of Nacl per mass of solution at 20°C (%) | Mass per unit volume of the solution at 15°C (kg/m³) | The same at 20°C | Specific heat at 20°C (J/kg K) |
|---|---|---|---|---|
| 0  | 0     | 999.13  | 998.23  | 4182 |
| 10 | 0.995 | 1006.30 | 1005.30 | 4127 |
| 20 | 1.976 | 1013.39 | 1012.29 | 4075 |
| 30 | 2.943 | 1020.41 | 1019.22 | 4024 |
| 40 | 3.898 | 1027.35 | 1026.07 | 3975 |
| 50 | 4.841 | 1034.25 | 1032.88 | 3929 |
| 60 | 5.772 | 1041.05 | 1039.60 | 3884 |
| 70 | 6.690 | 1047.83 | 1046.32 | 3841 |

## 2.4. *Properties of Soils and Rocks*

See the following page references for definitions of these properties.

Porosity, see p. 36; specific surface, p. 22; grain size, p. 21; suction potential, see p. 30; layer of adhesive water, see p. 23; hydraulic conductivity, see p. 78; hydraulic conductivity of a fracture, see p. 68; electro-osmotic permeability, see p. 83; compressibility, see p. 98, 102; compressibility of the solid grains, see p. 108; storage coefficient, see p. 111; tortuosity, see p. 233; dispersivity, see p. 239, 244–245, 247; heat conductivity, see p. 281; heat capacity, see p. 281; partition coefficient for sorption of organics, see p. 262–264.

# Bibliography

Abourifassi, M., and Marino, M. A. (1984). Cokriging of aquifer transmissivities from field measurements of transmissivity and specific capacity. *J. Int. Assoc. Math. Geol.* **16** (1), 19–35.

Ahlstrom, S. W., Foote, H. P., Arnett, R. C., Cole, C. R., and Serne, R. J. (1977). Multicomponent mass transport model. Theory and numerical implementation (discrete parcel random walk version). Battelle Rep. **BNWL 2127** for ERDA.

Allen, M. B., III. (1986). Numerical modeling of multiphase flow in porous media. *In* "Fundamentals of Transport Phenomena in Porous Media." *Proc. NATO-ASI* (Corapcioglu, M. V., Bear, J., eds.). Martinus Nijhoff, Dordrecht, The Netherlands (to be published).

Anderson, J. (1985). Predicting mass transport in fractured rock with the aid of geometrical field data. *Proc. Int. Symp. Stochastic Approach to Subsurface Flow, Montvillargenne, France, 1985,* Int. Assoc. Hydrol. Res., (Marsily, G. de, ed.). Fontainebleau, France.

Ardity, P. C. (1978). The earth tide effects on petroleum reservoirs. Ph.D. dissertation, Dept. of Petroleum Engineering, Stanford Univ., Stanford, Cal.

Aris, R. (1962). "Vector, Tensors and the Basic Equations of Fluid Mechanics." Prentice Hall, Englewood Cliffs, NJ.

Armstrong, M., and Jabin, R. (1981). Variogram models must be positive definite. *J. Int. Assoc. Math. Geol.* **13** (5), 455–459.

Armstrong, M. (1984). Common problems seen on variograms. *J. Int. Assoc. Math. Geol.,* **16** (3), 305–313.

Avogadro, A., and Marsily, G. de, (1984). The role of colloids in nuclear waste disposal. *Mater. Res. Soc. Symp. Proc.* **26,** 495–505. Elsevier, Amsterdam, The Netherlands.

Arnold, E. M., Gee, G. W., and Nelson, R. W. (1982). *Symp. Unsaturated Flow Transport Modeling.* US Nuclear Regulatory Commission Rep. **NUREG/CP-0030, PNL-SA-10325,** Washington DC.

Bachmat, Y., Bredehoeft, J., Andrews, B., Holts, D., and Sebastian, S. (1980). Groundwater management: the use of numerical models. *Am Geoph. Union, Water Resour.,* Monogr. 5, Washington DC.

Bakr, A. A., Gelhar, L. W., Gutjahr, A. L., and McMillan, J. R. (1978). Stochastic analysis of spatial variability in subsurface flow. Part I: comparison of one-and three-dimensional flows. *Water Resour. Res.* **14** (2), 263–271.

Barbreau, A., Sousselier, Y., Bonnet, M., Margat, J., Peaudecerf, R., Goblet, P., Ledoux, E., and Marsily, G. de (1980). Premières évaluations des possibilités d'évacuation des déchets radioactifs dans les roches cristallines. *In* "Underground Disposal of Radioactive Waste," *Proc. Symp. Otaniemi, 1979,* Vol. 1, 387–411. Int. Atomic Energy Agency, Vienna, Austria.

Barenblatt, G., Zheltov, I. P., and Kochina, I. N. (1960). Basic concepts in the theory of seepage of homogeneous liquids in fissured rocks. *Sov. Appl. Math. Mech. (Engl. transl.)* **24** (5), 852–864.

Batchelor, G. K. (1974). Transport properties of two-phase materials with random structure. *Ann. Rev. Fluid Mech.*, **6**, 227–255.

Bear, J. and Bachmat, Y. (1967). A generalized theory on hydrodynamic dispersion in porous media. *Symp. Artificial Recharge Management Aquifers, Haifa, Int. Assoc. Sci. Hydrol.*, **72**, 7–36.

Bear, J., Zaslavsky, D., and Irmay, S. (1968). "Physical Principles of Water Percolation and Seepage." UNESCO, Paris.

Bear, J. (1972). "Dynamics of Fluids in Porous Media." Amer. Elsevier, New York.

Bear, J. (1979). "Hydraulics of Groundwater." McGraw-Hill, New York.

Beran, M. J. (1968). "Statistical Continuum Theories." Wiley (Interscience), New York.

Berkaloff, E. (1966). Formulaire de l'hydrogéologue. Bur. Rech. Géol. Miniére, rep. **DS.66.A24**, Orléans, France.

Besbes, M., Ledoux, E., and Marsily, G. de (1976). Modeling of the salt transport in multilayered aquifers. *In* "System Simulation in Water Resources." (Vansteenkiste, G. C., ed.), pp. 229–246. North-Holland Publishing, Amsterdam, 1976.

Besbes, M., Delhomme, J. P., and Marsily, G. de (1978). Estimating recharge from ephemeral streams in arid regions: a case study at Kairouan (Tunisia). *Water Resour. Res.* **14** (2), 281–290.

Besbes, M., and Marsily, G. de (1984). From infiltration to recharge: use of a parametric transfer function. *J. Hydrol.* **74,** 271–293.

Binsariti A. A. (1980). Statistical Analysis and Stochastic Modeling of the Cortaro aquifer in Southern Arizona. Ph.D. dissertation, Dept. of Hydrology and Water Resources, Univ. of Arizona, Tucson, Arizona.

Biot, M. A. (1955). Theory of elasticity and consolidation for a porous anisotropic solid. *J. Appl. Phys.* **26** (2), 182–185.

Biot, M. A. (1956). General solution of the equations of elasticity and consolidation for a porous material. *J. Appl. Mech.* **78**, 91–96.

Bize, J., Bourguet, L. and Lemoine, J. (1972). "L'alimentation artificielle des nappes souterraines." Masson, Paris.

Bodelle, J. and Margat, J. (1980). "L'eau souterraine en France." Masson, Paris.

Bories, S. (1970). Sur les mécanismes fondamentaux de la convection naturelle en milieu poreux. Thèse d'Etat, Univ. Toulouse.

Boulton, N. S. (1963). Analysis of data from non-equilibrium pumping tests allowing for delayed yield from storage. *Proc. Inst. Civil. Eng.* **26** (6693), 469–482.

Boulton, N. S., and Streltsova, T. D. (1975). New equations for determining the formation constants of an aquifer from pumping test data. *Water Resour. Res.* **11** (1), 148–153.

Boulton, N. S., and Streltsova, T. D. (1977). Unsteady flow to a pumped well in a fissured water bearing formation. *J. Hydrol.* **35**, 257–269.

Boussinesq, J. (1904). Recherches théoriques sur l'écoulement des nappes d'eau infiltrées dans le sol et sur le débit des sources. *J. Math. Pure Appl.*, **10**, 5–78 and 363–394. (Originally published in *C. R. Acad. Sci, Paris*, June 22, 1903.)

Bouwer, H. (1964). Unsaturated flow in groundwater hydraulics. *J. Hydraul. Div.*, *Amer. Soc. Civil Eng.*, **90 HY5**, 121–127.

Braester, C. (1972). Simultaneous flow of immiscible liquids through porous fissured media. *Soc. Petr. Eng.*, *J.* **12** (4), 297–305.

Brebbia, C. A. (1978). "The Boundary Element Method for Engineers". Pentech Press, London.

Bredehoeft, J. D., and Pinder, G. F. (1970). Digital analysis of area flow in multiaquifer groundwater systems: a quasi three-dimensional model. *Water Resour. Res.* **6** (3), 883–888.

Bredehoeft, J. D., and Pinder, G. F. (1973). Mass transport in flowing groundwater. *Water Resour. Res.* **9** (1), 194–210.

Brenner, H., and Gaydos, L. J. (1977). *J. Colloid Interface Sci.* **58**, (2) 312–356.

Brochet, P., and Gerbier, N. (1974). L'évapotranspiration. Aspect agrométéorologique, évaluation pratique de l'évapotranspiration potentielle. Monogr. 65, Météorologie Nationale, Paris, France.

Brustkern, R. L., and Morel-Seytoux, H. J. (1970). Analytical treatment of the two-phase infiltration. *J. Hydraul. Div., Am. Soc. Civil Eng.* **96**, 2535–2548.

Buckley, S. E., and Leverett, M. C. (1942). Mechanism of fluid displacement in sands. *Trans. Amer. Inst. of Mining. Metall. Petr. Eng.* **146**, 107–116.

Budyko, M. I., Efimova, N. A., Zubenok, L. I., and Strokina, L. A. (1962). The heat balance of the earth's surface. *Akad. Nauk. URSS Izv. Ser. Geogr.* **1**, 6–16. Transl. 153M, US Dept. of Commerce, Office of Techn. Serv., doc. **63–19851**.

Burkholder, H. C. (1975). Incentives for partitioning high-level waste. Battelle, Pacific North West Lab., Rep. 1927, Richland, Wash.

Cady, R., and Neuman, S. P. (1986) (to be published). Advection–dispersion with adaptive Eulerian–Lagrangian finite elements. *In* "Fundamentals of Transport in Porous Media." *Proc. NATO-ASI*, (Corapcioglu, M. Y., Bear, J. eds.). Nijhoff, Dordrecht, The Netherlands.

Cambefort, H. (1966). "Forages et sondages." Eyrolles, Paris.

Carlsson, A. and Olson, T. (1977). Hydraulic properties of Swedish crystalline rocks. Hydraulic conductivity and its relation to depth. *Bull. Inst. Univ. Uppsala, Sweden.* 71–84.

Carrera, J. (1984). Estimation of aquifer parameters under transient and steady state conditions. Ph.D. thesis, Univ. of Arizona, Tucson (also to appear in 3 articles in *Water Resour. Res.*, by Carrera and Neuman, 1985–86).

Carslaw, H. S., and Jaeger, J. C. (1959). "Conduction of heat in solids," 2nd ed.. Oxford Univ. Press (Clarendon), London and New York.

Casagrande, L. (1952). Electro-osmotic stabilization of soils. *Boston Soc. Civ. Eng. Contrib. Soil. Mech.* 1941–1953, Boston, Massachussetts.

Casimir, H. B. G. (1945). Onsager's principle of microscopic reversibility. *Rev. Mod. Phys.* **17**, 343–350.

Castany, G. (1967). "Traité pratique des eaux souterraines," 2nd ed. Dunod, Paris.

Chauveteau, G. (1965). Essai sur la loi de Darcy. Thèse, Univ. Toulouse.

Chauveteau, G., and Thirriot, C. (1967). Régimes d'écoulement en milieu poreux et limite de la loi de Darcy. *La Houille Blanche* **1** (22), 1–8.

Chavent, G., Dupuy, M., and Lemonier, P. (1975). History matching by use of optimal theory, *Soc. Pet. Eng.* **15** (1) 74–86.

Chiles, J. P. (1977). Géostatistique des phénomènes non stationnaires. Thèse, Univ. of Nancy 1, France.

Choisnel, E. (1977). Le bilan d'énergie et le bilan hydrique du sol. *Météorologie* (special issue "Evapotranspiration") **11**, 103–159.

Chow, V. T. (1964). "Handbook of applied hydrology." McGraw-Hill, New York.

Christie, I., and Griffith, A. R., Mitchell, A. R., and Zienkiewicz, O. C. (1976). Finite element methods for second order differential equations with significant first derivatives. *Int. J. Numer. Methods* **10**, 1389–1396.

Clerc, J. P., Giraud, G., Roussenq, J., Blanc, R., Carton, J. P., Guyon, G., Ottavi, H., and Stauffer, D. (1983). La percolation. *Ann. Phys, Paris* **8** (1), 3–105.

Combarnous, M. A. (1970). Convection naturelle et convection mixte en milieu poreux. Thèse, Univ. de Paris, Technip.

Combarnous, M. A. and Bories, S. A. (1975). Hydrothermal convection in satured porous media. *Adv. Hydrosci* **10**, 232–307.

Cooley, R. L. (1977). A method of estimating parameters and assessing reliability for models of steady state groundwater flow, 1, Theory and numerical property. *Water Resour. Res.* **13** (2), 318–324.

Cooley, R. L. (1979). A method of estimating parameters and assessing reliability for models of steady state groundwater flow, 2, Application of statistical analysis. *Water Resour. Res.* **15**, (3), 603–617.

Cooley, R. L. (1982). Incorporation of prior information into non-linear regression groundwater flow models. 1. Theory. *Water Resour. Res.* **18** (4) 965–976.

Cooper, H. H., Jr. (1966). The equation of groundwater flow in fixed and deforming coordinates. *J. Geophys. Res.* **71** (20), 4785–4790.

Corapcioglu, M. Y., Abboud, N. M., and Haridas, A. (1986). Governing equations for particle transport in porous media. *In* "Fundamentals of Transport Phenomena in Porous Media." (Corapcioglu, M. Y., Bear, J., eds.), Martinus Nijhoff, Dordrecht, The Netherlands (to be published).

Cornet, F. H. (1979). Comparative analysis by the displacement discontinuity method of two energy criteria of fracture. *J. Appl. Mech.* **46** (2), 349–355.

Cornet, F. H. (1980). Analysis of hydraulic fracture propagation, a field experiment. *Int. Sem. Results E. C.*, 2nd "Geothermal Energy Research," Reidel, Dordrecht, The Netherlands.

Cornet, F. H., and Valette, B. (1984). *In situ* determination from hydraulic injection test data. *J. Geophys. Res* **89** (B13), 11,527–11,537.

Cottez, S. and Dassonville, G. (1965). Carte de la surface piézométrique de la nappe de la craie dans la région du Nord. Bur. Rech. Géol. et Min., map. Lille, France.

Dagan, G. (1967). A method of determining the permeability and effective porosity on unconfined anisotropic aquifers. *Water Resour. Res.* **3** (4), 1059.

Dagan, G. (1979). Models of groundwater flow in statistically homogeneous porous formations. *Water Resour. Res.* **15** (1), 47–63.

Dagan, G. and Bresler, E. (1980). Solute dispersion in unsaturated heterogeneous soil and field scale: theory. *Soil Sci. Soc. Am. J.* **43**, 461–467.

Dagan, G. (1981). Analysis of flow through heterogeneous random aquifers by the method of embedding matrix. 1. Steady flow. *Water Resour. Res.* **17** (1), 107–121.

Dagan, G. (1982a). Stochastic modelling of groundwater flow by unconditional and conditional probabilities. 2. The solute transport. *Water Resour. Res.* **18** (4), 835–848.

Dagan, G. (1982b). Analysis of flow through heterogeneous random aquifers. 2. Unsteady flow in confined formations. *Water Resour. Res.* **18** (5), 1571–1585.

Darcy, H. (1856) "Les fontaines publiques de la ville de Dijon". Dalmont, Paris.

Daubree, A. (1887). "Les eaux souterraines aux époques anciennes" et: "Les eaux souterraines à l'époque actuelle." Dunod, Paris.

Davis, S. N., and De Wiest, R. J. M. (1965). "Hydrogeology." John Wiley, New York.

Degallier, R., and Marsily, G. de (1977). Détermination des paramètres hydrodynamiques par interprétation dans des puits de variations brusques de niveau. Paris School of Mines, Bur. Rech. Géol. Min., rep. **78 SGN028-HYD**.

De Josselin De Jong, G. (1958) Longitudinal and transverse diffusion in granular deposits. *Trans. Am. Geophys. Union* **39**, 67–74.

Delfiner, P. (1976). Linear estimation of nonstationary spatial phenomena. *In* "Advanced Geostatistics in the Mining Industry," *NATO-ASI, Ser. C* **24**, (Guarascio, M., David, M., and Huijbregts, C., eds.). Reidel, Dordrecht, The Netherlands.

Delhomme, J. P. (1974). La cartographie d'une grandeur physique à partir de données de différentes qualités. *Proc. Meet., Int. Assoc. Hydrogeol.*, Mem. **10** (1), 185–194.

Delhomme, J. P. (1976). Application de la théorie des variables régionalisées dans les sciences de l'eau. Thèse, Univ. Paris VI.

Delhomme, J. P. (1978a). Kriging in hydroscience. *Adv. Water Resour.* **1** (5), 251–266.

Delhomme, J. P. (1978b). Application de la théorie des variables régionalisées dans les sciences de l'eau. *Bull. Bur. Rech. Géol. Min., Ser. 2, Sect. III*, No. 4–1978, 341–375.

Delhomme, J. P. (1979). Spatial variability and uncertainty in groundwater flow parameters: a geostatistical approach. *Water Resour. Res.* **15** (2), 269–280.

De Smedt, F., Belhadi, M., and Nurul, K. (1985). Study of the spatial variability of the hydraulic conductivity with different measurement techniques. *Proc. Int. Symp. Stochastic Approach Subsurface Flow, Montvillargenne, 1985, Int. Assoc. Hydraul. Res.* (Marsily, G. de, ed.). Fontainebleau, France.

De Swann, O. A. (1976). Analytical solutions for determining naturally fractured reservoir properties by well testing. *Trans. Am. Inst. Min. Eng.* **261**, 117–122.

Dhatt, G., and Touzot, G. (1981). "Une présentation de la méthode des éléments finis." Maloine, Paris.

Diamond, P., and Armstrong, M. (1984). Robustness of variograms and conditioning of kriging matrices. *J. Int. Assoc. Math. Geol.* **16** (8), 809–822.

Dieulin, A. (1980). Propagation de pollution dans un aquifère alluvial: l'effet de parcours. Thesis, Paris School of Mines-Univ. Paris VI.

Dieulin, A., Beaudoin, B., and Marsily, G. de (1980). Sur le transfert d'éléments en solution dans un aquifère alluvionnaire structuré. *C. R. Acad. Sci. Paris*, **291**, *Ser. D*, 805.

Dieulin, A., Marsily, G. de, and Beaudoin, B. (1981a). Sur l'existence d'un effet de parcours dans le transfert d'éléments en solution en milieu poreux. *C. R. Acad. Sci. Paris* **292**, *Ser. II*, 121.

Dieulin, A., Matheron, G., and Marsily, G. de (1981b). Growth of the dispersion coefficient with the mean travelled distance in porous media. *Sci. Total Environ.* **21**, 319–328.

Dieulin, A., Matheron, G., Marsily, G. de, and Beaudoin, B. (1981c). Time dependence of an "equivalent dispersion coefficient" for transport in porous media. *Proc. Euromech 143, Delft, 1981* (Verruijt, A. and Barends, F. B. J., eds.), 199–202. Balkema, Rotterdam, The Netherlands.

Dieulin, A. (1982). Filtration de colloïdes d'actinides par une colonne de sable argileux. Paris School of Mines, Rep. **LHM/RD/82/83**, Fontainebleau.

Domenico, P. (1972). "Concepts and Models in Groundwater Hydrology." McGraw-Hill, New York.

Dodds, J. (1982). La chromatographie hydrodynamique. *Analusis* **10**, 109–119.

Dupuit, J. (1848, 1863). "Etudes théoriques et pratiques sur le mouvement des eaux dans les canaux découverts et à travers les terrains perméables," 1st and 2nd Ed. Dunod, Paris.

Eagleson, P. S. (1970). "Dynamic Hydrology." McGraw-Hill, New York.

Einstein, A. (1905). Über die von der Molekularkinetischen Theorie der Wärme gefordete Bewegung von in ruhenden Flüssigkeiten suspendierten Teinchen. *Ann. Phys.* **17**, 539–560.

Emsellem, Y., and Marsily, G. de (1971). An automatic solution for the inverse problem. *Water Resour. Res.* **7** (5), 1264–1283.

Emsellem, Y., and Ledoux, E. (1971). Méthode de la simulation rapide. Intégration des systèmes différentiels linéaires du ler ordre à coefficients constants. Paris School of Mines Rep. **LHM/R71/8**, Fontainebleau.

Endo, H., Long, J. C. S., Wilson, C.R., and Witherspoon, P. A. (1984). A model for investigating mechanical transport in fracture networks. *Water Resour. Res.* **20** (10), 1390–1400.

Engelman, R., Gur, Y. and Jaeger, Z. (1983). Fluid flow through a crack network in rocks. *J. Appl. Mech.* **50**, 707–711.

Fairhurst, C. and Cornet, F. H. (1981). Rock fracture and fragmentation. *In* "Rock Mechanics from Research to Application." *22nd US Symp. on Rock Mechanics, Massachusetts Institute of Technology, 1981*, pp. 21–46.

Farengolts, Z. D., and Kolyada, M. N. (1969). Opyt primeneniyaé metodov matematicheskoy statistiki slya izutcheniya zakonov raspredeleniya gidrogeologichestikh parametrov. *Trudy Vsegingeo* **17**, 76–112.

Farmer, C. L. (1986). Moving point techniques. *In* "Fundamentals of Transport in Porous Media." *Proc. NATO-ASI, Ser. C* (Bear. J., and Corapcioglu, M. Y., eds.), Nijhoff, Dordrecht, The Netherlands (To be published.)

Fetter, C. W., Jr. (1980). "Applied Hydrogeology." Merrill, Columbus, Ohio.

Feuga, B. (1981). Détermination des directions principales et de l'anisotropie de perméabilité d'un milieu rocheux fracturé à l'aide de levés de fracturation. Approche théorique et premières applications. Rep. Bur. Rech. Géol. Min., **81 SGN 497 GEG**, Orléans.

Filliat, G. (ed.) (1981). "La pratique des sols et fondations." Editions du Moniteur, Paris.

Forscheimer, P. (1886). Über die Ergiebigkeit von Brunnen. Anlagen und Sickerschlitzen. *Zeitsch. Archit. Ing. Ver. Hannover* **32**, 539–563.

Forscheimer, P. (1901). Wasserbewegung durch Boden. *Z. Dtsch. Ing.* **45**, 1782–1788.

Fourmarier, (1939, 1958). "Hydrogéologie." 1st and 2nd ed., Masson, Paris.

François-Bongarçon, D. (1981). Les corégionalisations. Le cokrigeage. Paris School of Mines-Centre de Géostatistiques et de Morphologie Mathématique, note C-86, Fontainebleau, France.

Freeze, R. A., and Witherspoon, P. A. (1967). Theoretical analysis of regional groundwater flow. 2. Effect of watertable configuration and subsurface permeability variation. *Water Resour. Res.* **3**, 623–634.

Freeze, R. A. (1971). Three-dimensional, transient, saturated–unsaturated flow in a groundwater basin. *Water Resour. Res.* **7** (2), 347–366.

Freeze, R. A. (1975). A stochastic–conceptual analysis of one-dimensional groundwater flow in non-uniform, homogeneous media. *Water Resour. Res.* **11** (5), 725–741.

Freeze, R. A., and Cherry, J. C. (1979). "Groundwater." Prentice Hall, Englewood Cliffs, New Jersey.

Fried, J. J., and Combarnous, M. A. (1971). Dispersion in porous media. *Adv. Hydrosci.* **7**, 169–282.

Fried, J. J. (1972). Etudes théoriques et méthodologiques de la dispersion en milieu poreux naturel. Thèse, Univ. Bordeaux.

Fried, J. J. (1975). "Groundwater Pollution: Theory, Methodology, Modelling and practical rules." Elsevier, Amsterdam.

Fried, J. J., Munster, P., and Zilliox, L. (1979). Groundwater pollution by transfer of oil hydrocarbons. *Groundwater* **17** (6), 586–594.

Fried, Sh., Friedman, A. M., Atcher, R., and Hines, J. (1977). Retention of plutonium and americium by rocks. *Science* **196** (4294), 1087–1089.

Gale, J. E. (1975). A numerical field and laboratory study of flow in rocks with deformable fractures. Ph.D. dissertation, Univ. of California, Berkeley.

Gambolati, G. (1973). Equations for one-dimensional vertical flow of groundwater. 1) The rigorous theory. 2) Validity range of the diffusion equation. *Water Resour. Res.* **9** (4 and 5), 1022–1028 and 1385–1386.

Gambolati, G., and Freeze, R. A. (1973). Mathematical simulation of the subsidence of Venice. 1. Theory. *Water Resour. Res.* **9**, 721–733.

Gambolati, G., Gatto, P., and Freeze, R. A. (1974). Mathematical simulation of the subsidence of Venice. 2. Results. *Water Resour. Res.* **10**, 563–577.

Gambolati, G., and Perdon, A. (1984). The conjugate gradients in subsurface flow and land subsidence modelling. *In* "Fundamentals of Transport Phenomena in Porous Media," (Corapcioglu, M. Y. and Bear, J., eds.). Nijhoff, Dordrecht, The Netherlands.

Garder, A. O., Jr., Peaceman, D. W., and Pozzi, A. L., Jr. (1964). Numerical calculation of multidimensional miscible displacement by the method of characteristics. *Soc. Pet. Eng. J.* **4**, 26–36.

Gaudet, J. P., Jegat, H., Vachaud, G., and Wirenga, J. (1977). Solute transfer with exchange between mobile and stagnant water through unsaturated sand. *Soil Sci. Soc. Am. J.* **41** (4), 665–671.

Geerstma, J. (1957). The effect of fluid pressure decline on volumetric changes of porous rocks. *Trans. Am. Inst. Min. Metall. Pet. Eng.* **210**, 331–340.

Gelhar, L. W., Ko, P. Y., Kwai, H. H. and Wilson, J. L. (1974). Stochastic modelling of groundwater systems. R. M. Parsons Lab. for Water Resources and Hydrodynamics, Rep. 189, M.I.T., Cambridge, Mass.

Gelhar, L. W. (1976). Effects of hydraulic conductivity variations on groundwater flows. *Proc.*

*Int. Symp. Stochastic Hydraulics, 2nd, Int. Assoc. Hydraulic Res., Lund, Sweden.* (Hjort, P., Jönsson, L., and Larsen, P. eds.) Water Res. Pub., Fort Collins, Colorado, 1977, 409–428.

Gelhar, L. W., Bakr, A. A., Gutjahr, A. L., and McMillan, J. R. (1977). Comments on stochastic conceptual analysis of one dimensional groundwater flow in a non-uniform homogeneous medium, by Freeze, R. A., and reply by Freeze, R. A. *Water Resour. Res.* 13 (2), 477–480.

Gelhar, L. W., Gutjahr, A. L., and Naff, R. L. (1979a). Stochastic analysis of macrodispersion in a stratified aquifer. *Water Resour. Res.* 15 (6), 1387–1397.

Gelhar, L. W., Wilson, J. L., and Gutjahr, A. L. (1979b). Comments on "Simulation of groundwater flow and mass transport under uncertainty," by Tang, D. H., Pinder, G. F.—Reply. *Adv. Water Resour.* 2 (2), 101–102.

Gelhar, L. W., and Axness, C. L. (1983). Three-dimensional stochastic analysis of macrodispersion in aquifers. *Water Resour. Res.* 19 (1), 161–180.

Gelhar, L. W. (1986). Stochastic subsurface hydrology: from theory to applications. *Water Resour. Res.* (to be published).

Geze, B. Hydrogeology course. Given orally in 1967.

Girard, G., Ledoux, E., and Villeneuve, J. P. (1981). Le modèle couplé: simulation conjointe des écoulements de surface et des écoulements souterrains sur un système hydrologique. *Cah. ORSTOM, ser. Hydrol.* 18 (4), 195–280.

Goblet, P. (1981). Modélisation des transferts de masse et d'énergie en aquifère. Thesis, Paris School of Mines—Univ. Paris VI.

Graven, G., and Freeze, R. A. (1984). Theoretical analysis of the role of groundwater flow in the genesis of stratabound ore deposits. 1. Mathematical and numerical model. 2. Quantitative results. *Am. J. Sci.* 284, 1125–1174.

Green, W. H., and Ampt, G. A. (1911). Studies on soil physics. 1. The flow of air and water through soils. *J. Agri. Sci.* 4, 1–24.

Greenberg, J. A. (1971). Simultaneous flow of salt and water in soils. Ph.D. thesis, Univ. of Calif., Berkeley.

Gringarten, A. C., and Ramey, H. J. (1974). Unsteady-state pressure distribution created by a well with a single horizontal fracture, partially penetrating or restricted entry. *Trans. Am. Inst. Min. Eng.* 257, 413–426.

Gringarten, A. C., Ramey, H. J., and Raghavan, R. (1974). Unsteady-state pressure distribution created by a well with a single infinite conductivity vertical fracture. *Trans. Am. Inst. Min. Eng.* 257, 347–360.

Gringarten, A. C., Ramey, H. J., and Raghavan, R. (1975). Applied pressure analysis for fractured wells. *Trans. Am. Inst. Min. Eng.* 259, 887–892.

Gringarten, A. C., Landel, P. A., Menjoz, A., Sauty, J. P., and Thiery, D. (1979). Modélisation du fonctionnement d'un doublet hydrothermique sur le site de Bonnaud (Jura). Report Bur. Rech. Géol. Min., 79 SGN 063 GTH. Orléans, France.

Gringarten, A. C. (1982). Flow test evaluation of fractured reservoirs. *In* "Recent Trends in Hydrogeology," Geol. Soc. Am. Special Paper 189, (Narasimhan, T. N., ed.), 237–263.

Grund, A. (1903). Die Karsthydrographie, A. *Penck Geogr. Abhandl., Leipzig, Treubner* 7 (3), 3–200.

Gutjahr, A. L., Gelhar, L. W., Bakr, A. A., and Mc Millan, J. R. (1978). Stochastic analysis of spatial variability in subsurface flows. Part II: Evaluation and application. *Water Resour. Res.* 14 (5), 953–960.

Hammersley, J. M., and Welsh, D. J. A. (1980). Percolation. Theory and its ramifications. *Contemp. Phys.* 21, 593–605.

Hanshaw, B., and Back, W. (1974). Determination of regional hydraulic conductivity through use of 14C dating of groundwater. *Mém. Int. Assoc. Hydrogéol., Montpellier,* 10 (1), 195–198.

Hantush, M. S. (1966). Analysis of data from pumping tests in anisotropic aquifers. *J. Geophys. Res.* **71** (2), 421–426.

Harada, M., Chambé, P. L., Foglia, M., Higashi, K., Iwamoto, F., Leung, D., Pigford, D. H., and Ting, D. (1980). Migration of radionuclides through sorbing media. Analytical solutions. I. Rep. **LBL-10500**, Lawrence Berkeley Laboratory, Berkeley, Calif.

Harlan, R. L. (1973). Analysis of coupled heat-fluid transport in partially frozen soils. *Water Resour. Res.* **9** (5), 1314–1323.

Hartsock, J. H., and Warren, J. E. (1961). The effect of horizontal hydraulic fracturing on well performances. *J. Petrol. Technol.* **13**, 1050–1056.

Hennart, J. P., Yates, R., and Herrera, I. (1981). Extension of the integrodifferential approach to inhomogeneous multiaquifer systems. *Water Resour. Res.* **17** (4), 1044.

Herrera, I., and Yates, R. (1977). Integrodifferential equations for systems of leaky aquifers. 3. A numerical method of unlimited applicability. *Water Resour. Res.* **13** (4), 725.

Herrera, I. (1984). "Boundary Methods: Analgebraic Theory." Pitman Adv. Publ. Program, Boston, Mass.

Herrera, I. (1985). Unified formulation of numerical methods. 1. Green's formulas for operator in discontinuous fields. *Numer. Methods Partial Differential Eq.* **1**, 25–44.

Herzig, J. P., Leclerc, D. M., and Le Goff, P. (1970). Flow of suspension through porous media. Application to deep filtration. *Ind. Eng. Chem.*, **62** (5), 129–157.

Hillel, D. (1971). "Soil and Water: Physical Principles and Processes." Academic Press, New York.

Holtan, H. N. (1961). Concept for infiltration estimates in watershed engineering. US Dept Agr. Res. Ser. Pub., 41–51.

Houpeurt, A., Delouvrier, J., and Ifly, R. (1965). Fonctionnement d'un doublet hydraulique de refroidissement. *Mém. Trav. Soc. Hydrotech. Fr.*, No. 1.

Houpeurt, A. (1974). Eléments de mécanique des fluides dans les milieux poreux. Appeared in parts in *Rev. Inst. Fr. Pét.* from April 1955 to March 1957. Also Technip, 2nd ed.

Hubbert, M. K. (1940). The theory of groundwater motion. *J. Geol.* **48**, 785–944.

Hubert, P. (1984) Exploitation d'un Système de deux puits en situation de compétition et en situation de coordination. *J. Hydrol.* **70**, 353–367.

Huyakorn, P. S. (1976). An upwind finite element scheme for improved solution of the convection-diffusion equation. Rep. No. **76-WR-2**, Water Resources Program, Princeton University.

Huyakorn, P. S., and Pinder, G. I. (1983). "Computational Methods in Subsurface Flow." Academic Press, New York.

Illangasekare, T. H., and Morel-Seytoux, H. J. (1982). Stream-aquifer influence coefficients as tools for simulation and management. *Water Resour. Res.* **18** (1), 168–176.

Ilyin, N. I., Chernychev, S. N., Dzekster, E. S., and Zilberg, V. S. (1971). Ostenka tochnosti opredeleniya vodopronitsayemosti gornykh porod. Nedra, Moscow.

Jackson, R. E. (1980). Geochemical and biochemical attenuation processes. *In* "Aquifer Contamination and Protection." Studies and reports in hydrology, No. 30, UNESCO, Paris.

Jacob, C. E. and Lohman, S. W. (1952). Non-steady flow to a well of constant drawdown in an extensive aquifer. *Trans. Am. Geophys. Union* **33** (4), 559–569.

Jacquin, C. (1965a). Etude des écoulements et des équilibres de fluides dans les sables argileux. *Rev. Inst. Fr. Pet.* **20** (4), 1–49.

Jacquin, C. (1965b). Interactions entre l'argile et les fluides. Ecoulement à travers les argiles compactes. Etude bibliographique. *Rev. Inst. Fr. Pet.* **20** (10), 1475–1501.

James, R. B., and Rubin, J. (1979). Applicability of the local equilibrium assumption to transport through soils of solutes affected by ion exchange. *In* "Chemical Modelling in Aqueous Systems," *ACS Symp. Ser.* **93**, 225–235. Am. Chem. Soc., Washington, D. C.

Javandel, I., and Witherspoon, P. A. (1969). A method of analyzing transient fluid flow in multilayered aquifers. *Water Resour. Res.* **5** (4), 856–869.

Javandel, I., Doughty, C., Tsang, C. F. (1984). Groundwater transport: handbook of mathematical models. *Am. Geophys. Union.* Water Resources Monogr. 10, Washington DC.

Jenson, O. K. and Finlayson, B. A. (1978). Solution of the convection-diffusion equation using a moving coordinate system. *Proc. Int. Conf. Finite Elements Water Resour.*, (2nd ed.). Pentech Press, London.

Jetel, J. (1974). Complément régional de l'information sur les paramètres pétrophysiques en vue de l'élaboration des modèles de systèmes aquifères. *Proc. Meet. Int. Assoc. Hydrogeol., Mém.* **10** (1), 199–203.

Johnson, K. S. (1981). Dissolution of salt in the east flank of the Permian basin in the Southwestern USA. *J. Hydrol.* **54**, 75–93.

Journel, A., and Huijbregts, C. (1978). "Mining geostatistics." Academic Press New York.

Journel, A. G. (1980). The log-normal approach to predicting local distributions of selective mining unit grades. *J. Int. Assoc. Math. Geol.* **12** (4), 285–303.

Journel, A. G. (1984). The place of non-parametric geostatistics. *In* Geostatistics for Natural Resources Characterization. *NATO ASI, Ser. C*, 122, (Verly, G., David, M., Journel, A. G., and Maréchal, A. eds.), Vol. 1, 307–335. Reidel, Dordrecht, The Netherlands.

Karickhoff, S. W., Brown, D. S., and Scott, T. A. (1979). Sorption of hydrophobic pollutants on natural sediments. *Water Res.* **13**, 241–248.

Karickhoff, S. W. (1981). Semi-empirical estimation of sorption of hydrophobic pollutants on natural sediments and soils. *Chemosphere* **10** (8), 833–846.

Keller, J. B. (1964). Stochastic equations and wave propagation in random media. *Proc. Symp. Appl. Math., 16th, Am. Math. Soc.*, 145–170. Providence, R. I.

Kirkpatrick, S. (1973). Percolation and conduction. *Rev. Mod. Phys.* **45** (4), 574–588.

Kitanidis, P. K. (1983). Statistical estimation of polynomial generalized covariance functions and hydrologic applications. *Water Resour. Res.* **19** (4), 909–921.

Kolmogorov, A. N. (1931). Uber die analytischen Methoden in der Wahrscheinlichkeitsrechnung. *Math. Ann.* **104**, 415–458.

Konikow, L. F., and Bredehoeft, J. D. (1974). Modeling flow and chemical quality changes in an irrigated stream aquifer system. *Water Resour. Res.* **10** (3), 546–562.

Konikow, L. F., and Bredehoeft, J. D. (1978). Computer model of 2-dimensional solute transport and dispersion in groundwater. *In* "USGS Techniques of Water Resources Investigations," Book 7, Chap. C2. US Geol. Survey, Washington DC.

Krasny, J. (1974). Les différences de la transmissivité statistiquement significatives dans les zones de l'infiltration et du drainage. *Proc. Meet. Int. Assoc. Hydrogeol., Mem.* **10** (1), 204–211.

Krumbein, W. C. (1936). Application of the logarithmic moments to size frequency distributions of sediments. *J. Sediment. Petrol.* **6** (1), 35–47.

Kruseman, G. P., and De Ridder, N. A. (1970). Interpretation and discussion of pumping tests. *Bull.* No. 11/F, Int. Inst. Land Reclamation Improvement, Wageningen, Holland.

Lallemand-Barrès, A., and Peaudecerf, P. (1978). Recherche des relations entre les valeurs mesurées de la dispersivité macroscopique d'un milieu aquifère, ses autres caractéristiques et les conditions de mesure. Etude bibliographique. *Bull. Bur. Rech. Géol. Min. Sér. 2, Sec. III*, 4–1978, 277–284.

Lantz, R. B. (1971). Quantitative evaluation of numerical diffusion (truncation error). *Soc. Pet. Eng. J.* **11**, 315–320.

Law, J. (1944). A statistical approach to the interstitial heterogeneity of sand reservoirs. *Trans. Am. Inst. Min. Metall. Pet. Eng.* **155**, 202–222.

Ledoux, E., and Clouet d'Orval, M. (1975). Détermination in situ des paramètres de transfert de la chaleur dans les aquifères en écoulement monophasique. *Bull. Bur. Rech. Géol. Min., Ser. 2, Sect. III*, 1.

Ledoux, E., and Clouet d'Orval, M. (1977). Etude expérimentale d'un doublet hydrothermique. Interprétation des expériences "puits unique." Paris School of Mines-BURGEAP. Rep. for Dir. Gén. Rech. Sci. Tech., Fontainebleau, France.

Ledoux, E., Girard, G., and Villeneuve, J. P. (1984). Proposition d'un modèle couplé pour la simulation conjointe des écoulements de surface et des écoulements souterrains sur un bassin hydrologique. *La Houille Blanche*, **1** (2), 101–110.

Lefèvre du Prey, E. J. and Weill, L. M. (1974). Front displacement model in a fractured reservoir. *In* "Finite Elements Method in Flow Problems," (Oden, J. T., Zienkiewicz, O. C., Gallagher, R. H., and Taylor, C., eds.). Univ. of Alabama Press, Huntsville, Alabama.

Legait, B. (1983): Interprétation de certains types d'écoulements diphasiques en milieu poreux à partir des écoulements dans des capillaires. Thesis, Univ. Bordeaux I, France.

Levan, Ph., and Morel-Seytoux, H. J. (1972). Effect of soil air movement and compressibility on infiltration rates. *Soil. Sci. Soc. Am. Proc.* **36**, 237–241.

Liggett, J. A., and Liu, P. L. F. (1979). Unsteady flow in confined aquifers: a comparison of two boundary integral methods. *Water Resour. Res.* **15** (4), 861–866.

Liu, P. L. F., and Liggett, J. A. (1979). Boundary solutions to two problems in porous media. *J. Hydraul. Div., Am. Soc. Civil Eng.* HY3, 171–183.

Long, J. C. S., Remer, J. S., Wilson, C. R., and Witherspoon, P. A. (1982). Porous media equivalents for networks of discontinuous fractures. *Water Resour. Res.* **18** (3), 645–658.

Louis, C. (1974). Introduction à l'hydraulique des roches. *Bull. Bur. Rech. Géol. Min., Ser. 2, Sect. III*, No. 4.

Lumley, J. L., and Panofsky, H. A. (1964). "The structure of atmospheric turbulence." Wiley, New York.

Maddock, T., III. (1972). Algebraic technological function from a simulation model. *Water Resour. Res.* **8** (1), 129–134.

Maini, T. and Hocking, G. (1977). An examination of the feasibility of hydrologic isolation of a high level waste repository in crystalline rocks. Invited paper, Geologic Disposal of High Radioactive Waste Session, *Ann. Meet, Geol Soc. Am., Seattle,* Washington.

Mantoglou, A. and Wilson, J. L. (1982). The turning bands method for simulation of random fields using line generation by a spectral method. *Water Resour. Res.* **18** (5), 1379–1394.

Marle, C. (1967). Ecoulements monophasiques en milieu poreux. *Rev. Inst. Fr. Pét.* **22** (10), 1471–1509.

Marle, C. (1972). "Ecoulements polyphasiques en milieu poreux." Vol. IV of cours de production, 2nd ed. Technip. Paris.

Marsily, G. de (1971). Mathematical models for hydrologic processes. *In* "Computer Application in the Earth Sciences," (D. F. Merriam, ed.). Plenum Press.

Marsily, G. de (1978). De l'identification des systèmes hydrogéologiques. Thèse, Univ. Paris VI.

Marsily, G. de, Ledoux, E., Levassor, A., Poitrinal, D. and Salem, A. (1978). Modelling of large multilayered aquifer systems: theory and applications. *J. Hydrol.* **36**, 1–33.

Marsily, G. de, Lavedan, G., Boucher, M., and Fasanino, G. (1984). Interpretation of interference tests in a well field using geostatistical techniques to fit the permeability distribution in a reservoir model. *In* "Geostatistics for Natural Resources Characterization," Part 2, (Verly, G., Journel, A. C., Maréchal, A., and David, M. eds.), *Proc. NATO-ASI, Ser. C.,* **182**.

Martel, E. A. (1921). "Nouveau traité des eaux souterraines." Douin, Paris.

Matheron, G. (1965). "Les variables régionalisées et leur estimation." Masson, Paris.

Matheron, G. (1967). "Eléments pour une théorie des milieux poreux." Masson, Paris.

Matheron, G. (1970). La théorie des variables régionalisées et ses applications. Paris School of Mines, *Cah. Cent. Morphologie Math.*, **5**. Fontainebleau.

Matheron, G. (1971). The theory of regionalized variables and its applications. Paris School of Mines, *Cah. Cent. Morphologie Math.*, **5**. Fontainebleau.

Matheron, G. (1973). The intrinsic random functions and their applications. *Adv. Appl. Prob.* **5**, 438–468.

Matheron, G. (1974). Effet proportionnel et lognormalité ou le retour du serpent de mer. Paris School of Mines, Note Géostatistique No. 124, Centre Morphologie Math., Fontainebleau.

Matheron, G. (1976). A simple substitute for conditional expectation: disjunctive kriging. *In* "Advances in Geostatistics in the Mining Industry, "*NATO ASI Ser. C*, **24**, (Guarascio, M., David, M., Huijbregts, C. eds.), 221–236. Reidel, Dordrecht, The Netherlands.

Matheron, G., and Marsily, G. de (1980). Is transportation in porous media always diffusive? A counter example. *Water Resour. Res.* **16** (5), 901–917.

Meija, J. M., and Rodriguez-Iturbe I. (1974). On the synthesis of random field sampling from the spectrum: an application to the generation of hydrologic spatial processes. *Water Resour. Res.* **10** (4), 705–712.

Meinzer, O. E. (1923). The occurrence of groundwater in the United States, with a discussion of principle. US Geol. Survey Water Supply Paper 489.

Mercer, J. W., and Faust, Ch. R. (1981). "Groundwater modeling." National Water Well Assoc., Columbus, Ohio.

Miller, C. T., and Weber, W. J., Jr. (1984). Modelling of organic contaminant partitioning in groundwater systems. *Groundwater* **22** (5), 584–592.

Mitchell, A. R., and Wait, R. (1977). "The finite elements method in partial differential equations." Wiley, New York.

Morel, F. M. N. (1983). "Principles of Aquatic Chemistry." Wiley, New York.

Morel-Seytoux, H. J. (1973). Two-phase flow in porous media. *Adv. Hydrosci.* **9**, 119–202.

Morel-Seytoux, H. J., and Khanji, D. (1974). Derivation of an equation of infiltration. *Water Resour. Res.* **10** (4), 795–800.

Morel-Seytoux, H. J. (1984). From effective infiltration to aquifer recharge: a derivation based on the fluid mechanics of unsaturated porous media flow. *Water Resour. Res.* **20** (9), 1230–1240.

Morel-Seytoux, H. J., Illangasekare, T. H., and Simpson, A. R. (1981). Modeling for management of stream-aquifer systems. *Proc. Am. Soc. Civ. Eng. Water Forum, 1981, San Francisco, California.* 1342–1349.

Mualem, Y., and Dagan, G. (1972). Hysteresis in unsaturated porous media: a critical review and a new simplified approach. Technion, 2nd annual rep., part IV, Haïfa, Israël.

Mualem, Y. (1974). A conceptual model of hysteresis. *Water Resour. Res.* **10** (3), 514–520.

Muskat, M. (1946). "The Flow of Homogeneous Fluids through Porous Media," 2nd ed. Edwards, Ann Arbor, Michigan.

Myers, D. E. (1982). Matrix formulation of cokriging. *J. Int. Assoc. Math. Geol.* **14** (3), 249–257.

Myers, D. E. (1984). Estimation of linear combinations and cokriging, *J. Int. Assoc. Math. Geol.* **15** (5), 633–637.

Myers, D. E. (1984). Cokriging: new developments. *In* "Geostatistics for Natural Resources Characterization" (Verly, G., David, M., Journel, A. G., and Maréchal, A., eds.), *NATO-ASI, Ser. C*, *122*, **1**, 295–305. Reidel, Dordrecht, The Netherlands.

Nagy, D. J., Silebi, C. A., and McHugh, A. J. (1981). Hydrodynamic chromatography. An evaluation of several features. *J. Colloid Interf. Sci.* **79**, 264–267.

Narasimhan, T. N., Witherspoon, P. A. (1976). An integrated finite differences method for analyzing fluid flow in porous media. *Water Resour Res.* **12** (1), 57–64.

Narasimhan, T. N., and Neuman, S. P. (1977). Mixed explicit–implicit iterative finite element scheme for diffusion-type problems. *Int. J. Numer. Methods Eng.* **11**, 309–344.

Neretnieks, I. (1979). Transport mechanisms and rates of transport of radionuclides in the geosphere as related to the Swedish KBS concept. *In* "Underground Disposal of Radioactive Waste," *Proc. Int. Symp. Int. Atomic Energy Agency, July, 1979, Otaniemi, Finland* pp. 315–339. Int. Atomic Energy Agency, Vienna.

Neretnieks, I. (1980). Diffusion in the rock matrix: an important factor in radionuclide retardation. *J. Geophys. Res.* **85**, 4379–4397.

Neuman, S. P., and Witherspoon, P. A. (1968). Theory of flow in acquicludes adjacent to slightly leaky aquifers. *Water Resour. Res.* **4**(1), 103–112.

Neuman, S. P., and Witherspoon, P. A. (1969a). Theory of flow in a confined two aquifers system. *Water Resour. Res.* **5** (2), p. 803–816.

Neuman, S. P., and Witherspoon, P. A. (1969b). Applicability of current theories of flow in leaky aquifers. *Water Resour. Res.* **5** (4), p. 817–829.

Neuman, S. P., and Witherspoon, P. A. (1970). Finite element method of analysing steady seepage with a free surface. *Water Resour. Res.* **6** (3), 889–897.

Neuman, S. P., and Witherspoon, P. A. (1971). Analysis of non-steady flow with a free surface using the finite element method. *Water Resour. Res.* **7** (3), 611–623.

Neuman, S. P. (1972). Theory of flow in unconfined aquifers considering delayed response of the watertable. *Water Resour. Res.* **8** (4), 1031–1045.

Neuman, S. P., and Witherspoon, P. A. (1972). Field determination of the hydraulic properties of leaky multiple aquifer systems. *Water Resour. Res.* **8** (5), 1284–1298.

Neuman, S. P. (1973a). Calibration of distributed parameter groundwater flow models viewed as a multiple objective decision process under uncertainty. *Water Resour. Res.* **9** (4), 1006–1021.

Neuman, S. P. (1973b). Supplementary comments on "Theory of flow in confined aquifers considering delayed response of the water table." *Water Resour. Res.* **9** (4), 1102–1103.

Neuman, S. P. (1973c). Saturated-unsaturated seepage by finite elements. *J. Hydraul. Div. Amer. Soc. Civil Eng.* **99**, 2233–2251.

Neuman, S. P. (1974). Effect of partial penetration on flow in unconfined aquifers considering delayed gravity response. *Water Resour. Res.* **10** (2), 303–312.

Neuman, S. P. (1975a). Analysis of pumping tests data from anisotropic unconfined aquifers considering delayed gravity response. *Water Resour Res.* **11** (2), 329–342.

Neuman, S. P. (1975b). A computer program to calculate drawdown in an anisotropic unconfined aquifer with a partially penetraging well. Unpub., Dept. Hydrol. Water Resour., Univ. Arizona, Tucson, Arizona.

Neuman, S. P. (1975c). Galerkin's approach to saturated–unsaturated flow in porous media. *In* "Finite Elements in Fluids" (Taylor *et al.*, eds.), Vol. 1, Chap. 10. Wiley, New York.

Neuman, S. P. (1976). Wetting front pressure head in the infiltration model of Green and Ampt. *Water Resour. Res.* **12** (3), 564–566.

Neuman, S. P. (1979). Perspective on "delayed yield." *Water Resour. Res.*, **15** (4), 899–908.

Neuman, S. P., and Yakowitz, S. (1979). A statistical approach to the inverse problem of aquifer hydrology. 1. Theory. *Water Resour. Res.* **15** (4), 845–860.

Neuman, S. P., Fogg, G. E., and Jacobson, E. A. (1980). A statistical approach to the inverse problem of aquifer hydrology. 2. Case study. *Water Resour. Res.* **16** (1), 33–58.

Neuman, S. P. (1980). A statistical approach to the inverse problem of aquifer hydrology. 3. Improved solution method and added perspective. *Water Resour. Res.* **16** (2), 331–346.

Neuman, S. P. (1981). A Eulerian–Lagrangian numerical scheme for the dispersion convection equation using conjugate space-time grids. *J. Computat. Phys.* **41**, 270–294.

Neuman, S. P. (1982). Statistical characterization of aquifer heterogeneities: an overview. (P. A. Witherspoon's 60th birthday, Berkeley, 1979). *In* "Recent Trends in Hydrogeology" (Narasimhan, T. N., ed.), *Spec. Pap. Geol. Soc. Amer.*, **189**, 81–102. Boulder, Colorado.

Neuman, S. P., Walter, G. R., Bentley, H. W., Ward, J. J., and Gonzalez, D. D. (1984). Determination of horizontal aquifer anisotropy with three wells. *Groundwater* **22** (1), 66–72.

Neuman, S. P. (1984). Role of geostatistics in subsurface hydrology. *In* "Geostatistics for Nature Resources Characterization." *Proc. NATO-ASI*, (Verly, G., David, M., Journel, A. G., and Maréchal, A., eds.), Part 1, pp. 787–816. Reidel, Dordrecht, The Netherlands.

Nielsen, D. R., Biggar, J. M., and Erh, K. T. (1973). Spatial variabiility of field measured soil–water properties. *Hilgardia* **42**, 215–259.

Noblanc, A., and Morel-Seytoux, H. J. (1972). A perturbation analysis of two-phase infiltration. *J. Hydraul. Div. Am. Soc. Civil Eng.* **98**, 1527–1541.

Nordstrom, D. K., Plummer, L. N., Wigley, T. M. L., Wolery, T. J., Ball, J. W., Jenne, E. A., Basset, R. L., Crerar, D. A., Florence, T. M., Fritz, B., Hoffman, M., Holdren, G. R., Lafon, G. M., Mattigod, S. V., McDuff, R. E., Morel, F., Reddy, M. M., Sposito, G., and Thraikill, J. (1979). A comparison of computerized chemical models for equilibrium calculations in aqueous systems. *In* "Chemical Modelling in Aqueous Systems" (E. A. Jenne, ed.), *Am. Chem. Soc. Symp. Series* **93** (38), 857–892.

O'Neil, K., Pinder, G. F., and Gray, W. G. (1976). Simulation of heat transport in fractured, single-phase geothermal reservoirs. Note, Water Resources Program, Princeton University.

O'Neil, K. (1977). The transient three-dimensional transport of liquid and heat in fractured porous media. Ph.D. thesis, Princeton Univ., Princeton, NJ.

Onsager, L. (1931). *Phys. Rev.* **37**, 405–426; **38**, 2265–2279.

Oster, C. A. (1982). Review of groundwater flow and transport in the unsaturated zone. Rep. **NUREG/CR-2917, PNL-4427**, US Nuclear Regulatory Commission, Washington DC.

Papadopoulos, I. S., and Cooper, H. H., Jr. (1967). Drawdown in a well of large diameter. *Water Resour. Res.* **3** (1), 241–244.

Papadopoulos, I. S., Bredehoeft, J. D., and Cooper, H. H. (1973). On the analysis of slug test data. *Water Resour. Res.* **9** (4), 1087–1089.

Paramelle (l'abbé) (1926). "L'art de découvrir les sources," 2nd ed. Béranger, Paris.

Parlange, J. Y. (1971). Theory of water movement in soils. 2. One-dimensional infiltration. *Soil Sci.* **111**, 170–174.

Parlange, J. Y. (1972). Theory of water movement in soils. 8. One-dimensional infiltration with constant flux at the surface. *Soil Sci.* **114**, 1–4.

Parlange, J. Y. (1976). Capillary hysteresis and relationship between drying and wetting curves. *Water Resour. Res.* **12** (2), 224–228.

Pearson, F. J., Noronha, C. J., and Andrews, R. W. (1983). Mathematical modeling of the distribution of natural $^{14}C$, $^{234}U$, and $^{238}U$ in a regional groundwater system. *Radiocarbon* **25** (2), 291–300.

Peaudecerf, P., Gaillard, B., Lallemand-Barrès, P., Molinari, J., Guizerix, J., and Margat, J. (1975). Etude méthodologique des caractéristiques du transfert de substances chimiques dans les nappes. Etude expérimentale sur la parcelle de Bonnaud (Jura). Report ATP Hydrogéologie, Cent. Nat. Rech. Sci., Bur. Rech. Géol. Min., Commiss. Energ. Atom., Orléans, France.

Pfankuch, H. O. (1963). Contribution à l'étude des déplacements de fluides miscibles dans un milieu poreux. *Rev. Inst. Fr. Pét.* **2** (18), 215–270.

Philips, J. R. (1957). The theory of infiltration. *Soil Sci.* **83**, 345–357.

Pigford, T. H., Chambré, P. L., Albert, M., Foglia, M., Harada, M., Iwamoto, F., Kanki, T., Leung, D., Masuda, S., Muraoka, S., and Ting, D. (1980). Migration of radionuclides through sorbing media, analytical solutions. II. Rep. **LBL-11616**, 2 vol., Lawrence Berkeley Laboratory, Berkeley, Calif.

Pinder, G. F., and Bredehoeft, J. D. (1968). Application of a digital computer for aquifer evaluation. *Water Resour. Res.* **4** (5), 1069–1093.

Pinder, G. F., and Cooper, H. H., Jr. (1970). A numerical technique for calculating the transient position of a saltwater front. *Water Resour. Res* **6** (3) 875–882.

Pinder, G. F., and Gray, W. G. (1977). "Finite Element Simulation in Surface and Subsurface Hydrology." Academic Press, London.

Polubarinova-Kochina, P. Y. (1962). "Theory of Groundwater Movement." (Translated from Russian by R. J. M. De Wiest). Princeton Univ. Press, Princeton, New Jersey.

Potié, L. (1973). Etude et captage des résurgences d'eau douce sous-marine. *Int. Symp. Groundwater, 2nd*, Palermo, Sicily. 603–620.

Price, H. S., Varga, R. S. and Warren, J. R. (1966). Application of oscillation matrices to diffusion-convection equations. *J. Math. Phys.* **45**, 301–331.

Prickett, T. A. (1975). Modeling techniques for groundwater evaluation. *Adv. Hydrosci.* **10**, 1–143.

Prickett, T. A., Naymik, T. C., and Lonnquist, C. G. (1981). A random-walk solute transport model for selected groundwater quality evaluations. *Bull. Ill. State Water Survey, Champaign, Illinois.*

Remson, I., Hornberger, G. M., and Molz, F. J. (1971). "Numerical Methods in Subsurface Hydrology with an Introduction to the Finite Elements Method." Wiley (Interscience), New York.

Renard, D. Geffroy, F. Touffait, Y., and Seguret, S. (1985). Présentation de Bluepack 3D release 4. Paris School of Mines, Centre Géostatistique-Morphologie Mathématique, Service Informatique, Fontainebleau, France.

Roach, G. F. (1982). "Green's functions," 2nd ed. Cambridge Univ. Press. Cambridge.

Robinson, P. C. (1984). Connectivity, flow and transport in network models of fractured media. Ph.D. thesis, Oxford. Rep. **TP 1072**, Atomic Energy Research Authority, Harwell, United Kingdom.

Rouleau, A., and Gale, J. (1985). Fracture size interconnectivity and groundwater flow: a site specific stochastic analysis. *Proc. Int. Symp. Stochastic Approach Subsurface Flow, Int. Assoc. Hydraul. Res., Montvillargenne, France, 1985* (Marsily, G. de, ed.). Fontainebleau, France.

Rousselot, D. (1976). Proposition pour une loi de distribution des perméabilités ou transmissivités. Rep. Bur. Rech. Géol. Min., Serv. Géol. Jura-Alpes, Lyon.

Rumer, R. R. J., and Shiau, J. C. (1968). Salt water interface in a layered coastal aquifer. *Water Resour. Res.* **4** (6), 1235–1247.

Russo, D., and Bresler, E. (1980). Scaling soil hydraulic properties of a heterogeneous field. *Soil Sci. Soc. Amer J.* **44**, 681–684.

Sa Da Costa, A. (1981). Seawater intrusion in aquifers: the modeling of interface discontinuities. *Proc. Euromech* **143**, *Delft, 1981* (A. Verruijt and F. B. J. Barends, ed.), 109–117. Balkema, Rotterdam, The Netherlands.

Saffman, P. G. (1959). A theory of dispersion in porous media. *J. Fluid Mech.* **6**(3), 321–349.

Saffman, P. G. (1960). Dispersion due to molecular diffusion and macroscopic mixing in flow through a network of capillaries. *J. Fluid Mech.* **7**(2), 194–208.

Sagar, B. (1978). Galerkin finite elements procedure for analyzing flow through random media. *Water Resour. Res.* **14** (6), 1035–1044.

Saltelli, A., Avogadro, A., and Bidoglio, G. (1984). Americium filtration in glauconitic sand columns. *Nuc. Technol.* **67**, 245–254.

Sauty, J. P. (1977). Contribution à l'identification des paramètres de dispersion dans les aquifères par interprétation des expériences de traçage. Thesis, Univ. Grenoble, France.

Sauty, J. P. (1978a). Identification des paramètres du transport hydrodispersif dans les aquifères par interprétation de traçages en écoulement cylindrique convergent ou divergent. *J. Hydrol.* **39**, 69–103.

Sauty, J. P. (1978b). The effect of thermal dispersion on injection of hot water in aquifers. *Well Testing Symp., 2nd, Berkeley, California, 1978.*

Scheidegger, A. L. (1960). "The Physics of Flow through Porous Media." Univ. Toronto Press, Toronto.

Schellenberg, K., Leuenberger, C. and Schwartzenbach, R. P. (1984). Sorption of chlorinated phenols by sediments and aquifer materials. *Environ. Sci. Technol.* **18**, (9) 652–657.

Schneebeli, G. (1966). "Hydraulique souterraine." Eyrolles, Paris.

Schoeller, H. (1962). "Les eaux souterraines." Masson, Paris.

Schwartz, L. (1961). "Methodes mathematiques pour les sciences physique." Hermann, Paris.

Schwartz, F. W., Smith, L., and Crowe, A. S (1983). A stochastic analysis of macrodispersion in fractured media. *Water Resour. Res.* **19** (5), 1253–1265.

Schwartzenbach, R. P., and Westall, J. (1981). Transport of non-polar organic compounds from surface water to groundwater. Laboratory sorption studies. *Environ. Sci. Technol.* **15**, 1360–1375.

Schwartzenbach, R. P., and Westall, J. (1984). Sorption of hydrophobic trace organics in groundwater systems. *Proc. Symp. Degradation, Retention Dispersion Pollutants Groundwater, 1984*, (Arvin, E. ed.), p. 39–55. Univ. Copenhagen, Denmark.

Schweppe, F. (1973). "Uncertain dynamic systems." Prentice Hall, New York.

Schwille, F. (1967). "Petroleum Contamination of the Subsoil. A Hydrological problem. The Joint Problems of the Oil and Water Industry." Peter Hepple, Elsevier, Amsterdam.

Schwille, F. (1984). Migration of organic fluids immiscible with water in the unsaturated zone. *In* "Pollutants in porous media," (Yaron, B., Dagan, G., and Goldshmid, J., eds.) Ecol. Stud. **47**, p. 27–48, Springer-Verlag, Berlin and New York.

Schwydler, M. I. (1962). Les courants d'écoulement dans les milieux hétérogènes. *Izv. Akad. Nauk. SSSR* **3**, 185–190.

Seguin, B. (1980). Détermination de l'évapotranspiration réelle ETR dans les bilans hydrologiques par télédétection en thermographie infra-rouge. *Hydrol. Sci. Bull.* **25** (2), 143–153.

Serra, J., and Kolomenski, E. N. (1976). La quantification en pétrologie et trois études de morphologie mathématique en géologie de l'ingénieur. *Bull. Assoc. Int. Géol. Ing.* **13**, 83–97.

Shante, V. K. S., and Kirkpatrick, S. (1971). An introduction to percolation theory. *Adv. Phys.* **20**, 325–357.

Simpson, E. S. and Duckstein, L. (1975). Finite state mixing cells models. *Proc. US-Yugoslavian Symp. Karst Hydrology Water Resour., Dubrovnik, 1975*, 489–508.

Simpson, E. S. (1978). A note on the structure of the dispersion coefficient. *Geol. Soc. Amer. Symp. Dispersion Groundwater Flow Syst. Toronto, Canada, 1978.*

Slattery, J. C. (1972). "Momentum, energy, and mass transfer in continua." McGraw-Hill, New York.

Small, H. (1974). *J. Colloid Interface Sci.* **48**, 147–161.

Small, H. (1977). *Adv. Chromatogr.* **15**, 113–129.

Smith, R. E. (1972). The infiltration envelope: results from a theoretical infiltrometer. *J. Hydrol.* **17**, 1–22.

Smith, L., and Freeze, R. A. (1979). Stochastic analysis of steady state groundwater flow in a bounded domain. 1. One-dimensional simulations. *Water Resour. Res.* **15** (3), 521–528. 2. Two-dimensional simulations. *Water Resour. Res.* **15** (6), 1543–1559.

Smith, L., and Schwartz, F. W. (1980). Mass transport. 1. A Stochastic analysis of macrodispersion. *Water Resour. Res.* **16** (2), 303–313.

Smith, L., and Schwartz, F. W. (1981a). Mass transport. 2. Analysis of uncertainty in prediction. *Water Resour. Res.* **17** (2), 351–369.

Smith, L., and Schwartz, F. W. (1981b). Mass transport. 3. Role of hydraulic conductivity data in prediction. *Water Resour. Res.* **17** (5), 1463–1479.

Smith, L., and Schwartz, F. W. (1984). An analysis of the influence of fracture geometry on mass transport in fractured media. *Water Resour. Res.* **20** (9), 1241–1252.

Snow, D. T. (1968). Hydraulic character of fractured metamorphic rocks of the front range and implications to the Rocky Mountains Arsenal well. *Colo. Sch. Mines.* **63** (1), 201–244.

Snow, D. T. (1969). Anisotropic permeability of fractured media. *Water Resour. Res.* **5**, 1273–1289.

Strack, O. D. L. (1985). "Groundwater Mechanics." Prentice Hall, Englewood Cliffs, N. J. (To be published.)

Strang, G., and Fix, G. T. (1973). "An analysis of the finite element method." Prentice Hall, Englewood Cliffs, New Jersey.

Streltsova, T. D. (1976a). Hydrodynamics of groundwater flow in a fractured formation. *Water Resour. Res.* 12 (3), 405–414.

Streltsova, T. D. (1976b). Analysis of aquifer–aquitard flow. *Water Resour. Res.* 12 (3), 415–422.

Streltsova, T. D. (1984). Well hydraulics in heterogeneous porous media. *In* "Fundamentals of Transport Phenomena in Porous Media" (Bear, J. and Corapcioglu, M. Y., eds.) *NATO-ASI, Ser.* 82, 315–346. Martinus Nijhoff, Dordrecht, The Netherlands.

Sudicky, E. A., and Frind, E. O. (1982). Contaminant transport in fractured porous media: Analytical solutions for a system of parallel fractures. *Water Resour. Res.* 18 (6), 1634–1642.

Tal, A., and Dagan, G. (1983). Flow toward storage tunnels beneath a water table. *Water Resour. Res.* 19 (1), 241–250.

Talbot, A. (1979). The accurate numerical inversion of Laplace transforms. *J. Inst. Math. Appl.* 23, 97–120.

Tang, D. H., and Pinder, G. F. (1977). Simulation of groundwater flow and mass transport under uncertainty. *Adv. Water Resour.* 1(1), 25–30.

Taylor, Sir G. (1953). Dispersion of soluble matter in solvent flowing slowly through a tube. *Proc. R. Soc., London, Ser. A,* 219, 186–203.

Terzaghi, K., and Peck, R. B. (1967). "Soil mechanics in engineering practices," 2nd ed. Wiley, New York.

Theis, C. V. (1935). The relation between the lowering of the piezometric surface and the rate and duration of discharge of a well using groundwater storage. *Trans. Am. Geophys. Union, Ann. Meet., 16th,* 519–524.

Thiem, G. (1906). "Hydrologische Methoden." J. M. Gebhardt, Leipzig.

Thomas, R. G. (1973). Groundwater models. Irrigation and drainage. *Spec. Pap. Food Agricultural Organis.* No. 21, U. N., Rome.

Tien, C, Turian, R. M., and Pendse, H. (1979). Simulation of the dynamic behaviour of deep bed filters. *AICE J.* 25 (3), 385–395.

Todd, D. K. (1959). "Groundwater Hydrology." Wiley, New York.

Topp, G. C. (1971). Soil-water hysteresis: the domain theory extended to pore interaction conditions. *Soil Sci. Soc. Am. Proc.* 35, 219–225.

Toth, J. (1963). A theoretical analysis of groundwater flow in small drainage basins. *J. Geophys. Res.* 68, 4795–4812.

Townley, L. R., and Wilson, J. L. (1985). Computationally efficient algorithms for parameter estimation and uncertainty propagation in numerical models of groundwater flow. *Water Resour. Res.,* (To be published.)

Trescott, P. C., Pinder, G. F., and Carson, S. P. (1976). Finite difference model for aquifer simulations in two dimensions with results of numerical experiments. *In* "Technique of Water Resources Investigations of the USGS," Book 7, Chap. Cl: Automated data processing and computation. US Geol. Survey, Washington, DC.

Tyson, N. H., and Weber, E. M. (1964). Groundwater management for the nation's future. Computer simulation of groundwater basins. *Proc. Am. Soc. Civ. Eng.* 90 HY4, 59–77.

UNESCO (1972–1975). "Groundwater Studies (Studies and Reports in Hydrology)."

Vachaud, G. (1968). Contribution à l'étude des problèmes d'écoulement en milieux poreux non saturés. Thesis, Univ. Grenoble.

Valocchi, A. J., Street, R. L. and Roberts, P. W. (1981). Transport of ion exchanging solutes in groundwater: chromatographic theory and field simulation. *Water Resour. Res.* 17 (5), 1517–1527.

Van Genuchten, M. T. (1977). On the accuracy and efficiency of several numerical schemes for solving the convective–dispersive equation. *In* "Finite Elements in Water Resources," (W. G. Gray, G. F. Pinder, C. A. Brebbia, eds.). Pentech Press, London.

Varga, R. S. (1962). "Matrix Iterative Analysis." Prentice Hall, Englewood Cliffs, New Jersey.

Varoglu, E., and Finn, W. D. L. (1978). A finite element method for the diffusion–convection equation. *Proc. Int. Conf. Finite Elements Water Resour.,* 2nd ed. Pentech Press, London.

Vauclin, M., Haverkamp, R., and Vachaud, G. (1979a). "Résolution numérique d'une équation de diffusion non linéaire (application à l'infiltration de l'eau dans les sols non saturés)." Press. Univ. Grenoble.

Vauclin, M., Khanji, D. and Vachaud, G. (1979b). Experimental and numerical study of a transient two-dimensional unsaturated-saturated watertable recharge problem. *Water Resour. Res.* **15** (5), 1089–1101.

Vauclin, M., Vieira, S. R., Vachaud, G. and Nielsen, D. R. (1983). The use of cokriging with limited field soil observations. *Soil Sci. Proc. Am. J.*, **47** (2), 175–184.

Vauclin, M. and Vachaud, G. (1984). Transferts hydriques dans les sols non saturés non homogènes. *Ann. Min.* **191** (5–6), 63–74.

Vauclin, M. (1984). Infiltration in unsaturated soils. *In* "Fundamentals of Transport Phenomena in Porous Media" (Bear, J., and Corapcioglu, M. Y., eds.), *NATO-ASI Ser.* **82**, 257–313. Reidel, Dordrecht, The Netherlands.

Verruijt, A. (1968). A note on the Ghybben–Herzberg formula. *Bull. Int. Assoc. Sci. Hydrol.* **13** (4), 43–46.

Verruijt, A. (1970). "Theory of Groundwater Flow." MacMillan, New York.

Wackernagel, H. (1985). L'ajustement multivariable d'un modèle linéaire. Paris School of Mines, Centre de Géostatistique et de Morphologie Mathématique, Note 945, Fontainebleau, France.

Walton, W. C., and Neill, I. C. (1963). Statistical analysis of specific capacity data for a dolomite aquifer, *J. Geophys. Res.* **68**, (8) 2251–2262.

Walton, W. C. (1970). "Groundwater resources evaluation." McGraw-Hill, New York.

Wang, H. F., and Anderson, M. P. (1982). "Introduction to Groundwater Modeling. Finite Difference and Finite Element Methods." Freeman, San Francisco.

Warren, J. E., and Root, P. J. (1963). The behaviour of naturally fractured reservoirs. *Soc. Petr. Eng. J.*, 245.

Warrick, A. W., Mullen, G. J. and Nielsen, D. R. (1977). Scaling field measured soil hydraulic properties using similar-media concepts. *Water Resour. Res.* **13** (2), 355–362.

Wenzel, L. K. (1942). Methods of determining permeability of water bearing materials. US Geol. Survey Water Supply Pap. 887.

Westall, J. C. (1984). Properties of organic compounds in relation to chemical binding. *In* "Biofilm Processes in Groundwater Research," *Proc. Symp. Stockholm, 1983.* (Brinck, P. ed.), 65–90. Ecologic Res. Comm. of NFR, Stockholm, Sweden.

Wilke, S., Guyon, E., and Marsily, G. de (1985). Water penetration through rocks: test of a tridimensional percolation description. *J. Int. Assoc. Math. Geol* **17** (1), 17–24.

Willis, M. S. (1986). Theory of filtration. *In* "Fundamentals of Transport Phenomena in Porous Media." *Proc. NATO-ASI* (Corapcioglu, M. Y., Bear, J., eds.). Martinus Nijhoff, Dordrecht, The Netherlands (To be published.)

Winter, C. L., Newman, C. M., and Neuman, S. P. (1984). A perturbation expansion for diffusion in a random field. *SIAM J. Appl. Math.* **44** (2), 411–424.

Witherspoon, P. A., Gale, J. E., Taylor, R. L., and Ayatollahi, M. S. (1973). Investigation of fluid injection in fractured rock and effect on stress distribution. Rep. No. **TE 73-2**, Dept. of Civil Engineering, Univ. California, Berkeley.

Wnek, W. J., Gidaspow, D., and Wasan, D. T. (1975). The role of colloid chemistry in modelling deep bed liquid filtration. *Chem. Eng. Sci.* **30**, 1035–1047.

Yeh, W. W., Yoon, Y. S., and Lee, K. S. (1983). Aquifer parameter identification with kriging and optimum parametrization. *Water Resour. Res.* **19** (1), 225–233.

Zienkiewicz, O. C. (1977). "The Finite Elements Method," 3rd ed. McGraw-Hill, New York.

# Index